This book is a basic introduction to the principles of circular particle accelerators and storage rings, for scientists, engineers and mathematicians.

Particle accelerators used to be the exclusive province of physicists exploring the structure of the most fundamental constituents of matter. Nowadays, particle accelerators have also found uses as tools in many other areas, including materials science, chemistry, and medical science. Many people from these fields of study, as well as from particle physics, have learned about accelerators at various courses organised by CERN, the European Organisation for Nuclear Research, which has established a reputation as the world's top accelerator facility. Kjell Johnsen and Phil Bryant, the authors of this book, are distinguished accelerator physicists who have also run the CERN Accelerator School. The text they present here starts with a historical introduction to the field and an outline of the basic concepts of particle accelerators. It goes on to give more details of how the transverse and longitudinal motions of the particle beams can be analysed, including treatments of lattice design, compensation schemes, phase focusing, transition crossing, and other radio frequency effects. Operational and diagnostic techniques and the optimisation of luminosity are discussed in detail. One chapter is devoted to synchrotron radiation and the special features of synchrotron light sources. Although the book emphasises circular machines, much of the treatment applies equally to linear machines and transfer lines.

The book will be an essential reference for anyone working with particle accelerators as a designer, operator or user, as well as being a good preparation for those intending to go to the frontiers of accelerator physics.

The Principles of
Circular Accelerators and Storage Rings

The Principles of
Circular Accelerators
and Storage Rings

Philip J. Bryant

CERN, Geneva, Switzerland

Kjell Johnsen

Formerly of CERN, Geneva, Switzerland

CAMBRIDGE
UNIVERSITY PRESS

PHYSICS

Published by the Press Syndicate of the University of Cambridge
The Pitt Building, Trumpington Street, Cambridge CB2 1RP
40 West 20th Street, New York, NY 10011-4211, USA
10 Stamford Road, Oakleigh, Melbourne 3166, Australia

First published 1993

Printed in Great Britain at the University Press, Cambridge

A catalogue record for this book is available from the British Library

Library of Congress cataloguing in publication data
Bryant, Philip J.
The principles of circular accelerators and storage rings /
Philip J. Bryant, Kjell Johnsen.
p. cm.
ISBN 0-521-35578-8
1. Particle accelerators. 2. Storage rings. I. Johnsen, Kjell.
II. Title.
QC787.P3B79 1993
539.7′3—dc20   92-17737 CIP

ISBN 0 521 35578 8   hardback

KW

CONTENTS

QC787
P3
B79
1993
PHYS

FOREWORD

A dictionary definition of *acceleration* is *an increase in speed*[*] from which one understands that a *charged-particle accelerator* would increase the speed of charged particles – as indeed it does. However, today's accelerators work at ultra-relativistic energies and it is not so much the particle's speed that increases as its mass. For example, between 1 MeV and 1 GeV an electron gains speed modestly from approximately 95% of the speed of light to what is virtually the full value, but its mass leaps forward from approximately three times its rest value to around 2000 times. This anomaly led Ginzton, Hansen and Kennedy[†] to propose the names *mass aggrandiser* or *ponderator*, but neither became fashionable. More strictly one should speak of a *momentum aggrandiser*, but since this is sure to be as unfashionable as the others, we are left with the simple name *accelerator*. The accelerator family is, however, very large, so the authors will concentrate on *synchrotrons* and *storage rings* with only brief references to linear accelerators and many of the early circular machines.

Although universities often include some lectures on accelerators in their physics courses, there are very few courses which can claim to be principally about accelerators. The machines and the expertise in this field are mainly in national and international laboratories. Since these laboratories have a more mission-orientated approach than universities, relatively few books have been written and the accelerator community has relied heavily on a 'learning-by-working apprenticeship' for newcomers and on personal contacts and conferences for the dissemination of knowledge. The subject has not been stationary. Machine designs continuously develop, industrial applications increase and the need for more trained personnel is ever-present. The progress in this field has been such that even at an elementary level there are many differences between a text book written now and one produced 10 years ago.

This is a basic book on circular accelerators and storage rings, but as *basic*

---

[*] *The new national dictionary*, W. Collins Sons & Co. Ltd., London & Glasgow (1966).
[†] E. L. Ginzton, W. W. Hansen and W. R. Kennedy, *Rev. Sci. Instr.*, vol. **19**, No. 2 (1948), p. 89.

is a relative term we shall try to describe what is meant by this. It is assumed that the reader starts with either a first degree knowledge of physics or electrical engineering. Using simple mathematical tools, the book then aims to treat mainly the single particle and linear theory of accelerators for the transverse and longitudinal phase planes. However, some of the simpler collective effects are also included and in the Appendixes some effort is made to prepare the way for those readers who will wish to carry on to more advanced topics. For example, the basic equations for the transverse and longitudinal motions are derived in an intuitively straightforward manner in the main part of the book, but in Appendixes A and D these derivations are repeated using the Hamiltonian formalism. This adds little to the results of the elementary treatment for the newcomer to the field, but it is the stepping stone needed for analysing non-linear resonances, dynamic aperture, stochasticity, chaos, etc. The validity of certain commonly-made approximations is also discussed. The under-standing of these finer points is essential when preparing for more advanced work. Space charge and image forces and coherent instabilities in coasting beams are included, but the extremely complex and large field of coherent instabilities in bunched beams is omitted. The Vlasov equation is mentioned, but it is not widely applied. Collective processes with diffusion, such as the action of radiofrequency noise on full buckets and stochastic cooling, which can be treated by the Fokker–Planck equation, are also omitted as being beyond elementary.

We have written this book with the main idea of helping those who are just entering the field of accelerators and have no previous experience of the subject. However, it should also be helpful to those who have experience in some accelerator speciality and now wish to broaden their knowledge and much of the book should be useful to accelerator users. Whatever the situation may be, we hope that this book will be of interest and of use to you and that you will derive as much satisfaction from reading it as we have had from writing it.

In producing this book, we have been greatly helped by our colleagues in CERN and the many contacts we made through the CERN Accelerator School. We are especially grateful to W. Hardt for his thorough and critical reading of the manuscript and the many important improvements that he proposed. H. G. Hereward, in particular, contributed significantly to the clarification of specific problems and G. Guignard made many constructive suggestions. We also wish to acknowledge the helpful discussions we have had with O. Gröbner, A. Hofmann and G. Dôme. The labour of producing and labelling the diagrams was shared by S. M. Bryant to whom we give our thanks. We are indebted to CERN and its Directors General for the support given during the period of preparation of the manuscript and for permission to make generous use of some extracts from CERN Reports.

The authors

COORDINATE SYSTEM

The curvilinear coordinate system following the central orbit (x, s, z) is shown in the diagram below.

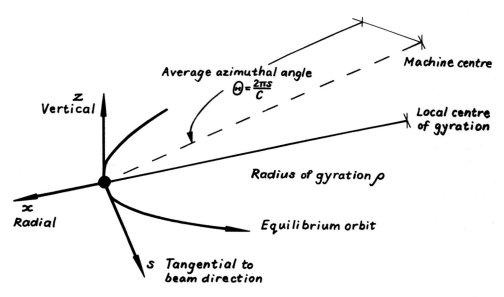

The particle motion is defined with respect to an equilibrium orbit by means of a right-handed curvilinear coordinate system x, s, z. The azimuthal coordinate s is directed along the tangent of the orbit. The local radius of curvature ρ and the bending angle s/ρ subtended by the equilibrium orbit are defined as positive for anticlockwise rotation when viewed from positive z. In the vertical plane, radii of curvature and bending angles are defined as positive for anticlockwise rotation viewed from positive x. Anticlockwise rotation is assumed throughout the book unless otherwise stated.

The above can be summarised as: ρ *is positive when bending to the left or upwards when looking along the beam direction*. This has the consequence that the local, radial coordinate x is positive to the *outside* of a positive bend, while the local, vertical coordinate z is positive to the *inside* of a positive bend. Special care is needed to account for the effect of this sign asymmetry in the transfer matrices of the accelerator elements. When relating the beam to an external reference system, or when defining a transfer line, the sign convention must be rigorously applied. However, when dealing with *only the optical parameters of a ring* the sense of rotation is unimportant and it is usual to use a simplified convention that can be expressed as: *ρ is positive when bending towards the ring centre or upwards when looking along the beam direction.*

The general transverse coordinate y is frequently used to represent x or z. In some applications, an average azimuthal angle $\Theta = 2\pi s/C$ is used where C is the circumference of the equilibrium orbit and Θ is defined as positive for anticlockwise rotation.

SYMBOLS

SPECIAL SYMBOLS AND CONVENTIONS

$\pm, \mp$ the upper sign corresponds to the horizontal (x, s) plane and the lower sign to the vertical (z, s) plane in expressions containing y the general transverse coordinate

$\cdot$ differentiation w.r.t. to time, t

$'$ differentiation w.r.t. a specified variable

$-$ average value

$\langle \; \rangle$ average over a distribution

$\wedge \; \vee$ maximum and minimum values

$\wedge$ also indicates the complex amplitude of a phasor

$\sim$ an extreme value (maximum or minimum), also used for the Fourier Transform, and in accordance with common usage for $\tilde{D}$ the lattice damping constant

$*$ complex conjugate

$\oplus \; \odot$ field vector or particle motion entering or leaving the plane of the paper

$(\quad)^{\mathrm{T}}$ rounded brackets denote a matrix and T denotes transpose

Tr   trace of a matrix

suffix $_0$ denotes a reference value

bold type   denotes vector or matrix

PI   particular integral

PV   Cauchy principal value of an integral

Re, Im   real, or imaginary, part of a parameter or function

xix

ARABIC SYMBOLS

a	chromatic variable, also used as horizontal beam radius for an elliptical beam [m]
A	magnetic vector potential $[\text{T}\cdot\text{m}]$ or $[\text{V}\cdot\text{s}\cdot\text{m}^{-1}]$
A	transverse acceptance [m], also used as an amplitude function [m], and as a constant
A_l	longitudinal acceptance $[\text{eV}/\text{c}]$ or $[\text{eV}\cdot\text{s}]$
b	chromatic variable, also used as vertical beam radius for an elliptical beam [m]
B	magnetic induction [T], also used unscripted as a constant
$B\rho$	magnetic rigidity $[\text{T}\cdot\text{m}]$
c	speed of light $[\text{m}\cdot\text{s}^{-1}]$
$C = C_q + jC_b$	complex coupling coefficient
$C = 2\pi R$	circumference of an equilibrium orbit [m], also used as capacitance [Farad]
$\mathscr{C}$	'Cosine-like' principal trajectory
$D_y = \dfrac{y}{\Delta p/p_0}$	transverse dispersion function [m]
$\tilde{D}$	the lattice damping constant
$D = (d_{nn})$	diagonal matrix
e	unit electronic charge $[\text{A}\cdot\text{s}]$
E	electric field strength $[\text{V}\cdot\text{m}^{-1}]$
E, E_0	total energy and rest energy of a particle [eV]
f	focal length of a lens [m]
$f(\Phi)$	dipole error distribution $\beta^{3/2}\Delta B/(B_0\rho_0)$ $[\text{m}^{\frac{1}{2}}]$
$f(p)$	beam distribution function in momentum space, $\int f(p)\,\mathrm{d}p = N$, number of particles in the beam
f_{rev}	revolution frequency [Hz]
F	force [N]
F_n, f_n	harmonic coefficients of the closed orbit distortion and the dipole error function $f(\Phi)$
$g = -1/f$	focal strength of a lens $[\text{m}^{-1}]$, also used as half-gap between magnet poles [m]
$G(t)$	wake potential $[\text{V}\cdot\text{A}^{-1}\cdot\text{s}^{-1}]$
h	radio-frequency (rf) harmonic number, also used as half-height of vacuum chamber [m], and height of a ribbon beam [m]
h_{eff}	effective beam height [m]
$\hbar$	Planck's constant normalised by 2π $[\text{J}\cdot\text{s}]$
H	dispersion or lattice invariant [m], also used as the Hamiltonian [J] or [eV], and a constant
i	integer index
I	identity matrix

I, I_w beam current and induced wall current [A]

$j = \sqrt{-1}$ denotes an imaginary quantity

$\boldsymbol{j}$ current density [$A \cdot m^{-2}$]

j integer index

J_x, J_z, J_l partition numbers

$$J = \begin{pmatrix} \alpha & \beta \\ -\gamma & -\alpha \end{pmatrix}$$

$$k = -\frac{1}{B_0\rho_0}\left[\frac{\partial B_z}{\partial x}\right]_0 \quad \text{normalised gradient } [m^{-2}]$$

$$k^{(n)} = \frac{\partial^n k}{\partial x^n} = -\frac{1}{B_0\rho_0}\left[\frac{\partial^{n+1} B_z}{\partial x^{n+1}}\right]_0 \quad n\text{th order normalised gradient } [m^{-n-2}]$$

$$k_s = -\frac{1}{B_0\rho_0}\left[\frac{\partial B_x}{\partial x}\right]_0 \quad \text{normalised skew gradient } [m^{-2}]$$

$k_n = -[K(\Delta p/p_0) - K(0)]/\Delta p/p_0$ normalised chromatic error [m^{-2}]

$K(s)$ general focusing constant (equivalent to the spring constant for simple harmonic motion) [m^{-2}]

$$K_s = -\frac{R^2}{2B_0\rho_0}\left[\frac{\partial B_x}{\partial x} - \frac{\partial B_z}{\partial z}\right]_0 \quad \text{generalised skew gradient}$$

l length of an element [m]

L half length of a FODO lattice cell [m], also used as the full length of an insertion or period [m], inductance [Henry], luminosity [$cm^{-2} s^{-1}$], and the Lagrangian [J] or [eV]

m, m_0 mass and rest mass of a particle [kg], also used as an integer index

$$M(\phi) = \frac{B_s(\phi)}{B_0\rho_0} \quad \text{normalised longitudinal field } [m^{-1}]$$

$$n = -\frac{\rho_0}{B_0}\left[\frac{\partial B_z}{\partial x}\right]_0 = k\rho_0{}^2 \quad \text{field index}$$

n also used as integer index

$n(u)$ number of quanta with energy u

N order of a non-linear resonance, also number of cells in a dispersion suppressor, also $(N + 1)$ is number of sextupole families, and number of particles in a beam

N_c number of cells in a full machine

N_γ number of quanta emitted

$\boldsymbol{p}$ kinetic momentum [eV/c]

P order of azimuthal harmonic of a field error

$\boldsymbol{P}$ generalised momentum used in Hamiltonian

$\boldsymbol{P} = (p_{mn})$ also used as a beam transfer matrix

P_γ total power radiated by a point charge [$J \cdot s^{-1}$]

q	fractional part of the tune Q
$\boldsymbol{q}$	position vector used in the Hamiltonian
$q = ne$	multiple electronic charge [As]
Q	number of betatron oscillations per revolution, also used as the quality factor of a cavity
$\boldsymbol{Q} = (q_{mn})$	beam transfer matrix
r_e, r_p	classical radii of electron and proton respectively [m]
$R = C/2\pi$	average machine radius [m]
R_m, R_{uv}	modulation of the squares of the betatron amplitude functions with coupling
R_s	shunt impedance [Ω]
s	distance along central orbit [m]
$\boldsymbol{S}$	generalised angular momentum [kg·m^2·s^{-1}], also used as similarity transform matrix
S	area enclosed by equilibrium orbit [m^2] also used for area enclosed by a single-particle, phase-space trajectory [m]
$\mathscr{S}$	'Sine-like' principal trajectory
$\boldsymbol{T} = (t_{mn})$	beam transfer matrix
t	time [s]
T	interchange period for betatron coupling [s]
T_0	revolution period of synchronous particle [s]
u, u_c	photon energy and critical photon energy [J] or [eV]
u	rf cavity voltage [V]
U, V	'nearly' horizontal and vertical normal modes
U	generalised position and velocity dependent potential [V]
U_γ	energy radiated by a particle during one turn [J] or [eV]
$\boldsymbol{v}$	velocity [m·s^{-1}]
V, V_w	voltage and induced wall voltage [V]
w	wake potential [V·A^{-1}·s^{-1}],
	also used as width of a ribbon beam [m]
$\boldsymbol{w} = b + ja$	chromatic vector
$W = \displaystyle\int_{E_0}^{E} \frac{\mathrm{d}E}{\varOmega(E)}$	action variable [kg·m^2·s^{-1}]
x, z, s	curvilinear coordinates, which follow the central orbit [m, m, m] (see Coordinate System)
y	general transverse coordinate replacing x or z [m]
X, Y, Z	general coordinate system [m, m, m], also used as amplitudes
$Y = y/\beta^{1/2}$	normalised betatron excursion
Z	impedance [Ω]
Z_0	impedance of vacuum [Ω]
$Z_\parallel, Z_\perp$	longitudinal and transverse coupling impedances [Ω]

GREEK SYMBOLS

$\alpha_y = -\dfrac{1}{2}\dfrac{d\beta_y}{ds}$ modified derivative of betatron amplitude function*

$\alpha_x, \alpha_z, \alpha_l$ also used for the damping constants in the three degrees of freedom $[s^{-1}]$

$\alpha = \dfrac{\Delta C/C}{\Delta p/p_0}$ momentum compaction function, also used as the beam crossing angle [radian]

β_y transverse betatron amplitude function [m]*

$\beta = v/c$ ratio of particle speed to that of light

$\boldsymbol{\beta}$ ratio of particle velocity to speed of light

$\gamma_y = (1 + \alpha_y{}^2)/\beta_y$ Courant and Snyder beam parameter* $[m^{-1}]$

$\gamma = E/E_0$ ratio of particle's total energy to its rest energy

γ_{tr} γ value at transition energy

Γ propagation coefficient for chromatic variables

δ angular deviation due to a dipole kick [rad], also used for the skin depth [m]

δ the delta function, $\displaystyle\int_{-\infty}^{\infty} \delta(t - t')\, dt = 1$ where $\delta(t - t') = 0$ for all t except $t = t'$; also used to denote a small change in a variable

$\Delta = Q_x - Q_z$ difference of uncoupled tunes

Δ small changes or increments

ε angle of end faces of a dipole with respect to the perpendicular to the equilibrium orbit in the plane of bending [rad]

ε_0 permittivity of vacuum $[F \cdot m^{-1}]$

$\varepsilon_{0,y}$ self-field space-charge coefficient

$\varepsilon_{1,y}, \varepsilon_{2,y}$ image space-charge coefficients for electric and magnetic images for incoherent motion

ε transverse geometrical emittance [m]

$\varepsilon_n = \beta\gamma\varepsilon$ normalised transverse emittance [m]

$\varepsilon_l(\Delta\theta, \Delta p), \varepsilon_l(\Delta\tau, \Delta E)$ longitudinal emittance $[eV/c]$ and $[eV \cdot s]$ respectively

$\eta = \gamma^{-2} - \gamma_{tr}^{-2}$ fractional revolution frequency spread per fractional momentum spread

θ rf phase seen by a particle crossing an rf cavity [rad]

$\Theta = s/R = 2\pi s/C$ average azimuthal angle [rad]

λ line charge density $[A \cdot s \cdot m^{-1}]$

λ_β wavelength of the betatron oscillation [m]

---

* When there is no risk of confusion the Courant and Snyder parameters α, β and γ are often used without suffix.

λ_c critical wavelength [m] or [Å]

$\lambda_\beta = \lambda_\beta/2\pi$ modified betatron wavelength [m]

Λ bunching factor (average current/peak current), also used as the charge in a bunch [A·s], and as a switch between sector ($\Lambda = 1$) and straight ($\Lambda = 0$) dipoles

$\mu_y(s_0) = \displaystyle\int_s^{s_0} \frac{ds}{\beta_y(s)}$ phase advance of the betatron oscillation [rad]

μ_0 betatron phase advance in a lattice period [rad], also used as the permeability of vacuum [H·m^{-1}]

$\xi_{1,y}, \xi_{2,y}$ image space-charge coefficients for the electric and magnetic images for the coherent motion

ρ local bending radius of trajectory [m], also used as resistivity [Ω·m]

σ distance (similar to s) [m], also used as standard deviation of a distribution

$$\sigma = \begin{pmatrix} \beta & -\alpha \\ -\alpha & \gamma \end{pmatrix}$$

Σ propagation coefficient for chromatic variables, also used as cross-section for a particle interaction

τ damping time constant [s], also used for lead or lag of a particle w.r.t. the rf phase [s]

ϕ electric scalar potential [V], or magnetic scalar potential [T·m], also used as a phase angle

$\Phi(s_0) = \displaystyle\int_s^{s_0} \frac{ds}{Q\beta(s)}$ normalised betatron phase advance [rad]

$\chi = (Y^2 + Y'^2)^{\frac{1}{2}}$ an amplitude in normalised coordinates, also used as an integration variable

ψ normalised distribution function, $\displaystyle\int_{-\infty}^{\infty} \psi(x)\,dx = 1$

ω general angular frequency [s^{-1}]

ω_{rf} rf angular frequency [s^{-1}]

ω_s synchrotron oscillation angular frequency [s^{-1}]

ω_c critical angular frequency [s^{-1}]

Ω general revolution angular frequency [s^{-1}], also used for solid angles [steradian]

Ω_c cyclotron angular frequency [s^{-1}]

Where symbols do not conform to the above lists they are defined in the text.

USEFUL CONSTANTS

$\pm 1.602\,177 \times 10^{-19}$ [A·s] e, unit electronic charge (upper sign protons, lower sign electrons)

$2.997\,924\,58 \times 10^{8}$ [m·s^{-1}] c, speed of light in vacuum (exact)

$9.109\,390 \times 10^{-31}$ [kg] m_0, rest mass of electron

$1.672\,623 \times 10^{-27}$ [kg] m_0, rest mass of proton

$0.510\,999$ [MeV]   equiv. energy of rest mass of electron

$938.272\,3$ [MeV]   equiv. energy of rest mass of proton

$2.817\,941 \times 10^{-15}$ [m] r_e, classical radius of electron

$1.534\,699 \times 10^{-18}$ [m] r_p, classical radius of proton $[r_{e,p} = e^2/(4\pi\varepsilon_0 m_0 c^2)]$

$4\pi \times 10^{-7}$ [H·m^{-1}] μ_0, permeability of vacuum

$8.854\,188 \times 10^{-12}$ [F·m^{-1}] ε_0, permittivity of vacuum $[\varepsilon_0 = 1/(\mu_0 c^2)]$

376.730 [Ω] Z_0, impedance of vacuum $[Z_0 = 1/(\varepsilon_0 c)]$

$1.054\,573 \times 10^{-34}$ [J·s] $\hbar$, Planck's constant divided by 2π

CHAPTER 1

Introduction

Nuclear physics research was the birth-place of charged-particle accelerators and for many decades their main 'raison d'être'. This has given them a somewhat specialised and academic image in the eyes of the general public and indeed accelerators and storage rings do provide an extremely rich field for the study of fundamental physics principles. However, this academic image is fast changing as the applications for accelerators become more diversified. They are already well established in radiation therapy, ion implantation and isotope production. Synchrotron light sources form a large and rapidly growing branch of the accelerator family. The spallation neutron source is based on an accelerator and there are many small storage rings for research around the world relying on sophisticated accelerator technologies such as stochastic and electron cooling. In time accelerators may be used for the bulk sterilisation of food and waste products, for the cleaning of exhaust gases from factories, or as the drivers in inertial fusion devices.

During the first third of our century, natural radioactivity furnished the main source of energetic particles for research in atomic physics. Let us mention a famous example. At McGill University, Montreal, Canada, in 1906, Rutherford bombarded a thin mica sheet with alpha particles from a natural radioactive source. He observed occasional scattering, but most of the alpha particles traversed the mica without deviating or damaging the sheet. He continued this work at Manchester University, UK, and in 1911 Geiger and Marsden, under Rutherford's guidance, verified Rutherford's theory for atomic scattering. The RBS technique (Rutherford Back-Scattering) is still a standard technique used at modern particle accelerators. In 1908 Rutherford received the Nobel Prize in chemistry for his investigations into the disintegration of elements and the chemistry of radioactive substances.

Natural sources are limited in energy and intensity and it is not surprising that in 1928, Cockcroft and Walton, encouraged by Rutherford, set about the task of building a particle accelerator (see Section 1.1.1) at the Cavendish Laboratory, UK. By 1932 the apparatus was finished and used to split lithium nuclei with 400 keV protons. This was the first fully man-controlled splitting of the atom. From the measurement

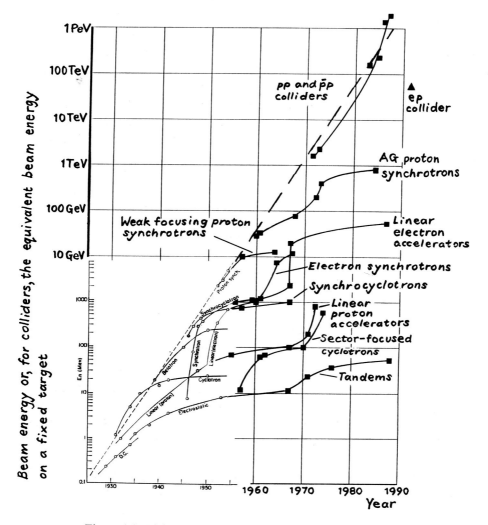

Figure 1.1. Livingston chart (1989). The insert in Figure 1.1 is a reproduction of the original chart first presented by Livingston in his book *High-energy accelerators* published in 1954 by Interscience Publishers Inc. New York (reprinted by permission of John Wiley & Sons, Inc.). It has since been updated by Livingston himself and many other workers, but the original trend-line has proved to be remarkably accurate.

of the binding energy, it provided the first experimental verification of Einstein's mass–energy relationship, known since 1905. They received the Nobel Prize for this work in 1951.

This was the start of the era of particle accelerators as the most important tool for experimental nuclear physics. Since then there has been tremendous progress in the construction of such accelerators with an increase of about one and a half orders of magnitude in beam energy per decade, as illustrated by the updated Livingston chart shown in Figure 1.1. While the energy increased eight orders of magnitude, the cost per GeV of a typical large accelerator installation has been drastically reduced, resulting in only modest inflation-adjusted cost increases of the largest accelerator facilities[1]. Just by looking at Figure 1.1 one notices a very important feature of the growth. The progress of each type of accelerator has saturated fairly quickly, but new ideas have been proposed regularly and have been the main contributors to the rapid advance. Two startling examples stand out: the invention of the alternating gradient focusing in the early fifties and the application of colliding beams in the sixties. In order to satisfy the far-reaching visions of high-energy physicists many more equally outstanding inventions and technological developments will be needed. The Livingston chart is a succinct representation of that challenge.

In this chapter the highlights in the history of accelerators and their operational principles will be briefly reviewed.

1.1 DIRECT-VOLTAGE ACCELERATORS

1.1.1 Cockcroft–Walton rectifier generator[2]

Using a voltage quadrupling circuit of capacitors and diodes powered by a step-up transformer, Cockcroft and Walton built the first proton accelerator (schematically shown in Figure 1.2). Three metal tubes were mounted vertically in an evacuated glass tube about 2 m long. The top tube was held at 400 kV, the middle tube at 200 kV and the bottom tube at earth potential. The protons were injected at the top and accelerated downwards onto the target.

This kind of rectifier generator is still widely used because of the high current it can deliver. In air, however, sparking limits the voltage to just over 1 MV.

1.1.2 Van de Graaff generator[3]

While in Oxford, UK, as a Rhodes scholar in 1928, Van de Graaff became aware of the need for a high-voltage source for nuclear physics research and later in Princeton, USA, he built his first machine, which reached a potential of 1.5 MV.

In this type of machine, the rectifier unit is replaced by an electrostatic charging belt. Corona discharge at the spraycomb shown in Figure 1.3 charges the belt, which then delivers its charge to the collector mounted inside the top terminal. In later versions the sparking threshold was raised by placing the electrical system and the accelerator tube in a high-pressure tank containing dry nitrogen or Freon at about 10 atmospheres.

The Van de Graaff accelerator is a high-precision instrument delivering a stable

3

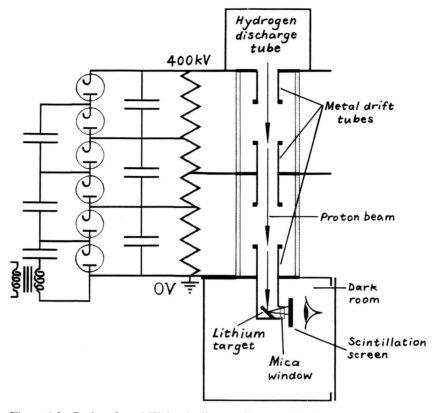

Figure 1.2. Cockcroft and Walton's direct-voltage accelerator.

voltage in the range 1–10 MV and beams of positive or negative ions or electrons with small energy spreads.

1.1.3   Tandem electrostatic accelerator

This is a further development of the Van de Graaff generator. The new feature is to use the electrostatic potential twice over. First an extra electron is attached to a neutral atom to create a negative ion. In recent years, there has been considerable development and it is now possible to obtain negative ion sources for almost all elements. The negative ion beam is injected at ground potential into the tandem (see Figure 1.4). At the high-voltage terminal the beam passes through a thin foil which strips at least two electrons from each ion converting them to positive ions. They are then accelerated a second time back to earth potential. The present-day record is held by the tandem at Oak Ridge, which operates with 24.5 MV on the high-voltage terminal, but a machine called the Vivitron is under construction at Strasbourg with a design voltage of 35 MV.

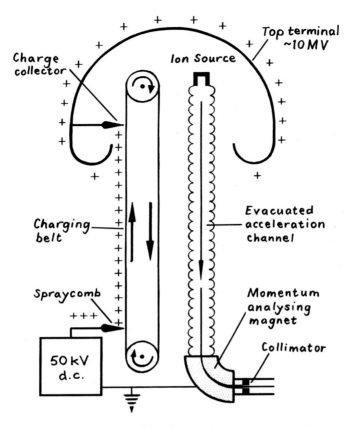

Figure 1.3. Van de Graaff particle accelerator (shown without pressure vessel and insulating support).

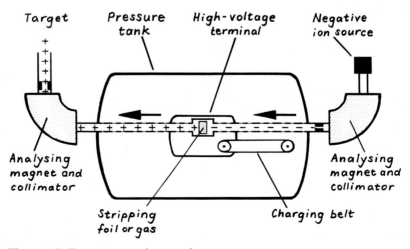

Figure 1.4. Two-stage tandem accelerator.

1.2   ACCELERATORS THAT USE TIME-VARYING FIELDS

Although a direct-voltage accelerator was the first to be exploited for nuclear physics research, an alternative was already being developed. In 1924 in Sweden, Ising proposed the first particle accelerator that would give particles more energy than the maximum voltage in the system[4]. He proposed, in fact, an electron linear accelerator with drift tubes, but did not build one. In 1928, Wideröe built the first linear accelerator[5], which produced 50 keV potassium ions. It used a radiofrequency (rf) voltage of 25 kV from a 1 MHz oscillator applied over two gaps. This was the first practical demonstration of Ising's principle. Jumping ahead a little, Sloan and Lawrence, in 1931, also built a linear accelerator for 1.2 MeV Hg ions[6] which was probably the first linear accelerator to be used for nuclear physics experimentation. Again it was an application of rf acceleration, which is sometimes referred to as *resonance acceleration*. In those days, the resonators were lumped circuits.

1.2.1   Principle of the Wideröe rf linear accelerator

Although the direct-voltage accelerators are geometrically straight, as are the accelerating sections in all accelerators except the betatron, the name linear accelerator or 'linac' is unambiguously applied to linear accelerators with time-varying fields.

The Wideröe-type linac comprises a series of conducting drift tubes in a glass envelope (see Figure 1.5). Alternate drift tubes are connected to the same terminal of an rf generator. The frequency is chosen so that when a particle traverses a gap it sees an electric field in the direction of its motion and by the time this field reverses the particle is shielded inside a drift tube. As the particle gains energy and velocity, the structure periods must be made longer in order to maintain synchronism.

Clearly, as the velocity increases the drift tubes become inconveniently long, unless the frequency can be increased, as was done by Sloan and Lawrence who used a 7 MHz oscillator[6]. At still higher frequencies, the open drift-tube structure becomes lossy, which makes it necessary to enclose the structure to form a cavity, or series

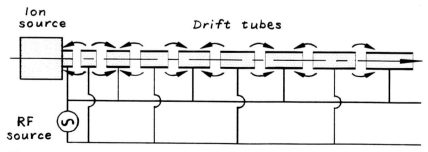

Figure 1.5. Wideröe-type linear accelerator.

of cavities. The underlying principle remains unchanged of course, although there are several variants of the accelerating structure.

1.2.2 Fixed-frequency cyclotron

Technologically the linear accelerator was rather difficult to build and, during the remarkable thirties, it was pushed into the background by a simpler idea conceived by Lawrence in 1929; the fixed-frequency cyclotron[7]. Livingston, who was Lawrence's research student, demonstrated the principle of the cyclotron by accelerating hydrogen ions to 80 keV in 1931. Lawrence's first machine worked in 1932[8] and he also produced nuclear reactions just a few weeks after Cockcroft and Walton. Lawrence's cyclotron was less than a foot in diameter and could accelerate protons to 1.25 MeV. This was quickly followed by $27\frac{1}{2}$ inch and 37 inch versions and by 1939, the year in which Lawrence received the Nobel Prize for his invention, the University of California, USA, had a 60 inch cyclotron (the Crocker cyclotron) capable of delivering 20 MeV protons, twice the energy of the most energetic alpha particles emitted from radioactive sources.

For the invention of the cyclotron Lawrence applied Ising's resonant principle, as demonstrated by Wideröe, but now inside a homogeneous magnetic field so that the particle would be bent back to the same rf gap twice for each rf period (see Figure 1.6).

The resonance condition in the cyclotron is obtained by choosing the rf period equal to the cyclotron period, which is independent of the particle's energy (and radius of its orbit) as long as the particle remains non-relativistic. Thus particles that pass the gap near the peak of the rf voltage will continue to pass near the peak at every half turn, moving in ever-increasing half-circles. With a fixed frequency the particles should continue to spiral until they reach the edge of the magnetic field or until they become relativistic and slip back with respect to the gap voltage. This intrinsic limit was confronted in the late thirties[9] at about 25 MeV for protons and 50 MeV for deuterons and alpha particles.

1.2.3 Synchro-cyclotron or frequency-modulated cyclotron

The remedy for the relativistic limit on the cyclotron was to modulate the applied rf in order to maintain synchronism with the decreasing cyclotron frequency. Although necessary, this is not sufficient by itself, since the natural energy spread in a bunch of relativistic ions causes a spread in their cyclotron frequencies and some longitudinal focusing is required to maintain the bunch intact. This problem was overcome by McMillan[10] and independently by Veksler[11], who both discovered the principle of phase stability in 1944. Phase stability is general to all rf accelerators, except the fixed-frequency cyclotron, and will be discussed in detail in Chapter 7. The effect is that a bunch of charged particles, with an energy spread, can be kept bunched throughout the acceleration cycle by simply injecting them at a suitable phase of

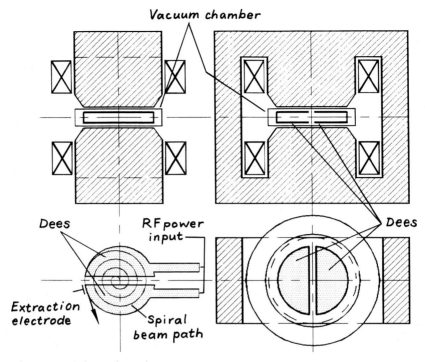

Figure 1.6. Schematic cyclotron.

the rf cycle. The longitudinal focusing effect proved strong enough that the frequency modulation did not have to be specially tailored and could conveniently be made sinusoidal. Synchro-cyclotrons can be used to accelerate protons in the range 100–1000 MeV, a great improvement on the simple cyclotron. The highest energy was reached at Berkeley, USA, where 720 MeV was achieved in 1957 in the '184 inch'. In the same year, the synchro-cyclotron at CERN reached 600 MeV and operated as a fine nuclear physics tool until its closure in December 1990. The international laboratory in Dubna, Russia, also made one of 680 MeV. These higher energies had, however, been obtained at the expense of intensity as the beam in a synchro-cyclotron is pulsed and thus gives far less average intensity than a continuous-beam, or fixed-frequency, cyclotron.

Transverse weak focusing was also understood by this time and incorporated in synchro-cyclotron designs[12]. The focusing of the transverse oscillations was achieved by making the field decrease with radius according to an inverse power law of the radius between zero and unity. This will be discussed more fully in Chapter 2. When the very first fixed-frequency cyclotrons were made, however, accelerator theory was very rudimentary, even close to non-existent. Livingston tells how, at Berkeley, they discovered the really hard way that the magnetic field has to decrease slightly with increasing orbit radius to prevent the particles from getting lost. They shimmed and adjusted the magnet for each small step they took in energy (and radius), thus ending up with a field shape that, years afterwards, theory predicted was necessary for transverse stability[13].

Another incident was the invention by Thomas in 1938 of a cyclotron made of sectors with alternating strong and weak fields arranged in such a way that the rotational frequency again became independent of energy[14]. He even showed that the restoring forces at the edges of the sectors were such that they compensated the defocusing forces of the average field. This was, therefore, a forerunner of alternating gradient focusing for a kind of accelerator that in the fifties, was re-invented as the iso-synchronous cyclotron. The outbreak of World War II was probably the reason why no Thomas-type cyclotron was ever built and after the war it was forgotten with the success of synchronous acceleration.

1.2.4   Betatron

The cyclotron could accelerate protons and heavier ions, but for electrons, which quickly become relativistic, it was useless. In 1923, Wideröe had already conceived an accelerator adapted to electrons. He referred to this as a 'beam transformer' since the circulating beam fulfilled the role of a secondary winding. He had worked out the condition for keeping the radius of the electron orbit constant and in 1925 he also found the condition for the electrons to be stable on this orbit, but neither result was published[15]. This concept was what later became known as the betatron.

It was left to Kerst in 1940 to construct the first betatron, which produced 2.3 MeV electrons[16]. The acceleration in this device is achieved by the electric field induced by the change in magnetic flux going through the circular electron orbit. By that time, the theory of transverse stability of a particle was sufficiently well developed that the guide field was carefully shaped and given a radial gradient in order to provide vertical and horizontal stability.

In 1941, Kerst and Serber published a paper[17] on particle oscillations in the betatron, after which the name 'betatron oscillation' was universally adopted, although these oscillations are common to all devices. Kerst also built the world's largest betatron (300 MeV) in 1950.

The betatron is a robust device, and for this reason it has found lasting applications in laboratories and hospitals. It is shown schematically in Figure 1.7. The magnet is fed with an alternating current at 50 to 200 Hz. The yoke must, of course, be laminated. The magnetic field not only guides the particles in a circular orbit, but, by virtue of its rate of change, it induces a circumferential voltage which accelerates the particles. It is easily shown that the average field, which is inducing the voltage, has to increase at twice the rate of the guide field in order to keep the beam on an orbit of constant radius. This is well known as the 2-to-1 rule (also called the Wideröe condition). Betatrons are limited by saturation in the iron yoke.

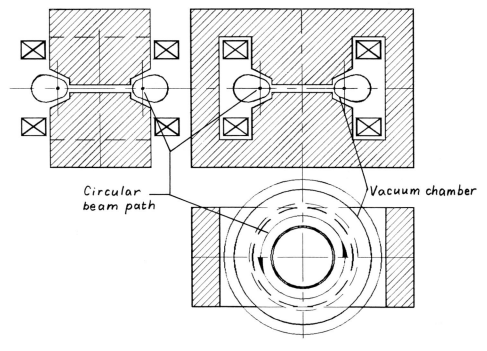

Figure 1.7. Schematic betatron.

1.2.5 Synchrotron

The betatron was soon overshadowed by the synchrotron, which almost put the betatron out of business for nuclear physics research. In connection with the cyclotron we have already mentioned synchronous acceleration invented by McMillan[10] and Veksler[11]. In their original papers they both proposed the electron synchrotron as an application of their principle. The principle, however, is equally valid for protons and Figure 1.8 schematically shows the Cosmotron, a 3 GeV proton synchrotron built at the Brookhaven National Laboratory (BNL), USA. This machine was unusual at that time (1952) in having straight sections. The guide field is similar to that of the betatron, with a bending field to keep the particles on a circular orbit and an appropriate radial gradient to achieve vertical and horizontal stability. Acceleration is by rf voltage operated at the revolution frequency or a higher harmonic. As the particle energy increases, the field is also increased at a rate that makes the particle orbit approximately the same at all energies and the rf frequency is changed in synchronism with the revolution frequency. For synchrotrons this means an upward modulation of the frequency, in contrast to the downward modulation in the synchro-cyclotron.

Several laboratories started at once to build electron synchrotrons. Goward and Barnes in the UK were the first to make a synchrotron work using a converted

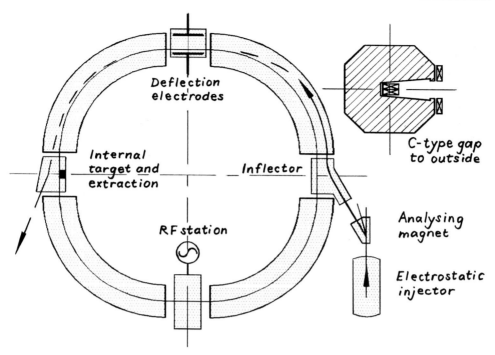

Figure 1.8. Schematic synchrotron.

betatron magnet early in 1946[18]. The synchrotron has been tremendously important for nuclear and elementary particle physics and, ever since its invention, larger and larger machines have been built. The highest energy electron synchrotron built for fixed-target operation is the 12 GeV machine at Cornell, USA. The first analysis of a proton synchrotron was published by Oliphant and Hyde in 1947[19] and the construction of a 1 GeV machine was started soon afterwards in Birmingham, UK. However, in the meantime, this group was overtaken by one at BNL, which completed the 3 GeV Cosmotron in 1952, just one year ahead of the Birmingham group. Many larger machines followed all over the world, the largest now being the 1000 GeV superconducting synchrotron (under R. R. Wilson and later L. M. Lederman) at the Fermi National Accelerator Laboratory (Fermilab), USA and the 500 GeV SPS (under J. B. Adams) at the European Organisation for Nuclear Research (CERN).

In all accelerators whose construction started before 1952, and several later ones, the maintenance of transverse stability depended upon what is now called weak focusing in the magnet system. This will be analysed in Chapter 2. In this case, the guide field decreases slightly with increasing radius in the vicinity of the particle orbit and this gradient is the same all round the circumference of the magnet. However, the tolerance on the gradient is severe and sets a limit to the size of such an accelerator. The aperture needed to contain the beam becomes very large and the magnet correspondingly bulky and costly. In the early fifties this limit was believed to be around 10 GeV.

The invention of the alternating-gradient principle by Christofilos[20], and independently by Courant, Livingston and Snyder in 1952[21], changed the situation completely and made it possible to go up in energy by two orders of magnitude compared with weak focusing machines. In fact, from a technical point of view, machines in the range of 10–100 TeV (1 TeV $\equiv 10^{12}$ eV) seem quite feasible. Cost is more likely to be the limiting factor.

Alternating-gradient focusing is directly analogous to a well-known result in geometrical optics, that the combined focal length of two appropriately spaced lenses of equal strength will be focusing overall when one lens is focusing and the other is defocusing. In fact such a doublet will remain focusing over quite a large range of values for the focal strengths of the two lenses. It will be shown later that quadrupole lenses focus in one plane while defocusing in the orthogonal plane. Thus in a similar way a pair of quadrupole lenses can still be focusing overall in both planes and it can further be shown that within certain limits an infinite series of alternating lenses will also focus. Structures based on this principle are referred to as alternating-gradient (AG) structures.

1.2.6   The linear accelerator revisited

The invention of circular machines took almost all interest away from linear accelerators for some years. However, the advances in ultra-high frequency technology during World War II opened up new possibilities and renewed interest in linac structures.

Again Berkeley was first, with a proton linear accelerator of 32 MeV built by Alvarez in 1946[22]. Alvarez went to the newly developed radar technology and chose the frequency of 200 MHz. This choice made it convenient to enclose the entire drift tube structure in a single resonant cavity, so reducing losses. Since then, this kind of accelerator has become very popular as an injector for large proton and heavy-ion synchrotrons all over the world with injection energies in the range of 50–200 MeV; that is, essentially non-relativistic particles. Alvarez's choice of 200 MHz has also become a *de facto* standard, which was again recently adopted for the HERA project at the Deutsches Elektronen Synchrotron (DESY) at Hamburg in Germany. The largest proton linear accelerator to date is the 800 MeV 'pion factory' (LAMPF) at Los Alamos, USA.

The first electron linear accelerators were studied at Stanford, USA, and at the Massachusetts Institute for Technology (MIT), USA, in 1946[23] and the Telecommunications Research Establishment (TRE), UK, in 1947[24]. This type of accelerator has also had a spectacular development, up to the largest now in operation, the 50 GeV linear accelerator at the Stanford Linear Accelerator Centre

(SLAC). Like betatrons they have become very popular in fields outside nuclear physics, particularly for medicine.

1.2.7 Other accelerators

There are still other accelerators that have not been mentioned, such as the microtron, a kind of electron cyclotron, the principle of which was put forward by Veksler in his original paper on phase stability[11]. A number of these accelerators have been constructed with energies up to about 30 MeV. Although this type of accelerator has not played a large role in high-energy physics, it is frequently found in the injector chains of synchrotron light sources.

The radio-frequency quadrupole (RFQ) is a relatively new type of accelerator suggested in 1970 by Kapchinski and Teplyakov[25]. The proof-of-principle experiment, however, was even more recent and was carried out at Los Alamos in 1979[26]. RFQs are useful at low energies and are rapidly replacing the Cockcroft–Walton generators as pre-injectors for larger accelerators. In this device the same rf field is used for both focusing and accelerating.

1.3   STORAGE RINGS

The next class of accelerator facilities to be briefly described is storage rings. In physics experiments only the centre-of-mass energy is 'useful' in an interaction. When using an accelerator beam on a fixed target only a fraction of the particle's energy is available in the centre-of-mass system, whereas for two equal particles in a head-on collision all of the particles' energy is available. This fundamental drawback of fixed-target accelerators becomes more punitive as the energy increases. The purpose of a storage ring is to make head-on collisions possible with a useful interaction rate.

The history of colliding-beam devices begins in 1956, when the group at the Midwestern Universities Research Association (MURA) in the USA put forward the idea of particle stacking in circular accelerators[27,28]. Of course, people who worked with particle accelerators had speculated earlier about the possibilities of obtaining high centre-of-mass energies with colliding beams, but such ideas were unrealistic with the particle densities then available in normal accelerator beams. The invention of particle stacking changed the picture significantly. This opened up the possibility of making two intense beams collide with sufficiently high interaction rates that experimentation would be feasible in an energy range otherwise unobtainable by known techniques, except at enormous cost.

In electron storage rings, the characteristics of the magnets can be determined in a way such that the synchrotron radiation will cause the beam to shrink in all dimensions and by this means intense electron beams can be accumulated.

In proton rings the synchrotron radiation is so small that it is neither a disturbance nor a help and momentum stacking, as proposed in the original MURA papers, has been used for stacking very high intensities. Later more sophisticated accumulation processes were invented, such as electron cooling and stochastic cooling.

The first storage rings to operate were the 2 × 250 MeV electron ring, Anelli di Accumulazione (AdA) at Frascati, Italy, and the 2 × 500 MeV Princeton–Stanford ring. Both of these rings started operating in 1961. Novosibirsk, Russia, followed shortly afterwards with VEPP-1 in 1963 and later with several others. Orsay, France, made ACO and Frascati constructed ADONE, which was the first storage ring above 1 GeV in each beam. The most successful electron storage rings have been SPEAR at SLAC (1972) and DORIS at DESY (1974), both at around 2 × 4 GeV. These two machines have had a tremendous impact on high-energy physics, first through the discovery of the J/ψ particle at SPEAR* and then through the study of the properties of this new particle at both SPEAR and DORIS. Several larger storage-ring projects have been constructed since: 2 × 15 GeV PEP at SLAC; 2 × 19 GeV PETRA at DESY; 2 × 9 GeV at Cornell and 2 × 33 GeV TRISTAN in Japan. The latest, and at present the largest, is the 2 × 55 GeV e^+e^- collider LEP at CERN. LEP was put into operation in 1989 and after one year reached a luminosity of $\sim 10^{31}$ cm^{-2} s^{-1}. It is planned to upgrade LEP to 2 × 100 GeV.

The first colliding beam facility for protons was the 2 × 31 GeV Intersecting Storage Rings (ISR). The authorisation for construction was given by the CERN Council in 1965, after an enthusiastic promotion by Professor V. Weisskopf (assisted by M. G. N. Hine) during his term as the CERN Director General. The facility started operation in 1971 and improved its performance steadily, and sometimes spectacularly, until it was closed in 1984 to provide some financial relief for LEP. From the performance of this facility and the related accelerator and detector development, a general confidence in the predictability of hadron colliders has emerged and a considerable change in attitude has taken place. Before the ISR, the physics community considered only fixed-target facilities as practical for hadron physics, whereas since the ISR, all plans for future hadron facilities are for colliders like UNK in Russia, LHC at CERN and SSC in USA (see Section 12.2.6 for more details).

A special development in the hadron collider family was the proton–antiproton collider. There were some early speculations of such collisions in the ISR, but the estimates for the luminosity had always been discouragingly low. With the invention of stochastic cooling by van der Meer[29] at CERN, it became possible to reduce the beam dimensions and thus increase the beam density. This is particularly effective on low-intensity beams, which led CERN to adopt this technique to implement the proposal made by Rubbia et al.[30] to accumulate antiprotons over long periods, then to accelerate them to 270 GeV in the SPS and to make them collide with protons of the same energy. This became a spectacular success both from an accelerator point

* At the same time, Ting and his group at BNL discovered the J/ψ and the 1976 Nobel Prize in physics was shared between Ting and Richter from SLAC.

of view and as a physics tool. The W and Z particles were produced and identified in 1982/83 and as a result Rubbia and van der Meer shared the Nobel Prize in 1984. A similar facility is now running at Fermilab. Commissioning of this facility started in summer 1985, the first 2×800 GeV events were seen on 14 October 1985 and it has since operated routinely at 2×900 GeV.

The HERA machine at DESY is an important deviation from the mainstream of colliders as it is constructed for colliding electrons against protons. The electron ring will reach 30 GeV with superconducting cavities while the superconducting proton ring is designed for 820 GeV, but will probably reach 1 TeV after a running-in period. This facility was first operated as a collider in 1991.

For hadron rings, there has been no severe technological limitations to-date, although financial limits are more and more acutely felt. At the very highest energies talked about nowadays the synchrotron radiation is no longer negligible and puts a bothersome heat load on the superconducting magnet system. There may also be limits related to the experimental possibilities. As one goes up in energy the cross-sections for interesting events decrease sharply, but not the total cross-section. This means that detectors may be swamped by uninteresting events that may make it impossible to single out the few important ones.

1.4   LINEAR COLLIDERS

In general, particle physicists prefer electron colliders to proton colliders, but the electron–positron storage ring at high energies is crippled by synchrotron radiation losses. In 1965, Tigner suggested that electron linear colliders would be more cost effective at very high energies[31]. The principle has been tested using the SLAC linac by deflecting consecutive bunches of electrons and positrons via opposing arcs into head-on collision. This project, called the Stanford Linear Collider (SLC), started operation for physics in 1989[32] and is the forerunner of a possible new generation of accelerators. However, the large linear colliders of the future will need a new technology for their linacs and the search is now on for methods of increasing acceleration gradients in order to reduce size and cost. There are many new ideas such as superconducting accelerating structures, the wakefield accelerator, the laser-plasma beat-wave accelerator, the switch-mode linac and many others which are just starting to be tested in various laboratories around the world. At present a promising approach is a 'semi-conventional' disc-loaded waveguide run at extremely high frequencies, say 10–30 GHz.

1.5   CONCLUDING REMARKS

This has been a brief review of the history of particle accelerators, which hopefully has transmitted some of the vitality of this field. The development has been spectacular with, on the average, an increase of one and a half orders of magnitude in energy per decade since the 1 MeV barrier was breached in the early thirties until the Fermilab collider came into operation in 1985. The development has been equally spectacular in the sophistication of these machines, in their technological demands and in the experimental particle physics results obtained. The knowledge in sub-nuclear physics has increased quite in step with the accelerators that have been put at the disposal of the experimentalists. Some may feel that this knowledge has not simplified our picture of nature, but that is another matter and one that makes this field perhaps even more fascinating.

The diversification into industrial, medical and general research applications has also increased and is now making accelerator physics a more 'complete' field of activity.

Since the subject of accelerators is an extremely wide one, this book will mainly concentrate on aspects related to circular machines and, in particular, synchrotrons and storage rings. However, it is believed that the knowledge thus gained will nevertheless be useful and a good starting point for those who also wish to understand the behaviour of other types of accelerators.

CHAPTER 2

Basic concepts and constant-gradient focusing

This chapter introduces the important concepts of

- betatron oscillations
- transverse stability
- adiabatic damping
- acceptance/emittance, and
- momentum compaction

by using the example of *constant-gradient focusing*, or *weak focusing* as it is also known. This approach is simple, yet fairly rigorous, while also having the advantage of respecting the historical order of events. The equations of motion that will be developed for a constant-gradient field will be used directly in the next chapter as the 'building blocks' for the *alternating-gradient*, or *strong-focusing*, *theory*.

Constant-gradient focusing was rapidly abandoned in favour of *alternating-gradient focusing*, after the latter was published in 1952[1]. This transition was rapid, since weak focusing had few advantages over its successor. Although weak focusing may now appear as something of an anachronism, there are in fact still some examples to be found among small machines.

2.1   CYCLOTRON MOTION

The force $\boldsymbol{F}$ acting on a charged particle in a magnetic field has the general form, $\boldsymbol{F} = e(\boldsymbol{v} \times \boldsymbol{B})$, so that the rate of change of momentum of the particle is given by,

$$\frac{\mathrm{d}\boldsymbol{p}}{\mathrm{d}t} = \frac{\mathrm{d}}{\mathrm{d}t}(m\boldsymbol{v}) = e(\boldsymbol{v} \times \boldsymbol{B}). \tag{2.1}$$

The mass has been kept inside the differential, so that the equation is still rigorous for relativistic mass changes.

The well-known relationships for the three components of this force in cylindrical co-ordinates (ρ, Θ, z) for a uniform magnetic field B_0 aligned with the z-axis are quoted below (see also Figure 2.1).

$$F_r = \frac{\mathrm{d}}{\mathrm{d}t}(m\dot{\rho}) - m\rho\dot{\Theta}^2 = e\rho\dot{\Theta}B_0 \tag{2.2}$$

$$F_\Theta = \frac{1}{\rho}\frac{\mathrm{d}}{\mathrm{d}t}(m\rho^2\dot{\Theta}) = -e\dot{\rho}B_0 \tag{2.3}$$

$$F_z = \frac{\mathrm{d}}{\mathrm{d}t}(m\dot{z}) = 0. \tag{2.4}$$

Apart from rectilinear motion parallel to the magnetic field, the simplest solution to (2.2), (2.3) and (2.4) is a circular motion perpendicular to the magnetic field (see Figure 2.1), which is obtained by putting $\rho = \rho_0 =$ constant, so that from (2.2),

$$\dot{\Theta} = -\frac{e}{m}B_0 = \Omega_c. \tag{2.5}$$

Equation (2.5) gives the angular revolution frequency, which is called the *cyclotron frequency*, Ω_c. At a given field level, this frequency is constant for non-relativistic particles. For a positive charge rotating in an anticlockwise (i.e. positive) direction the field is downwards (i.e. negative). If the integration of (2.4) yields a finite velocity along the z-axis, then the circular motion is stretched into a spiral. Without focusing, therefore, the smallest vertical angle error in a cyclotron would cause the beam to spiral up or down into the poles.

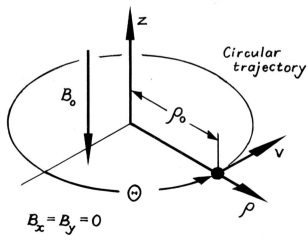

Figure 2.1. Cylindrical coordinates and cyclotron motion.

Replacing $\dot{\Theta}$ by v_0/ρ_0 in (2.5) and re-arranging gives,

$$\frac{mv_0}{e} = \frac{p}{e} = -B_0\rho_0. \tag{2.6}$$

This leads directly to the widely used and well-known numerical expression which relates the radius of gyration to the momentum,

$$|B[\text{T}]\rho[\text{m}]| = (10^9/c)p[\text{GeV}/c] \simeq 3.3356p[\text{GeV}/c]. \tag{2.7}$$

$|B\rho|$ is known as the *magnetic rigidity* and is a property of the beam; it is a measure of its 'stiffness' in the bending field.

If the relativistic mass increase with energy is taken into account,

$$m = m_0(1 - \beta^2)^{-1/2}, \tag{2.8}$$

where β is the relativistic variable v/c, then it can be seen that in the synchrocyclotron the revolution frequency decreases with the increasing energy and the frequency of the accelerating voltage must also be correspondingly decreased.

2.2   TRANSVERSE MOTION

First take a very simple example of a circular accelerator with a magnet gap in which the field does not change azimuthally, i.e.

$$\partial/\partial\Theta = 0$$

and in which the field configuration is symmetric about the $z = 0$ plane, known as the *median plane*. Then choose a circle of constant radius ρ_0 in the median plane with the constant field B_0 on this circle. A particle of momentum p_0 given by (2.6),

$$p_0 = -eB_0\rho_0 \tag{2.9}$$

will then follow this circular orbit if injected tangentially onto it. This orbit is called the *equilibrium orbit* for p_0, often also called the *central orbit* or the *closed orbit**. It is important to study the motion of particles deviating slightly from this orbit to see if the system is stable, i.e. focusing (see Figure 2.2).

2.2.1   Radial motion

The equation for the radial motion comes directly from (2.2) by replacing $\dot{\Theta}$ by v_0/ρ. This approximation assumes that the deviations from a circular orbit are small and

---

* Equilibrium orbit is generally the most correct and is equally applicable to circular machines, transfer lines and segments of structures. Central orbit and closed orbit refer to secondary properties of the orbit, but they are however the most often used terms.

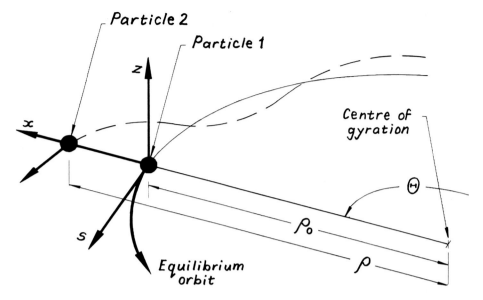

Figure 2.2. Motion of a particle close to the equilibrium orbit.

effectively makes the magnetic deflection look like a central force, which is balanced by the radial acceleration, so that

$$F_r = \frac{d}{dt}\left[m\frac{d\rho}{dt}\right] - \frac{mv_0{}^2}{\rho} = ev_0B_0. \tag{2.10}$$

The final equations are of more practical use if the independent variable t is changed to s, the distance measured along the trajectory,

$$\frac{d}{dt} \equiv v_0\frac{d}{ds}. \tag{2.11}$$

The introduction of the deviations,

$$\Delta B_z = (B_z - B_0) \tag{2.12}$$

and

$$x = (\rho - \rho_0) \tag{2.13}$$

with the assumption that $x \ll \rho_0$, which is always a good approximation, these expressions give with (2.10),

$$\frac{1}{mv_0}\frac{d}{ds}\left[mv_0\frac{d}{ds}(\rho_0 + x)\right] - \frac{1}{(\rho_0 + x)} = \frac{e}{mv_0}(B_0 + \Delta B_z)$$

which reduces to,

$$\frac{1}{\rho_0}\frac{d}{ds}\left[\rho_0\frac{dx}{ds}\right] + \frac{x}{\rho_0{}^2} + \frac{1}{\rho_0}\frac{\Delta B_z}{B_0} = 0. \tag{2.14}$$

By the use of the first two terms of the Taylor expansion for the magnetic field, i.e.

$$B_z = B_0 + \left[\frac{\partial B_z}{\partial x}\right]_0 x + \cdots, \tag{2.15}$$

the equation is linearised and by the introduction of the *normalised gradient k*, defined as

$$k = -\frac{1}{B_0 \rho_0}\left[\frac{\partial B_z}{\partial x}\right]_0, \tag{2.16}$$

the following equation is obtained for small radial deviations in a linear field,

$$\frac{1}{p_0}\frac{\mathrm{d}}{\mathrm{d}s}\left[p_0 \frac{\mathrm{d}x}{\mathrm{d}s}\right] + (\rho_0^{-2} - k)x = 0. \tag{2.17}$$

Some readers may be perturbed by the lack of rigour in this derivation so they are asked to turn to Appendix A where the problem is further discussed. However, the resulting equation is valid and sufficient for most purposes.

2.2.2 Vertical motion

The vertical motion is found in an analogous way from the equation of motion in the vertical plane, i.e.

$$F_z = \frac{\mathrm{d}}{\mathrm{d}t}\left[m \frac{\mathrm{d}z}{\mathrm{d}t}\right] = ev_0 B_x. \tag{2.18}$$

B_x will be zero on the median plane, since the field configuration was earlier restricted to being symmetric. In addition, $(\partial B_x/\partial z)_0$ will equal $(\partial B_z/\partial x)_0$, since $\nabla \times \boldsymbol{B} = 0$.

From the first term in the Taylor expansion, B_x is given by

$$B_x = -zB_0 \rho_0 k. \tag{2.19}$$

The combination of (2.18) and (2.19) then yields,

$$\frac{1}{p_0}\frac{\mathrm{d}}{\mathrm{d}s}\left[p_0 \frac{\mathrm{d}z}{\mathrm{d}s}\right] + kz = 0. \tag{2.20}$$

2.3 SOLUTIONS

The radial and vertical motions (2.17) and (2.20) can be represented by one simple expression valid for both planes,

$$\frac{1}{p_0}\frac{\mathrm{d}}{\mathrm{d}s}\left[p_0 \frac{\mathrm{d}y}{\mathrm{d}s}\right] + Ky = 0 \tag{2.21}$$

where, for the horizontal motion,

$$y \equiv x \quad \text{and} \quad K = K_x = (\rho_0^{-2} - k) \tag{2.22}$$

and, for the vertical motion,

$$y \equiv z \quad \text{and} \quad K = K_z = k. \tag{2.23}$$

The coefficients in (2.21) change slowly with time and the adiabatic (or WKB) solution can be applied. The solution can be written as

$$y(s_0) \simeq H(p_0\sqrt{K})^{-\frac{1}{2}} \cos\left[\int_0^{s_0} \sqrt{K}\, ds + \psi\right] \tag{2.24}$$

where H and ψ are integration constants.

Equation (2.24) is valid for weak focusing cyclotrons, betatrons and synchrotrons. For the last two, ρ_0 and k are kept constant during acceleration and the solution can be simplified to give,

$$y(s) \simeq H_0 p_0^{-\frac{1}{2}} \cos(\sqrt{K}s + \psi). \tag{2.25}$$

2.4 STABILITY

Stability can be secured only if the solutions are oscillatory, i.e. K must be positive for both planes:

Horizontally, K_x is positive if $(\rho_0^{-2} - k) > 0$ or $k\rho_0^2 < 1$,

Vertically, K_z is positive if $k\rho_0^2 > 0$.

This means there are oscillatory solutions for both planes if,

$$0 < k\rho_0^2 < 1 \tag{2.26}$$

or in a slightly different form

$$0 < -\frac{1}{B_0}\left[\frac{\partial B_z}{\partial x}\right]_0 < \frac{1}{\rho_0}. \tag{2.27}$$

It should be noted that overall stability is achieved by a small overlap of the stability regions of the two planes and this overlapping is entirely due to a focusing effect in the horizontal plane arising from the dipole field (see also Section 3.1.1). Since this means working near the limit of stability in both planes, the focusing is weak and this is why such focusing is called *weak focusing*.

In the second formulation of the stability criterion (2.27), it can be seen that as the size ρ_0 of the machine increases the fractional field gradient is reduced and hence the mechanical tolerance on the shape of the magnet pole becomes tighter and tighter. This sets a practical limit on the size of constant-gradient machines.

In the region of oscillatory motion it can be seen from the solutions that the wavelength of the oscillation is given by,

$$\lambda_\beta = 2\pi/\sqrt{K} \tag{2.28}$$

with K as given by the expressions (2.22) and (2.23). This oscillation is for historical reasons called the *betatron oscillation* and the wavelength is called the *betatron wavelength*. The number of oscillations per revolution is called the *betatron tune*,

$$Q = 2\pi\rho_0/\lambda_\beta. \tag{2.29}$$

It can be seen from (2.27) and (2.28) that in a weak focusing machine $Q < 1$. The amplitude of the betatron oscillation in (2.24) is

$$\hat{y} \propto (\lambda_\beta/p_0)^{\frac{1}{2}}, \tag{2.30}$$

which in the case of constant λ_β (i.e. betatrons and synchrotrons) is

$$\hat{y} \propto 1/\sqrt{p_0}. \tag{2.31}$$

Thus the amplitude decreases with increasing momentum. This important property is called *adiabatic damping*.

2.5   ACCEPTANCE AND EMITTANCE

From the general equation (2.21), the solution for the slope of the particle orbit can be found to the same approximation, so that

$$\frac{dy}{ds} \simeq -H\left[\frac{\sqrt{K}}{p_0}\right]^{\frac{1}{2}} \sin\left[\int_0^{s_0} \sqrt{K}\,ds + \psi\right]. \tag{2.32}$$

The combination of (2.24) and (2.32) gives,

$$y^2/\lambda_\beta + y'^2\lambda_\beta = H^2/p_0 \tag{2.33}$$

where for convenience and symmetry the modified wavelength has been introduced,

$$\lambda_\beta = \lambda_\beta/2\pi = 1/\sqrt{K}. \tag{2.34}$$

This means that in (y, y') space the particles trace elliptical orbits (see Figure 2.3).

The area of this ellipse is $\pi H^2/p_0$ and the ellipticity is λ_β. In a given machine there is a maximum possible amplitude a (e.g. given by the vacuum chamber or other restrictions). The *area* of the corresponding ellipse is called the *acceptance*,

$$A = \pi a^2/\lambda_\beta. \tag{2.35}$$

It should be noted that the acceptance is a property of the accelerator.

The particles in the beam oscillate with different amplitudes and phases, but they

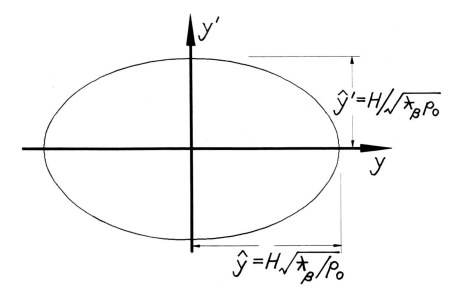

Figure 2.3. Phase-space trajectory.

all follow ellipses of the same form centred on the same point. The particle density across the aperture (i.e. the projection of the phase space population onto the y-axis) will, in most cases, be close to gaussian and it is convenient to take the ellipse corresponding to a particle oscillating with an amplitude equal to the standard deviation of this distribution to characterise the beam. The area of this 'characteristic' ellipse is a measure of the beam size and is defined as the *beam emittance*,

$$\mathcal{E} = \pi \hat{y}^2 / \lambda_\beta, \tag{2.36}$$

where $\hat{y}$ is made equal to the standard deviation of the particle density distribution. Taking into account (2.31) it is clear that

$$\mathcal{E} \propto p_0^{-1},$$

which means that the emittance diminishes with increasing momentum (i.e. the beam shrinks away from the vacuum chamber).

Instead of y', the variable for the corresponding transverse momentum could have been chosen as,

$$p_y = p_0 \frac{\mathrm{d}y}{\mathrm{d}s}. \tag{2.37}$$

In this case an invariant emittance is obtained, which is very often written in the normalised form as,

$$\mathcal{E}_n = \beta\gamma\mathcal{E} = \text{invariant}. \tag{2.38}$$

Correspondingly for the acceptance,

$$A_n = \beta\gamma A \propto p_0. \tag{2.39}$$

One of the practical consequences of the expressions for acceptance and emittance is that it is desirable to choose a high energy for the injection into the accelerator. Depending on the mode of operation, the acceptance will need to be larger or very much larger than the emittance. Transfer channels may operate with acceptances of only a few standard deviations, where colliding beam machines such as the CERN ISR required ± 10 standard deviations in the interaction regions to suppress the background from particle losses from the beam tails.

In the above, one standard deviation has been used for the definition of emittance. Unfortunately the experts do not all agree on this value, nor how the factor π should be treated. Electron machine designers usually use *one standard deviation*, while proton machine designers often use *two*, which results in a factor of four difference in the numerical values. The factor π is usually written separately but it may also be included in the numerical value or even omitted entirely. Be prepared for all combinations!

2.6 MOMENTUM COMPACTION

So far only deviations in position and slope of the orbits for particles of the same momentum p_0 have been discussed. It is also important to look at the effect on the equilibrium orbit for a particle with a deviation Δp in its momentum, assuming the same simple magnetic field satisfying $\partial/\partial\Theta = 0$. A very simple relation comes from (2.9), i.e.

$$\frac{\Delta B}{B_0} = \frac{\Delta p}{p_0} - \frac{\Delta\rho}{\rho_0}. \tag{2.40}$$

The orbit separation (see Figure 2.4) is characterised by the *momentum compaction*, α which is defined as,

$$\alpha = \frac{\Delta\rho}{\rho_0}\bigg/\frac{\Delta p}{p_0} = \frac{\Delta C}{C_0}\bigg/\frac{\Delta p}{p_0} \tag{2.41}*$$

and gives, with (2.40),

$$\frac{\Delta B}{B_0} = \frac{\Delta\rho}{\rho_0}\left(\frac{1}{\alpha} - 1\right). \tag{2.42}$$

By making use of (2.16) it follows that

---

* The ρ and C forms of the definition of α are fully equivalent in a constant-gradient machine, but the use of C allows the definition to be extended in a consistent way to machines with straight sections. This second definition will be used in later chapters.

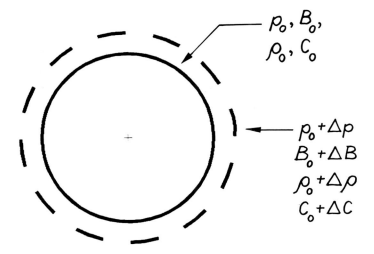

Figure 2.4. Orbit separation in a constant-gradient machine.

$$\alpha = 1/(1 - \rho_0^2 k) \qquad (2.43)$$

and from (2.22) and (2.29) that

$$\alpha = Q_x^{-2}. \qquad (2.44)$$

In the alternating-gradient systems discussed in later chapters it will be more difficult to calculate α, but somewhat surprisingly (2.44) will still prove to be a useful approximation, unless the designer of the lattice has made rather special efforts to ensure the contrary.

2.7 HISTORICAL NOTE

As a field of study develops it often happens that new variables are adopted, which have a simple, fixed relationship to earlier ones. Many readers will know that a variable n called the *field index* and defined as,

$$n = -\frac{\rho_0}{B_0}\left[\frac{\partial B_z}{\partial x}\right]_0, \qquad (2.45)$$

was used in the early literature in place of the normalised gradient k given in (2.16). The minus sign in the definition of n was introduced as a convention to make the n positive in a constant-gradient machine[2]. The constant-gradient focusing stability criterion (2.26) was then very simple and was expressed as,

$$0 < n < 1. \qquad (2.46)$$

This old sign convention has been carried over into the definition of k. The field index had the desirable property of being dimensionless, which k does not, but both variables have the important property of being normalised by momentum.

As was shown in Section 2.4, the tunes in a constant-gradient machine are less than unity. As different machine designs were developed, this characteristic was adopted as a general definition for weak focusing and it was applied independently of how the focusing was achieved. The largest weak-focusing machine to be built used a uniform dipole field and edge focusing. This was the 12 GeV Zero Gradient Synchrotron (ZGS) at the Argonne National Laboratory (ANL), USA. Towards the end of its life, it exploited the one advantage that weak focusing has, by accelerating polarised protons. The absence of depolarising resonances in a weak focusing structure made the ZGS an excellent source[3].

CHAPTER 3

Alternating-gradient focusing

Machine designers now think almost exclusively in terms of *alternating-gradient focusing*, or *strong focusing* as it is also known. Their conversion from weak to strong focusing was rapid and decisive following the publication of Ref. 1 by Courant, Livingston and Snyder from the Brookhaven National Laboratory in 1952. CERN, for example, immediately abandoned its already-approved project for a 10 GeV/c weak-focusing synchrotron in favour of a 25 GeV/c strong-focusing machine, which it was estimated could be built for the same price. Strong focusing had broken through a cost–size–tolerance barrier. Both the betatron amplitude and the momentum compaction functions are compressed and the required aperture is typically reduced from tens of centimetres to centimetres. The momentum compaction is the more strongly affected, but in accelerators the momentum spread is usually small and the betatron amplitude reduction dominates. The stronger gradients alleviate the tolerance problem and the alternating structure leads naturally to a modular design for the lattice. It took several years, however, before this latter point was fully exploited for special optics modules.

Alternating-gradient focusing was an attractive idea, but it was not entirely new. It is based on a long-known result in classical optics and, in fact, an American-born engineer, Christofilos, living in Athens, had already filed a USA patent on the same scheme in 1950[2].

3.1   A SEGMENT OF A MAGNET AS A FOCUSING ELEMENT

Instead of considering the whole circumference of a machine, consider only a segment, which is normally relatively short, and over this segment assume that the gradient is constant as before so that the equation of motion (2.21) will still be valid within the segment. In order to simplify the analysis (2.21) is re-written with acceleration disregarded (i.e. p_0 is constant) so that,

$$\frac{\mathrm{d}^2 y}{\mathrm{d}s^2} + K(s)y = 0. \tag{3.1}*$$

Note that in (3.1) K is shown with an s-dependence, so as to take account of changes in K from one segment to the next, i.e. from one element in the accelerator lattice to the next. This means that (3.1) can be applied to the whole machine and is therefore more general than (2.21) from which it is derived. When $K(s)$ is positive the element is focusing, whereas if it is negative the element is defocusing. The overall stability of a machine made from a series of focusing and defocusing elements will be investigated a little later. The solution for transmission through a single segment or element is the well-known solution to the equation of simple harmonic motion and is most conveniently written in matrix form as,

$$\begin{pmatrix} y \\ y' \end{pmatrix}_2 = \begin{pmatrix} \cos(K^{\frac{1}{2}}l) & K^{-\frac{1}{2}}\sin(K^{\frac{1}{2}}l) \\ -K^{\frac{1}{2}}\sin(K^{\frac{1}{2}}l) & \cos(K^{\frac{1}{2}}l) \end{pmatrix} \begin{pmatrix} y \\ y' \end{pmatrix}_1. \tag{3.2}$$

This formulation is quite general and includes the case of negative K, in which the '$\sin(\mathrm{i}x)$' terms transform to '$\mathrm{i}\sinh(x)$' and the '$\cos(\mathrm{i}x)$' terms to '$\cosh(x)$'. However, mainly for the ease of computer programming, it is usual to write down the focusing and defocusing solutions separately as given below,

$K > 0$ Focusing:

$$\begin{pmatrix} y \\ y' \end{pmatrix}_2 = \begin{pmatrix} \cos(|K|^{\frac{1}{2}}l) & |K|^{-\frac{1}{2}}\sin(|K|^{\frac{1}{2}}l) \\ -|K|^{\frac{1}{2}}\sin(|K|^{\frac{1}{2}}l) & \cos(|K|^{\frac{1}{2}}l) \end{pmatrix} \begin{pmatrix} y \\ y' \end{pmatrix}_1 \tag{3.3}$$

$K < 0$ Defocusing:

$$\begin{pmatrix} y \\ y' \end{pmatrix}_2 = \begin{pmatrix} \cosh(|K|^{\frac{1}{2}}l) & |K|^{-\frac{1}{2}}\sinh(|K|^{\frac{1}{2}}l) \\ |K|^{\frac{1}{2}}\sinh(|K|^{\frac{1}{2}}l) & \cosh(|K|^{\frac{1}{2}}l) \end{pmatrix} \begin{pmatrix} y \\ y' \end{pmatrix}_1 \tag{3.4}$$

$K = 0$ Drift space:

$$\begin{pmatrix} y \\ y' \end{pmatrix}_2 = \begin{pmatrix} 1 & l \\ 0 & 1 \end{pmatrix} \begin{pmatrix} y \\ y' \end{pmatrix}_1 \tag{3.5}$$

where $l = (s_2 - s_1)$ the length of the segment.

It should be noted that the above transfer matrices always have unit determinant. When there is a series of elements, the transfer matrix for the whole system is obtained by multiplying the individual matrices together in beam order and the resultant matrix will also have unit determinant. This fact can be used as a useful check when doing calculations.

Permissible forms of $K(s)$ are given in Table 3.1 using the definitions (2.22) and

---

* When $K(s)$ is periodic, equation (3.1) becomes of a type known as Hill's equation.

Table 3.1. *Forms of K(s)*

Magnet	K_x	K_z
Combined-function[a]:		
quadrupole plus vertical dipole field	$\rho_x^{-2} - k$	k
Combined-function:		
quadrupole plus horizontal dipole field	$-k$	$\rho_z^{-2} + k$
Pure quadrupole	$-k$	k
Vertical field dipole	ρ_x^{-2}	0
Horizontal field dipole	0	ρ_z^{-2}
Field-free drift space	0	0

[a] The first strong-focusing machines, like their constant-gradient antecedents, used hybrid dipole–quadrupole magnets, or *combined-function magnets* as they are known. Pure dipole and multipole lenses, or *separated-function magnets*, are now used almost exclusively.

(2.23) for K and (2.16) for k. In general, no restriction need be put on the magnitude of the normalised field gradient k. In all but very small machines, the $1/\rho^2$ term is less important than the gradient term and is frequently omitted in approximate calculations. Thus K and k are very similar, but have *opposite* signs for the horizontal motion; K_x is positive when the motion is stable (i.e. focusing) while k is negative. k has this sign for historic reasons (see Section 2.7).

3.1.1 Relation between vertical and horizontal focusing

Maxwell's equations state that in the current-free aperture of a magnet,

$$\mathbf{\nabla} \times \mathbf{B} = 0.$$

For a purely transverse field this reduces to,

$$\frac{\partial B_z}{\partial x} = \frac{\partial B_x}{\partial z}.$$

Figure 3.1 shows the special case of a pure quadrupole, where it can be seen from the symmetry and field pattern that this relation is true. Inspection of Figure 3.1 will also reveal that the forces on a moving charge will be towards the axis (i.e. focusing) in one plane, while being away from the axis (i.e. defocusing) in the other plane. Since the above formulation is quite general, it follows that *a transverse field that is focusing in one plane is defocusing in the other*. Common usage has defined a lens that focuses horizontally as a *focusing or F-type lens* and one that focuses vertically as a *defocusing or D-type lens*.

Table 3.1 shows this anti-symmetric focusing effect clearly for the pure-quadrupole magnet where K_x and K_z have opposite signs, but the combined-function magnet

(a)

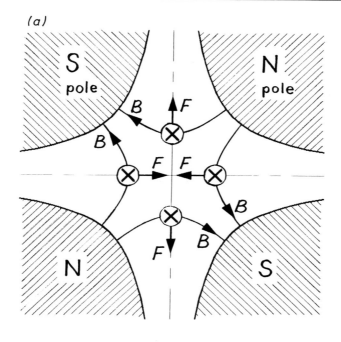

(b)

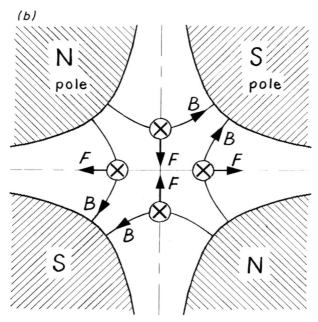

Figure 3.1. Fields and forces inside a quadrupole. (*a*) Focusing or F-type quadrupole, (*b*) defocusing or D-type quadrupole (drawn for positive particles entering the paper).

appears to be an exception by being focusing in both planes when $0 < k\rho_0^2 < 1$, i.e. for the constant-gradient focusing condition from (2.26). This arises from the rather special focusing action of a dipole field. Of course, a dipole field satisfies Maxwell's equations and the gradient condition mentioned earlier, but, in the case of the motion equations, its focusing action relies on the geometry of intersecting circles (see Figure 3.2(a)) and *not* on field gradients across the aperture. This is, for example, the principle behind the focusing action of a 180° dipole spectrometer magnet (see Figure 3.2(b)). *Geometric focusing* happens to have the opposite sign to *gradient focusing* and this is the reason for the small overlap of the otherwise separate stability regions exploited by the weak-focusing machines.

3.1.2 Point lens

The transfer matrices (3.3) and (3.4) can be considerably simplified if the argument

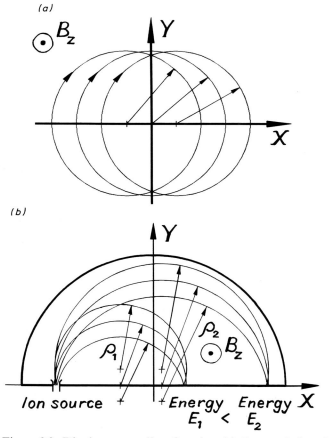

Figure 3.2. Dipole zero-gradient focusing. (*a*) Geometric focusing, (*b*) 180° spectrometer.

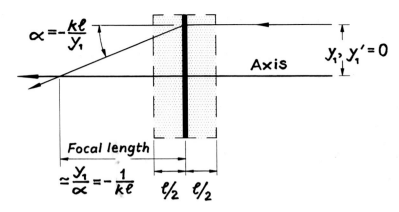

Figure 3.3. Focal length of a thin lens.

$|K|^{\frac{1}{2}}l$ is much less than unity*. In this case the lens can be replaced to a good approximation by an infinitely thin quadrupole lens of strength kl placed at the centre of the original lens (see Figure 3.3). The transfer matrix of this point lens will be,

$$\begin{pmatrix} y \\ y' \end{pmatrix}_2 = \begin{pmatrix} 1 & 0 \\ \pm kl & 1 \end{pmatrix}\begin{pmatrix} y \\ y' \end{pmatrix}_1, \tag{3.6}$$

where the upper sign $(+)$ corresponds to the horizontal plane and the lower sign $(-)$ to the vertical plane.

The approximations of replacing sin and sinh by their arguments and cos and cosh by unity are well known and the accuracy is evident. It is less evident why l has been replaced by zero in the top righthand position in the matrix. In order for the point lens approximation to hold, the length l must be small compared with the focal length of the lens. This element can then be replaced by zero so that the determinant of the matrix remains exactly unity. All transfer matrices must satisfy this requirement rigorously unless there is a dissipative mechanism such as energy loss by synchrotron radiation. The unit determinant is an expression of Liouville's theorem on the conservation of phase space (see Appendix A). The true length of the lens re-appears, of course, in the drift spaces on either side.

A slightly different viewpoint can be found in Ref. 3, where the thick lens is replaced by an infinitely thin lens of the same integrated strength set at one principal plane. The calculation of the drift space matrices needed to correctly position the lens on the principal plane is not simple, but when the principal plane is set arbitrarily to the centre of the thick lens the simplicity is regained. The quality of this approximation has been shown to be good if $|k|l^2$ is much less than unity, which is equivalent

---

* In Chapter 2, the general focusing constant K was related to the local wavelength of the betatron oscillation in (2.28) so that, $\sqrt{K(s)} = 2\pi/\lambda_\beta$. Thus the above criterion can be physically interpreted as the length of the lens l must be small compared to the local betatron wavelength λ_β.

to the condition set at the beginning of this section. This criterion can be physically interpreted as the length of the lens l must be short compared to its focal length.

It is common to replace the lens strength kl by a single symbol defined as,

$$\text{Lens strength,} \quad g = kl = \frac{-1}{\text{Focal length}}. \tag{3.7}$$

In the optical analogy, the focal length is the reciprocal of minus the lens strength when applying the 'real is positive' sign convention. Thus when g is negative the lens has a positive focal length in the horizontal plane and is focusing in that plane. The focal length is a useful concept for physical insight, but the lens strength g will be used here, as this simplifies the algebra in later chapters.

In separated-function, strong-focusing lattices, the $1/\rho^2$ focusing term in the dipoles can be neglected for simple calculations on the grounds that it is small. This is usually a good approximation. Thus dipoles are often represented in simple calculations by the drift space matrix (3.5) and combined-function magnets by the pure-quadrupole, point-lens matrix (3.6). With these approximations, many linear optics problems for circular machines and transfer lines can literally be solved on the back of an envelope. In fact, as machines progress towards TeV energies, these approximations steadily improve, since ρ increases with beam energy while the fields and gradients are quasiconstant being limited by technology. In this small way, today's big machines are becoming simpler.

3.1.3 Doublet

The point lens formulation can be used to make a simple analysis of the quadrupole doublet. Figure 3.4* shows a combination of two quadrupoles of equal but opposite strength g_0. Let the first magnet in this doublet be focusing in the horizontal plane, i.e. an F-type, and the second be defocusing, i.e. a D-type. In the vertical plane, their roles will be revised as described in Section 3.1.1.

Consider now the upper half of Figure 3.4 for the horizontal plane, where the transfer matrix from the first lens to the point P_1 will be,

$$T_x = \begin{pmatrix} 1 & s_1 \\ 0 & 1 \end{pmatrix} \begin{pmatrix} 1 & 0 \\ -g_0 & 1 \end{pmatrix} \begin{pmatrix} 1 & L \\ 0 & 1 \end{pmatrix} \begin{pmatrix} 1 & 0 \\ g_0 & 1 \end{pmatrix}$$

$$T_x = \begin{pmatrix} 1 + g_0 L - g_0{}^2 L s_1 & L + s_1 - g_0 L s_1 \\ -g_0{}^2 L & 1 - g_0 L \end{pmatrix}. \tag{3.8}$$

In the vertical plane, the transfer matrix is simply (3.8) with the sign of g_0 changed throughout and s_1 replaced by s_2, the distance to P_2.

---

* All figures which refer to the multiplication of matrices, will be drawn with the beam direction from right to left to correspond to the convention for matrix multiplication.

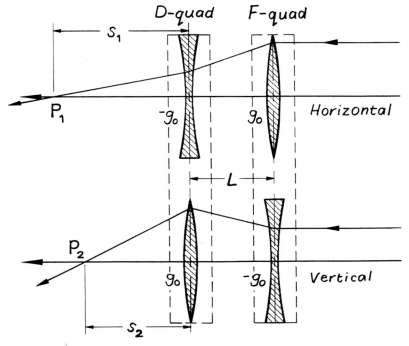

Figure 3.4. Symmetric doublet.

$$T_z = \begin{pmatrix} 1 - g_0 L - g_0{}^2 L s_2 & L + s_2 + g_0 L s_2 \\ - g_0{}^2 L & 1 + g_0 L \end{pmatrix}. \qquad (3.9)$$

If the doublet is to be focusing, then a particle travelling parallel to the axis must be brought to a focal point after traversing both lenses, i.e.

$$\begin{pmatrix} 0 \\ y' \end{pmatrix}_{\text{exit}} = T \begin{pmatrix} y \\ 0 \end{pmatrix}_{\text{entry}}.$$

For this to be true there must be values of s_1 and s_2 such that,

Horizontally: $\quad 1 + g_0 L - g_0{}^2 L s_1 = 0$, so that $s_1 = (1 + g_0 L) g_0{}^{-2} L^{-1}$ $\qquad (3.10)$

Vertically: $\quad 1 - g_0 L - g_0{}^2 L s_2 = 0$, so that $s_2 = (1 - g_0 L) g_0{}^{-2} L^{-1}.$ $\qquad (3.11)$

Equation (3.11) shows that s_2 is always positive (g_0 was defined as representing an F-type lens, i.e. g_0 must be numerically negative) and hence this doublet is focusing overall in the vertical plane, but in the horizontal plane (3.10) shows that focusing only occurs when $L < -g_0{}^{-1}$. Thus the condition for an anti-symmetric doublet to be focusing in both planes is that the lenses must be separated by less than their individual focal length.

From this starting point it would be natural to investigate combinations of more than two positive and negative lenses but, as so often is the case, it is easier and of rather fundamental importance to go straight to the case of an infinite number of positive and negative lenses, which is close to the practical case of an alternating-gradient accelerator.

3.2 SIMPLE DESCRIPTION OF AN ALTERNATING-GRADIENT ACCELERATOR

An alternating-gradient accelerator consists of magnetic elements that bend the particles 360° and that focus them by alternating between strong positive and negative gradients. The focusing gradients are either superimposed on the bending field, which is known as a *combined-function lattice*, or alternatively concentrated in separate elements, which is known as a *separated-function lattice*. To a first approximation these elements are arranged into a periodic structure around the machine (see Figure 3.5). The transfer matrix T for the *basic period* of this structure, which is often referred to as a *cell*, is found by multiplying the matrices of the individual elements. In practical machines, there will also be superperiodicity, e.g. a number of identical multi-cell arcs separated by identical straight sections. The unit of an arc and a straight section is then called a *superperiod*, but this refinement makes no difference to the stability criterion to be derived in the next section, so it will be ignored for the moment.

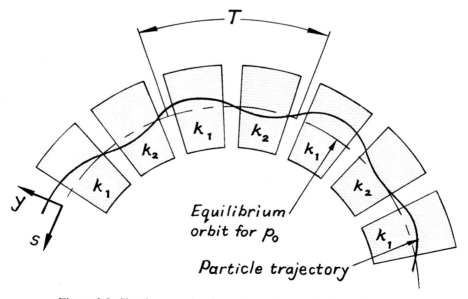

Figure 3.5. Simple example of an alternating-gradient accelerator.

3.2.1   Stability criterion

Particles going through an alternating gradient accelerator see what is effectively an infinite series of transfer matrices.

$$\begin{pmatrix} y \\ y' \end{pmatrix}_2 = T \begin{pmatrix} y \\ y' \end{pmatrix}_1 \dots \text{for 1 cell,} \qquad \begin{pmatrix} y \\ y' \end{pmatrix}_n = T^n \begin{pmatrix} y \\ y' \end{pmatrix}_1 \dots \text{for } n \text{ cells.} \qquad (3.12)$$

Suppose now that a transformation S is applied to (3.12) so that

$$S \begin{pmatrix} y \\ y' \end{pmatrix}_2 = \underbrace{STS^{-1}}_{D} S \begin{pmatrix} y \\ y' \end{pmatrix}_1. \qquad (3.13)$$

The new transformation STS^{-1} is a so-called *similarity transformation*[4], which can be shown to have the property of maintaining the same determinant and trace as the original matrix T. Now, S can be chosen such that it makes STS^{-1} diagonal. If D is defined as this diagonal matrix, then

$$D = STS^{-1} = \begin{pmatrix} d_{11} & 0 \\ 0 & d_{22} \end{pmatrix} \qquad (3.14)$$

and

$$\text{Tr } T = (d_{11} + d_{22}) \qquad \text{and} \qquad |T| = d_{11}d_{22} = 1. \qquad (3.15)$$

Consequently, d_{11} and d_{22} are the solutions of the equation

$$d^2 - d \text{ Tr } T + 1 = 0,$$

so that

$$d = \left\{ \frac{\text{Tr } T}{2} \right\} \pm \left[\left\{ \frac{\text{Tr } T}{2} \right\}^2 - 1 \right]^{\frac{1}{2}}. \qquad (3.16)$$

The two solutions of d are in fact the eigenvalues of the transfer matrix T.

The transfer matrix for the passage through n cells can now be written as,

$$T^n = S^{-1}DS \ S^{-1}DS \ S^{-1}DS \dots = S^{-1}D^nS$$

$$\begin{pmatrix} y \\ y' \end{pmatrix}_n = S^{-1}D^nS \begin{pmatrix} y \\ y' \end{pmatrix}_1. \qquad (3.17)$$

For the particle's excursion y to remain finite, however large n is made, requires that,

$$|d_1| \leq 1 \qquad \text{and} \qquad |d_2| \leq 1. \qquad (3.18)$$

Since from (3.15) $d_1 d_2 = 1$, this means that,

$$|d_1| = |d_2| = 1, \qquad (3.19)$$

which is only satisfied if

$$\left| \frac{\text{Tr } T}{2} \right| \leq 1. \qquad (3.20)$$

This criterion for alternating-gradient stability is of fundamental importance and must be satisfied both horizontally and vertically in order to assure full stability of the transverse particle motion. Once the criterion is satisfied the particles will oscillate about the equilibrium orbit with a pseudo-harmonic motion. This will be discussed further in Chapter 4.

3.2.2   'Necktie' stability plot

Through a very simple example it is now possible to demonstrate the importance of the findings so far. Consider a system of focusing (F-type) and defocusing (D-type) point lenses separated by drift spaces or dipoles (O) of length L (see Figure 3.6). This model is a good approximation for a structure with separate elements for focusing and bending and a fair approximation when the focusing and bending are combined. The basic cell of this structure is called a *FODO* cell and the whole structure is known as a *FODO lattice*.

The horizontal transfer matrix will be given by,

$$T_x = \begin{pmatrix} 1 & L \\ 0 & 1 \end{pmatrix}\begin{pmatrix} 1 & 0 \\ g_2 & 1 \end{pmatrix}\begin{pmatrix} 1 & L \\ 0 & 1 \end{pmatrix}\begin{pmatrix} 1 & 0 \\ g_1 & 1 \end{pmatrix} \tag{3.21}$$

and simple algebra yields the trace, which when inserted into the stability criterion (3.20), gives the condition,

$$\left|\frac{\mathrm{Tr}\,T_x}{2}\right| = |1 + g_1 L + g_2 L + \tfrac{1}{2}g_1 g_2 L^2| \le 1. \tag{3.22}$$

It is worth noticing that this condition cannot be satisfied with both g_1 and g_2

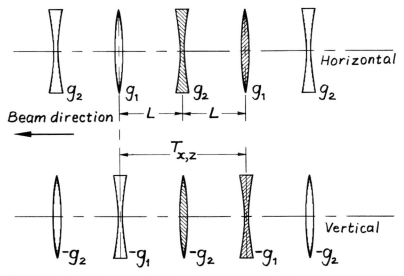

Figure 3.6. Point lens model of a periodic FODO lattice.

positive. The expression is symmetric in g_1 and g_2 and it is convenient to choose g_1 as negative to conform with Figure 3.6. The two limiting conditions are obtained by first putting the expression (3.22) equal to $+1$ and then equal to -1. This yields,

$$g_2 L = -g_1 L/(1 + \tfrac{1}{2}g_1 L) \qquad \text{and} \qquad g_1 L = -2. \tag{3.23}$$

These two curves are drawn in Figure 3.7. Horizontal stability is assured between the curves and instability outside.

The vertical motion similarly yields,

$$T_z = \begin{pmatrix} 1 & L \\ 0 & 1 \end{pmatrix}\begin{pmatrix} 1 & 0 \\ -g_2 & 1 \end{pmatrix}\begin{pmatrix} 1 & L \\ 0 & 1 \end{pmatrix}\begin{pmatrix} 1 & 0 \\ -g_1 & 1 \end{pmatrix} \tag{3.24}$$

$$\left|\frac{\operatorname{Tr} T_z}{2}\right| = |1 - g_1 L - g_2 L + \tfrac{1}{2}g_1 g_2 L^2| \le 1. \tag{3.25}$$

This condition cannot be satisfied with both g_1 and g_2 negative. Together with the corresponding horizontal condition, this means that for stability it is necessary (but not sufficient) to have opposite signs for g_1 and g_2. With g_2 made positive the vertical stability condition gives the following limits,

$$-g_1 L = g_2 L/(1 - \tfrac{1}{2}g_2 L) \qquad \text{and} \qquad g_2 L = 2. \tag{3.26}$$

The corresponding curves are also drawn on Figure 3.7. Vertical stability is assured between these curves and instability outside. It is interesting to note that the condition

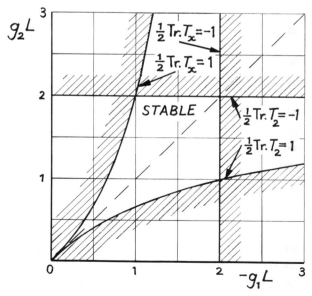

Figure 3.7. 'Necktie' stability diagram.

39

$L < 2/g$ means that a series of equally-strong and equally-spaced F- and D-lenses are stable for separations up to twice their individual focal length, whereas for the simple doublet studied earlier in Section 3.1 this separation had to be less than one focal length.

Similar diagrams can be calculated and drawn for any alternating-gradient structure, except that normally they require more work than this simple example. The important conclusion can be drawn that the accelerator parameters must be chosen such that the *working point* of the machine is well inside the region that is stable for both planes.

However, there are further complications. Most accelerators do not have the simple periodicity of the above example. Instead the magnets are arranged in superperiods, i.e. a number of regular multi-cell arcs separated by straight sections. This structure introduces resonance stopbands into the 'necktie' diagram, which are called *structure resonances*. The driving term for such a resonance has an azimuthal distribution at a harmonic of the superperiodicity or circumference (see Chapter 5 for a fuller description of resonances). Large local errors and special insertions can lead to the dominant periodicity being the whole machine circumference, which then enhances the first order, or integer, resonances.

Circular machines in fact have an infinite number of resonance conditions involving the horizontal tune or vertical tune or both. These resonances further divide the 'necktie' into many small regions. The resonances are driven by multipole field components in the magnets and by the field of the counter-rotating beam in colliders.

The practical consequences are that tight tolerances have to be imposed on the field quality of the magnets to keep the resonance stopbands sufficiently narrow and in colliders unused beam crossings often have to be avoided by separating the beams.

3.3   EDGE FOCUSING

So far, any problems that may arise at the boundaries between magnetic elements have been tacitly ignored. In mathematical terms, it has been assumed that all boundaries were planes perpendicular to the equilibrium orbit and that the uniform fields inside each element started or stopped discontinuously at these boundaries.

It is, however, relatively easy to extend the theory to account for the field boundaries of dipoles that are inclined with respect to the perpendicular to the equilibrium orbit in the plane of bending, which is usually the horizontal plane. In this plane, there is a focusing effect because particles will cross the boundary into the dipole field at different axial positions depending on their radial position. Perpendicular to the plane of bending there is also a focusing effect, but this time it is due to the angle at which the particles cross the longitudinal end field of such a boundary. These effects are known as *edge focusing* and are relatively important, especially in small machines. Indeed the weak focusing ZGS at the Argonne National

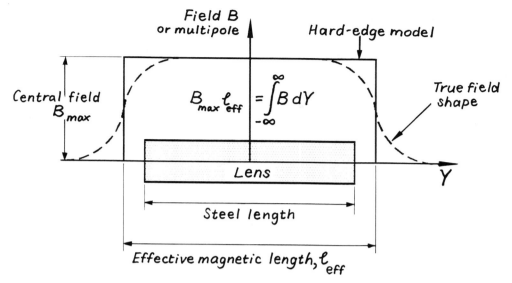

Figure 3.8. Hard-edge magnet model.

Laboratory relied on this type of focusing. If the boundary between elements is perpendicular to the equilibrium orbit, as is normally the case for quadrupoles and higher order lenses, there is no effect.

3.3.1 Hard-edge model

In order to render optics computations tractable, it is necessary to replace the real-world magnets by so-called *hard-edge models* (see Figure 3.8). In this model the field or multipole distribution of interest is replaced by a uniform block whose area equals the *central value times an effective length*, which ensures that the field integral through the magnet remains unchanged. Some more details about magnets are given in Appendix B.

The hard-edge model engenders very small errors in the beam geometry and occasionally it may be necessary to track through the field plots of dipole magnets or a stray field in order to satisfy the alignment tolerances. Likewise there are also very small errors in the focusing. Steffen[3] has evaluated this using the transfer matrices, Autin[5] has evaluated the perturbation to the betatron amplitude function and Hedin[6] proposes a modified effective length for the special case of quadrupoles whose lengths are not far larger than their apertures. Of course, a valid, but computationally expensive, way to model magnets more accurately is to cut them into many slices.

Longitudinal fields are not described by this model at all, but this omission is partly remedied by including them in the vertical edge focusing in Section 3.3.3.

Longitudinal end fields also affect non-linear coupling resonances, but fortunately this effect is rarely of importance[7].

3.3.2 Edge focusing in the plane of bending

An inclined boundary at a dipole can be modelled by a hard-edge field wedge superimposed on the ideal perpendicular boundary. The field in the wedge is made equal in magnitude to that in the dipole and the wedge is arranged so as to add the total field integral on one side of the equilibrium orbit and subtract on the other (see Figure 3.9). Since the angular deviation α in the wedge will be proportional to the distance of the particle from the equilibrium orbit, the wedge acts as a lens, so that

$$\alpha = \rho^{-1} x \tan \varepsilon$$

and

$$g = \rho^{-1} \tan \varepsilon. \tag{3.27}$$

This effect is very local and rather weak, so it is safe to use the point lens approximation (see Section 3.1.2) to describe the action of the wedge.

$$\begin{pmatrix} x \\ x' \end{pmatrix}_2 = \begin{pmatrix} 1 & 1 \\ (\tan \varepsilon)/|\rho| & 0 \end{pmatrix} \begin{pmatrix} x \\ x' \end{pmatrix}_1 . \tag{3.28}$$

It is now necessary to resolve any ambiguity in the sign of ε. The *wedge is focusing when it increases the field integral on the outside of the bend and decreases it to the inside*. Thus 'focusing' corresponds to a negative ε (see Figure 3.10 for examples). This convention agrees with (3.28), which represents a focusing lens when $\tan \varepsilon$ is

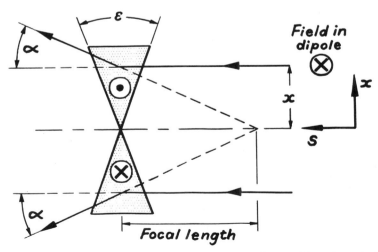

Figure 3.9. Hard-edge magnetic wedge (drawn for positive charge). (Diagram applies equally to both ends of the magnet.)

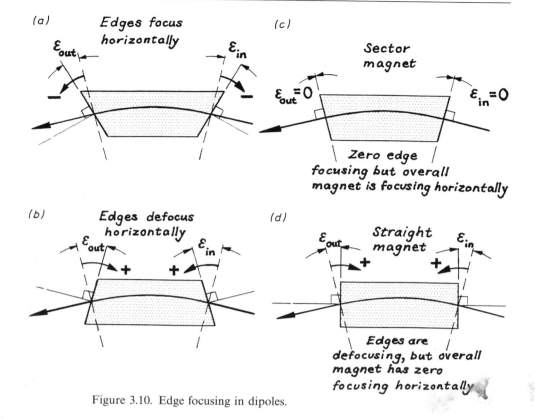

Figure 3.10. Edge focusing in dipoles.

negative and a defocusing lens when $\tan \varepsilon$ is positive. In the defocusing case, the focal length of the wedge is virtual and negative, which agrees with (3.27) and Figure 3.9. The above criterion is independent of the sign of ρ. If a dipole is providing reverse bending, then the sign of its edge focusing is established by the change in the field integral to the outside of the bend, *not* to the outside of the machine.

Unless the dipole ends are shaped with the express intention of adding exact amounts of focusing, there are normally only two cases in Figure 3.10 that are of interest: the sector magnet and the rectangular magnet.

(i) *Sector magnet*

This design eliminates edge focusing by putting $\varepsilon_{in} = \varepsilon_{out} = 0$. If there is no gradient in the block then only the geometric focusing remains in the horizontal plane and the transfer matrix (3.2) simplifies to,

$$T_x(\text{sector dipole}) = \begin{pmatrix} \cos|l/\rho| & |\rho|\sin|l/\rho| \\ -|\rho|^{-1}\sin|l/\rho| & \cos|l/\rho| \end{pmatrix}$$

and the vertical matrix to that of a simple drift space. Sector dipoles and sector combined-function magnets are typically found in small machines and especially in the older ones.

(ii) *Rectangular magnet*

The second case is the rectangular or straight magnet. This is easier to build and is often used in large machines. In very large machines, however, the distinction between sector and rectangular dipoles becomes of little practical importance.

Since the ends of a rectangular magnet are parallel, $\varepsilon_{in} = \varepsilon_{out} =$ half the bending angle. Figure 3.10 specifies that the angles are positive, so the edges are therefore horizontally defocusing. However, intuition indicates that there should be no net horizontal focusing in such a dipole, since two particles of the same momentum which are moving parallel to each other will be deflected equally by a uniform, parallel-sided block of dipole field. It is left as an exercise to the reader to multiply the horizontal matrix of a pure dipole with its edge-focusing matrices and to verify that in this particular case the edge focusing exactly compensates the geometric focusing in the body of the magnet to give the very simple transfer matrix,

$$T_x(\text{Rect. dipole}) = \begin{pmatrix} 1 & |\rho| \sin|l/\rho| \\ 0 & 1 \end{pmatrix}$$

which is equivalent to a drift space of length $|\rho| \sin|l/\rho|$. In the vertical plane the geometric focusing is not present, but the edge focusing remains.

3.3.3   Edge focusing perpendicular to the plane of bending

This effect invokes the focusing action of the longitudinal components in the fringe fields at the dipole's ends. Figure 3.11 illustrates how these fields give rise to a focusing action. As in the previous case, the sign and magnitude depends on the slope of the ends.

Equation (3.29) is the line integral of the field round a closed path in the plane perpendicular to the bending. The path starts deep inside the dipole, i.e. in the constant-field region, and passes well beyond the magnet's fringe field along the beam trajectory. This integral must give a zero result (see Figure 3.11(*a*)).

$$\oint \boldsymbol{B} \cdot \mathrm{d}\boldsymbol{l} = B_0 z + \int_{\text{fringe field}} B_s \, \mathrm{d}s = 0. \tag{3.29}$$

In the general case shown in Figure 3.11(*b*), the beam crosses the end face of the magnet at an angle ε. Since the longitudinal component of the fringe field will be perpendicular to the magnet's end face, the beam will see not only the axial component B_s but also a transverse component B_x perpendicular to its path. The

44

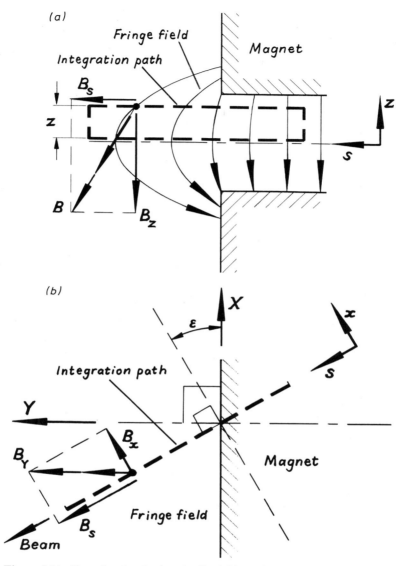

Figure 3.11. Focusing by the longitudinal fringe field in dipoles. (a) (s, z) plane perpendicular to bending, (b) (x, s) plane of bending. $[(X, Y)$ are local coordinates of the magnet face.]

integral of B_x is simply related to the integral of B_s by,

$$\int_{\text{fringe field}} B_x \, ds = \int_{\text{fringe field}} B_s \tan \varepsilon \, ds. \tag{3.30}$$

The combination of (3.29) and (3.30) gives an expression for the angular kick α in the plane perpendicular to the bending and this expression has the very nice feature of being independent of the shape of the fringe field.

$$\alpha = \frac{1}{B_0 \rho_0} \int_{\text{fringe field}} B_x \, ds = -\frac{\tan \varepsilon}{\rho} z.$$

This kick is proportional to the height above the plane of bending and therefore behaves as a linear lens of strength,

$$g = -\rho^{-1} \tan \varepsilon. \qquad (3.31)$$

Equation (3.31) is directly analogous to (3.27) and since the fields in Figures 3.9 and 3.11 have been made of the same sign it can be directly appreciated that while the edge focusing is positive in one plane it is negative in the other. This result can be summarised by the transfer matrix,

$$\begin{pmatrix} z \\ z' \end{pmatrix}_2 = \begin{pmatrix} 1 & 0 \\ -(\tan \varepsilon)/|\rho| & 1 \end{pmatrix} \begin{pmatrix} z \\ z' \end{pmatrix}_1, \qquad (3.32)$$

where ε is the slope of the end face as defined in Figure 3.10, i.e. it is positive when the integral of the bending field decreases on an outer orbit.

3.3.4   General formulation of edge focusing

In order to take into account both horizontal and vertical bending magnets, such as would be needed in vertically stacked colliding rings, (3.28) and (3.32) can be expressed in a more general form as,

$$\begin{pmatrix} y \\ y' \end{pmatrix} = \begin{pmatrix} 1 & 0 \\ \pm(\tan \varepsilon)/|\rho| & 1 \end{pmatrix} \begin{pmatrix} y \\ y' \end{pmatrix}_1, \qquad (3.28)' \, \& \, (3.32)'$$

where the upper sign $(+)$ is used when the bending ρ is in the plane y, s and the lower sign $(-)$ is used when the bending ρ is perpendicular to the plane y, s. The edge face angles are always assumed to be in the plane of bending and are negative when they increase the field integral on the outside of the bend.

3.4   MOTION WITH A MOMENTUM DEVIATION

So far only the equations of motion for charged particles with small angle and position deviations from the equilibrium orbit have been derived, but beams are never perfectly mono-energetic, so the corresponding equations must also be found for small deviations in momentum.

3.4.1   Equations of motion

The procedure is essentially that applied in Chapter 2, Section 2.2 and starts with the situation as shown in Figure 2.2. Equations (2.10) and (2.18) are rewritten with an increase Δm in the particle's mass and a corresponding increase Δv in its velocity. The increments are assumed constant for any given particle, i.e. acceleration is disregarded as in Section 3.1, so that the mass can be moved to the left of the differential operator d/dt.

$$(m + \Delta m)\frac{d^2}{dt^2}(x + \rho_0) - \frac{(m + \Delta m)(v + \Delta v)^2}{(x + \rho_0)} = e(B_0 + \Delta B_z)(v + \Delta v) \tag{3.33}$$

$$(m + \Delta m)\frac{d^2 z}{dt^2} = -eB_x(v + \Delta v). \tag{3.34}$$

The same assumption is made as before that $x \ll \rho_0$ and in addition that $\Delta m \ll m$ and $\Delta v \ll v$. After expansion of B_x in a Taylor series, rearrangement and the omission of second order and higher terms in the small quantities Δm, Δv, x and z, exactly the same equations can be obtained only with some new terms on the righthand side, i.e.

$$\frac{d^2 x}{ds^2} + \left(\frac{1}{\rho_0{}^2} - k\right)x = \frac{1}{\rho_0}\left(\frac{\Delta m}{m} + \frac{\Delta v}{v}\right) = \frac{1}{\rho_0}\frac{\Delta p}{p_0} \tag{3.35}$$

$$\frac{d^2 z}{ds^2} + kz = 0. \tag{3.36}$$

The z–s plane is therefore unchanged to the first order, while the x–s plane is modified by a term on the righthand side which is constant in any given magnetic element. A more rigorous derivation is given in Appendix A using the Hamiltonian formalism. It should be noted that the level of approximation applied above neglects the source of many chromatic errors and a more detailed analysis of chromaticity will be made in Chapter 6.

Equations (3.35) can be generalised by replacing the coefficients $(\rho_0{}^{-2} - k)$ and k by $K(s)$ as described earlier. Table 3.1 remains valid for the detailed forms of $K(s)$.

$$\frac{d^2 y}{ds^2} + K(s)y = \frac{1}{\rho_0(s)}\frac{\Delta p}{p_0} \tag{3.37}$$

The s-dependence of K and ρ_0 is shown in order to account for changes in these parameters between magnetic elements. Thus (3.37) applies to a complete structure whereas (3.35) and (3.36) are restricted to a single element. The general transverse coordinate y indicates the plane in which the bending occurs, i.e. x, as above, or z, or both.

The solution to (3.35), the motion through a single element, will be of the form,

$$y = A \cos(K^{\frac{1}{2}}l) + B \sin(K^{\frac{1}{2}}l) + \frac{1}{\rho_0} \frac{\Delta p}{p_0} \frac{1}{K} \left.\right\}$$
$$y' = -K^{\frac{1}{2}}A \sin(K^{\frac{1}{2}}l) + K^{\frac{1}{2}}B \cos(K^{\frac{1}{2}}l) \left.\right.$$

(3.38)

where $l = (s_2 - s_1)$, the length of the element, and the integration constants are determined from the initial conditions as,

$$A = \left(y_1 - \frac{1}{\rho_0} \frac{\Delta p}{p_0} \frac{1}{K} \right), \qquad B = y_1' K^{-\frac{1}{2}}.$$

(3.39)

This result can be expressed in matrix form by treating the $\Delta p/p_0$ as a quasi-variable. As was done in Section 3.1 for the motion without a momentum error, the focusing and defocusing solutions are given separately below. The matrices (3.40) to (3.42) are thus the logical generalisation of (3.3) to (3.5). As before, the complete solution for a lattice is found by multiplying the individual matrices for each element in beam order.

$K > 0$ Focusing: (3.40)

$$\begin{pmatrix} y \\ y' \\ \dfrac{\Delta p}{p} \end{pmatrix}_2 = \begin{pmatrix} \cos(|K|^{\frac{1}{2}}l), & |K|^{-\frac{1}{2}}\sin(|K|^{\frac{1}{2}}l), & \pm\rho_y^{-1}|K|^{-1}[1 - \cos(|K|^{\frac{1}{2}}l)] \\ -|K|^{\frac{1}{2}}\sin(|K|^{\frac{1}{2}}l), & \cos(|K|^{\frac{1}{2}}l), & \pm\rho_y^{-1}|K|^{-\frac{1}{2}}\sin(|K|^{\frac{1}{2}}l) \\ 0 & 0 & 1 \end{pmatrix} \begin{pmatrix} y \\ y' \\ \dfrac{\Delta p}{p} \end{pmatrix}_1$$

$K < 0$ Defocusing: (3.41)

$$\begin{pmatrix} y \\ y' \\ \dfrac{\Delta p}{p} \end{pmatrix}_2 = \begin{pmatrix} \cosh(|K|^{\frac{1}{2}}l), & |K|^{-\frac{1}{2}}\sinh(|K|^{\frac{1}{2}}l), & \pm\rho_y^{-1}|K|^{-1}[\cosh(|K|^{\frac{1}{2}}l) - 1] \\ -|K|^{\frac{1}{2}}\sinh(|K|^{\frac{1}{2}}l), & \cosh(|K|^{\frac{1}{2}}l), & \pm\rho_y^{-1}|K|^{-\frac{1}{2}}\sinh(|K|^{\frac{1}{2}}l) \\ 0 & 0 & 1 \end{pmatrix} \begin{pmatrix} y \\ y' \\ \dfrac{\Delta p}{p} \end{pmatrix}_1$$

$K = 0$ Drift space: (3.42)

$$\begin{pmatrix} y \\ y' \\ \dfrac{\Delta p}{p} \end{pmatrix}_2 = \begin{pmatrix} 1 & l & 0 \\ 0 & 1 & 0 \\ 0 & 0 & 1 \end{pmatrix} \begin{pmatrix} y \\ y' \\ \dfrac{\Delta p}{p} \end{pmatrix}_1$$

where the upper sign $(+)$ in the dispersion terms is used when the bending is in the horizontal plane $(\rho = \rho_x)$ and the lower sign $(-)$ is used when the bending is in the vertical plane $(\rho = \rho_z)$. The dispersion terms are zero when considering the plane perpendicular to the bending because ρ is then infinite. The need to introduce this sign change when bending vertically is due to the choice of coordinates (see Coordinate system).

3.4.2   Local dispersion function

The solution of the generalised equation (3.37) can be regarded as being the sum of two parts,

$$y(s) = y_\beta(s) + \text{PI}. \tag{3.43}$$

In the theory of differential equations, $y_\beta(s)$ would be called the *complementary function* (i.e. the full solution of (3.37) with the righthand side set to zero) and PI would be the *particular integral* (i.e. any particular solution of (3.37) complete with the righthand-side term). In more descriptive terms, $y_\beta(s)$ is the betatron motion that was studied in earlier sections and it is superimposed on a new equilibrium orbit PI, which is driven by the momentum deviation.

In a mathematical sense PI can be any valid solution of (3.37), but since there are many possibilities, the question of how it should be chosen arises and uncovers a distinction between transfer lines and circular machines. In a transfer line, it is the designer who chooses PI when he sets the boundary conditions at the entry to the line, usually to match the beam as it leaves the upstream accelerator. The subsequent forms of both terms are then fully determined by the lattice. In a circular machine, however, the periodicity forces the choice of PI to be a *closed orbit* and hence it is unique. This orbit will be such that the integrated bending over one turn matches the particle's momentum*. This distinction between transfer lines and circular machines will be further developed in Section 4.5 of the next chapter.

The PI that either the transfer line designer needs or the periodicity imposes will be called $y_D(s)$ and is usually re-expressed as,

$$y_D(s) = D(s) \frac{\Delta p}{p_0}, \tag{3.44}$$

where $D(s)$ is referred to as the *local dispersion function*, which must, of course, satisfy the equation,

$$\frac{\mathrm{d}^2 D}{\mathrm{d}s^2} + K(s)D = 1/\rho_0(s). \tag{3.45}$$

Both $K(s)$ and $\rho_0(s)$ are shown with an s-dependence to indicate that (3.45) applies to the whole lattice. The local dispersion function can be found from the 3×3 transfer matrices as follows.

---

* The condition for a stable closed orbit is examined in Appendix D.

(i) *For transfer lines*

This is the straightforward case in which the vector $(D, D', \Delta p/p = 1)$ is mapped through the line using the input values determined by the designer.

$$
\begin{pmatrix} D(s) \\ D'(s) \\ 1 \end{pmatrix} = \begin{pmatrix} t_{11}(s) & t_{12}(s) & t_{13}(s) \\ t_{21}(s) & t_{22}(s) & t_{23}(s) \\ 0 & 0 & 1 \end{pmatrix} \begin{pmatrix} D(0) \\ D'(0) \\ 1 \end{pmatrix}
$$

(3.46)

(ii) *Circular machines*

In this case the periodicity imposes a closed orbit for D, so that $D(s_0) = D(s_0 + C)$ and $D'(s_0) = D'(s_0 + C)$, where s_0 is any point around the machine. In matrix form this becomes,

$$
\begin{pmatrix} D(s_0) \\ D'(s_0) \\ 1 \end{pmatrix} = \begin{pmatrix} t_{11} & t_{12} & t_{13} \\ t_{21} & t_{22} & t_{23} \\ 0 & 0 & 1 \end{pmatrix} \begin{pmatrix} D(s_0) \\ D'(s_0) \\ 1 \end{pmatrix}.
$$

(3.47)

$D(s_0)$ and $D'(s_0)$ can be solved from the two simultaneous equations coming from (3.47) to give,

$$
D(s_0) = \frac{(1 - t_{22})t_{13} + t_{12}t_{23}}{(2 - t_{11} - t_{22})}
$$

(3.48)

$$
D'(s_0) = \frac{(1 - t_{11})t_{23} + t_{21}t_{13}}{(2 - t_{11} - t_{22})}.
$$

(3.49)

The values for the dispersion parameters at other positions can then be found by tracking from s_0, or by stepping round the machine element by element, finding the full turn matrix and applying (3.48) and (3.49)*. An alternative analytic expression for D will be derived in the next chapter in Section 4.6.2.

3.4.3   Momentum compaction in alternating-gradient lattices

In Chapter 2 the momentum compaction was defined in (2.41) in terms of the separation of the equilibrium orbits for different momenta and in terms of the equivalent difference in orbit length. In an alternating gradient machine the latter

---

* Once the full turn matrix T has been found at one point, it can be found at the position of the next element by applying $\Delta T \, T \, \Delta T^{-1}$ where ΔT is the step through the next element.

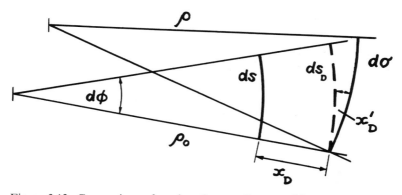

Figure 3.12. Comparison of arc lengths on adjacent orbits.

definition is applicable, i.e.

$$\alpha = \frac{\Delta C}{C_0} \bigg/ \frac{\Delta p}{p_0} \tag{3.50}$$

where,

$$C_0 = \oint ds \quad \text{and} \quad (C_0 + \Delta C) = \oint d\sigma. \tag{3.51}$$

Let ds, which is on the equilibrium orbit for the design momentum, and $d\sigma$, which is on an adjacent equilibrium orbit, subtend the same infinitesimally small sector of the machine $d\phi$. The lengths of ds and $d\sigma$ are compared in Figure 3.12.

Simple geometry gives,

$$d\phi = \frac{ds}{\rho_0} = \frac{ds_D}{\rho_0 + x_D} \quad \text{and} \quad d\sigma = (1 + \tfrac{1}{2}x_D'^2 + \cdots)(1 + x_D/\rho_0)\, ds.$$

After the introduction of the local dispersion function from (3.44), the arc length $d\sigma$ is given to first order by,

$$d\sigma = \left(1 + \frac{1}{\rho_0} D_x(s) \frac{\Delta p}{p_0}\right) ds. \tag{3.52}$$

The substitution of (3.52) into (3.51) yields,

$$\Delta C = \frac{\Delta p}{p_0} \oint \frac{1}{\rho_0} D_x(s)\, ds,$$

and finally the momentum compaction (3.50) can be expressed as

$$\alpha = \frac{1}{C_0} \oint \frac{1}{\rho_0} D_x(s)\, ds. \tag{3.53}$$

3.5   GENERAL REMARKS

This chapter has provided the basic tools for a numerical analysis of the single-particle, uncoupled motion in a linear lattice. The matrix formalism introduced for expressing the motion through single elements is ideal for computer codes and is in fact the basis upon which lattice programs are built. Two contributions to the motion were identified, that due to the momentum deviation from an equilibrium orbit and the so-called betatron motion that occurs independently of the momentum deviation. The matrix formalism makes it easy to separate these motions. The stability criterion, formulated for an infinite, periodic series of elements, is also based on the matrix formalism and fits naturally into a lattice program.

While three-dimensional lattices are common in transfer lines, the use of vertical bending in circular machines is much rarer. Lattice designers prefer to avoid vertical bending, since the resultant vertical dispersion is an unwanted complication, but this is not possible in the cases of vertically-stacked, colliding rings and vertical by-passes. In principle, all that is needed for a three-dimensional lattice has been included providing care is taken not to introduce coupling (weak coupling will be treated in Section 5.3). If horizontal bends are avoided while the equilibrium orbit is sloping up or down, then the normal modes, which are aligned with the (x, s) plane and the (z, s) plane, remain vertical and horizontal at any plane transverse to the equilibrium orbit. If, however, horizontal bends are made while the equilibrium orbit is inclined, the normal modes become tilted. The following magnet units must then also be tilted transversely to match the (x, s, z) system of the beam, if coupling is to be avoided[8]. In transfer lines, this is not always done for reasons of simplicity. The same situation occurs when a tilted dipole is used to combine a horizontal and vertical bend into a single unit with all other units untilted. In these cases, the coupling is either considered as negligible or, for electrons, as of no consequence when the beam will be damped by radiation in a later machine[8].

Although numerical computations are indispensable when designing and operating a machine, they are not always very helpful. A numerical analysis answers precise questions without providing much physical insight into the global picture. An attempt will be made in the next chapter to respond to this criticism by developing a more analytical description of the particle motion.

CHAPTER 4

Parameterisation of the transverse motion

In the previous two chapters it was shown how to build up a stepwise solution for an alternating-gradient lattice by solving the motion through each element and then multiplying the matrices together in the order seen by the beam. This is valid and appears quite satisfactory, but it is somewhat disjointed and lacking the usefulness and beauty of a smooth analytic function. In a circular machine the particles oscillate about the equilibrium orbit in a pseudo-harmonic manner. It is possible to exploit this pseudo-harmonic behaviour by choosing a rather special set of parameters by means of which the solution is transformed into simple harmonic motion, often facilitating optics calculations. Many of the results in the first part of this chapter can be found in Ref. 1.

4.1 PARAMETERISATION

The equation for the particle motion through the segments of an accelerator lattice was given in Chapter 3 as,

$$\frac{\mathrm{d}^2 y}{\mathrm{d}s^2} + K(s)y = 0. \tag{3.1}$$

The form of $K(s)$ determines whether a particular segment is a drift space, a quadrupole, a bending magnet or a combined-function element. The motion through an arbitrary lattice can be solved by stepping piecewise through the structure while adapting $K(s)$ to each segment in turn, as has already been demonstrated in Chapter 3. Since (3.1) is true anywhere within the lattice, it is in fact a generalised motion equation.

As a trial solution to this equation, consider the function,

$$y(s) = A\beta(s)^{\frac{1}{2}} \cos\left[\int_0^s \frac{\mathrm{d}\sigma}{\beta(\sigma)} + B \right] \tag{4.1}$$

in which the parameter $\beta(s)$ is made dependent on s to account for the different elements in the structure and A and B are real constants. This choice of function is partly guided by the form of the solution (2.24) for the weak focusing case, but with acceleration disregarded.

The substitution of the trial function (4.1) into (3.1) yields the result that $\beta(s)$ must satisfy the equation

$$\frac{d^2}{ds^2}\beta^{\frac{1}{2}} + K(s)\beta^{\frac{1}{2}} = \beta^{-3/2}. \tag{4.2}$$

With a solution for $\beta(s)$ that satisfies this equation, (4.1) becomes a solution for the general equation (3.1).

Since (4.2) is more complicated than the original equation (3.1), this transformation seems at first sight to be a rather poor exchange and indeed the function $\beta(s)$ must be calculated with a computer for all but the very simplest lattices. This, however, is not too difficult and for circular machines the method will be demonstrated in Section 4.1.2 using the full-period or full-turn matrix (4.14), whilst for transfer lines the input boundary conditions with the transfer matrix (4.26) or (4.30), which are both derived in the Section 4.3, will determine $\beta(s)$ at all points.

The main virtue of this approach is that once $\beta(s)$ is determined the transverse motion*, or *betatron motion* as it is known, closely resembles a harmonic oscillation when expressed in the form (4.1) and can in fact be expressed as a simple harmonic oscillator by the introduction of a *normalised excursion* as follows:

Normalised excursion, $Y(s) = y(s)\beta(s)^{-\frac{1}{2}} = A\cos[\mu(s) + B]$ (4.3)

with the *betatron phase,* $\mu(s) = \int_0^s \frac{d\sigma}{\beta(\sigma)}.$ (4.4)

This is valid for both circular machines and transfer lines. For circular machines, however, it is convenient for certain applications to use a *normalised phase*, which varies between 0 and 2π for one turn,

Normalised phase, $\Phi(s) = \int_0^s \frac{d\sigma}{Q\beta(\sigma)},$ (4.5)†

where Q is the tune of the machine as defined in (2.29) and has the following simple relation to μ,

$$Q = \mu(C)/(2\pi). \tag{4.6}$$

Although with the substitution of these fairly simple normalised parameters the motion has been transformed into that of a simple harmonic oscillator, in real space,

---

* In later sections other contributions to the transverse motion will be considered due to field errors and momentum deviations.

† Equation (4.5) leads to the useful approximation $\bar{\beta} = R/Q$.

the particle travels round the machine, or along the transfer line, with a distorted sine wave whose amplitude varies as $\sqrt{[\beta(s)]}$ and whose betatron phase advances unevenly as the integral $1/\beta(s)$. Furthermore, the number of oscillations in one full turn of the machine can now be much greater than unity, unlike the weak focusing machines in Chapter 2, which were restricted to $Q < 1$. This again illustrates that alternating-gradient focusing can be much stronger than constant-gradient, or weak, focusing.

From (4.1) it is natural that $\beta(s)$ has become known as the *betatron amplitude function*, although the amplitude is only proportional to the square root of this function.

4.1.1 Generalised transfer matrix

For this derivation, it is convenient to expand (4.1) into two terms to give

$$y(s) = A\beta^{\frac{1}{2}} \cos \mu + B\beta^{\frac{1}{2}} \sin \mu \tag{4.7}$$

where A and B are new constants. The derivative of $y(s)$ is easily obtained with the help of the differential form of (4.4), $d\mu/ds = 1/\beta$, and for convenience it is also common usage to define a new variable,

$$\alpha(s) = -\frac{1}{2}\frac{d}{ds}\beta(s) \tag{4.8}$$

so that

$$y'(s) = -A\beta^{-\frac{1}{2}}(\alpha \cos \mu + \sin \mu) + B\beta^{-\frac{1}{2}}(\cos \mu - \alpha \sin \mu). \tag{4.9}$$

The initial conditions at $s = s_1$ can be used to replace the constants A and B:

$$A = y_1\beta_1^{-\frac{1}{2}}; \qquad B = (y_1'\beta_1^{\frac{1}{2}} + y_1\alpha_1\beta_1^{-\frac{1}{2}}). \tag{4.10}$$

The substitution of (4.10) in (4.7) and (4.9) yields a solution at $s = s_2$ whose two terms are known as the *principal trajectories*.

$$y(s_2) = (\beta_2/\beta_1)^{\frac{1}{2}}(\cos \mu + \alpha_1 \sin \mu)y_1 + (\beta_1\beta_2)^{\frac{1}{2}} \sin \mu\, y_1' \tag{4.11}$$

$$y'(s_2) = -(\beta_1\beta_2)^{-\frac{1}{2}}[(1 + \alpha_1\alpha_2) \sin \mu + (\alpha_2 - \alpha_1) \cos \mu]y_1$$
$$+ (\beta_1/\beta_2)^{\frac{1}{2}}(\cos \mu - \alpha_2 \sin \mu)y_1'. \tag{4.12}$$

Equations (4.11) and (4.12) are more frequently seen in the form of the generalised transfer matrix,

$$T(s_2|s_1)$$

$$= \begin{pmatrix} (\beta_2/\beta_1)^{\frac{1}{2}}(\cos \Delta\mu + \alpha_1 \sin \Delta\mu), & (\beta_1\beta_2)^{\frac{1}{2}} \sin \Delta\mu \\ -(\beta_1\beta_2)^{-\frac{1}{2}}[(1 + \alpha_1\alpha_2) \sin \Delta\mu + (\alpha_2 - \alpha_1) \cos \Delta\mu], & (\beta_1/\beta_2)^{\frac{1}{2}}(\cos \Delta\mu - \alpha_2 \sin \Delta\mu) \end{pmatrix} \tag{4.13}$$

where $\Delta\mu = \mu(s_2) - \mu(s_1)$.

This form of the transfer matrix is completely general and applies equally to transfer lines or sections of circular machines.

4.1.2 The transfer matrix for a periodic lattice

In a circular machine the general focusing function $K(s)$ will be periodic, i.e. $K(s) = K(s + L)$, where L is the length of a period or cell. Floquet's theorem states that equations of the form (3.1) with periodic coefficients will have a particular solution of the form $\exp(as) \cdot \phi(s)$, where $\phi(s)$ is a periodic function with the same period as the coefficients. This is intuitively the solution one would expect to find in a circular accelerator for the beam envelope. This solution is found from (4.13) by imposing the conditions of periodicity: $\beta_1 = \beta_2 = \beta$ and $\alpha_1 = \alpha_2 = \alpha$. Thus for one period in a circular machine (4.13) reduces to,

$$T(s + L|s) = \begin{pmatrix} \cos \mu_0 + \alpha \sin \mu_0 & \beta \sin \mu_0 \\ -\gamma \sin \mu_0 & \cos \mu_0 - \alpha \sin \mu_0 \end{pmatrix} \tag{4.14}$$

where

$$\gamma = (1 + \alpha^2)/\beta. \tag{4.15}$$

This matrix provides a simple way of calculating $\beta(s)$ throught the period by comparing the matrix elements of (4.14) with the full-period matrices evaluated with the origin at the entry to each element in turn. It should be noted that in a periodic structure the phase advance in a period, μ_0, is independent of the origin as can be seen from (4.4).

The parameters α, β and γ are known as the *Courant and Snyder parameters* and sometimes as the *Twiss parameters**.

By inspection it can be seen that the matrix (4.14) can be split into two parts, such that,

$$T(s + L|s) = I \cos \mu_0 + J \sin \mu_0 \tag{4.16}$$

where

$$I = \begin{pmatrix} 1 & 0 \\ 0 & 1 \end{pmatrix} \quad J = \begin{pmatrix} \alpha & \beta \\ -\gamma & -\alpha \end{pmatrix}. \tag{4.17}$$

It is easily verified that $J^2 = -I$, so that the algebra of $T(s)$ is the same as that of a complex number. Thus De Moivre's formula[2] provides a quick way of finding the transmission through n identical periods of a lattice, each period with the phase advance μ_0.

$$T^n = (I \cos\mu_0 + J \sin \mu_0)^n = I \cos(n\mu_0) + J \sin(n\mu_0) \tag{4.18}$$

---

* Although these parameters are often referred to as the Twiss parameters it appears that they did not in fact originate with him.

and the inverse of T can be simply written as,

$$T^{-1} = I \cos \mu_0 - J \sin \mu_0. \tag{4.19}$$

The simplest single-period matrix can be found by choosing the origin to be at a symmetry plane, in which case $\alpha = 0$. Thus the matrix for such a period would be,

$$T(\text{symmetry}) = \begin{pmatrix} \cos \mu_0 & \beta \sin \mu_0 \\ -\beta^{-1} \sin \mu_0 & \cos \mu_0 \end{pmatrix}. \tag{4.20}$$

In practical machines the basic periodicity of the cells will be broken by long straight sections and other insertions to form a number of superperiods. In this case, L becomes the length of a superperiod. If errors are taken into account, only the full machine circumference can be taken as periodic and L is then replaced by C and μ_0 by $2\pi Q$ in (4.14) and (4.20).

4.2 INVARIANT OF THE UNPERTURBED MOTION

It is now convenient to return to the basic equation of the betatron motion (4.1), to introduce the phase advance μ and then to differentiate this equation to obtain

$$y(s) = A\beta^{\frac{1}{2}} \cos(\mu + B) \tag{4.1}'$$

$$y'(s) = -A\alpha\beta^{-\frac{1}{2}} \cos(\mu + B) - A\beta^{-\frac{1}{2}} \sin(\mu + B). \tag{4.21}$$

The phase terms can be extracted and eliminated between these two equations to yield an invariant for the motion.

$$\left. \begin{aligned} A \cos(\mu + B) &= y\beta^{-\frac{1}{2}} \\ A \sin(\mu + B) &= (y'\beta^{\frac{1}{2}} + y\alpha\beta^{-\frac{1}{2}}) \\ A^2 &= y^2\beta^{-1} + (y'\beta^{\frac{1}{2}} + y\alpha\beta^{-\frac{1}{2}})^2 \end{aligned} \right\} \tag{4.22}$$

which is often written as,

$$A^2 = \gamma y^2 + 2\alpha yy' + \beta y'^2 \ (=\text{constant}). \tag{4.23}$$

Equation (4.23) is in fact the equation of an ellipse in (y, y') space or *phase space* as it is known. The constant A^2 is equal to the area of this ellipse divided by π. The form of this ellipse will vary from point to point in the lattice but its area will remain unchanged*.

---

* The area of the ellipse, $S = A^2\pi = \pi(\gamma y^2 + 2\alpha yy' + \beta y'^2)$, is frequently referred to as the *single particle emittance* (as opposed to the beam emittance). This can be a useful concept, but strictly speaking *emittance* is defined as a phase-space area and single particles are singularities which have no definition in terms of area.

4.3   PROPAGATION OF THE COURANT AND SNYDER PARAMETERS

Use can be made of the invariant (4.23) to derive transfer matrices for the Courant and Snyder parameters through a lattice. Two solutions are given below.

4.3.1 3×3 matrix solution[3]

The invariant states that between two points,

$$\gamma_2 y_2{}^2 + 2\alpha_2 y_2 y_2{}' + \beta_2 y_2{}'^2 = \gamma_1 y_1{}^2 + 2\alpha_1 y_1 y_1{}' + \beta_1 y_1{}'^2. \qquad (4.24)$$

The trajectory at two different points is related by the transfer matrix $T(s_2|s_1)$, which on this occasion is more conveniently written in the inverse form, i.e. from point 2 to point 1,

$$\begin{pmatrix} y \\ y' \end{pmatrix}_1 = \begin{pmatrix} t_{22} & -t_{12} \\ -t_{21} & t_{11} \end{pmatrix} \begin{pmatrix} y \\ y' \end{pmatrix}_2. \qquad (4.25)$$

Equation (4.25) is used to substitute for y_1 and $y_1{}'$ in (4.24). The terms on the righthand side of (4.24) are then regrouped and the expressions given below for α_2, β_2 and γ_2 can be found in terms of α_1, β_1 and γ_1 by the direct comparison of terms on each side of the equality. These results are usually written in the form of a 3×3 matrix,

$$\begin{pmatrix} \beta \\ \alpha \\ \gamma \end{pmatrix}_2 = \begin{pmatrix} t_{11}{}^2 & -2t_{11}t_{12} & t_{12}{}^2 \\ -t_{11}t_{21} & [t_{11}t_{22} + t_{12}t_{21}] & -t_{12}t_{22} \\ t_{21}{}^2 & -2t_{21}t_{22} & t_{22}{}^2 \end{pmatrix} \begin{pmatrix} \beta \\ \alpha \\ \gamma \end{pmatrix}_1. \qquad (4.26)$$

In a drift space for example, $t_{11} = t_{22} = 1$, $t_{12} = l$ and $t_{21} = 0$ from (3.5), so that,

$$\beta_2 = \beta_1 - 2l\alpha_1 + l^2\gamma_1. \qquad (4.27)$$

Thus β varies parabolically and if the origin is chosen at a 'waist' in the beam where $\alpha = 0$ then (4.27) simplifies to $\beta(l) = \beta(0)[1 + l^2/\beta(0)^2]$, which is useful for calculating apertures in low-β insertions.

It is also interesting to note that γ is constant within a drift space.

4.3.2 2×2 matrix solution[4]

The above method is perhaps the better known way of transferring the Courant and Snyder parameters through a lattice, but the second method given below is more useful for analytical manipulations.

The reader can quickly verify that the motion invariant (4.23) can be written in a matrix form as

$$A^2 = Y^T \sigma^{-1} Y,$$ (4.28)

where,

$$\sigma = \begin{pmatrix} \beta & -\alpha \\ -\alpha & \gamma \end{pmatrix} \quad \text{and} \quad Y = \begin{pmatrix} y \\ y' \end{pmatrix}.$$

Thus the invariant states that between two points,

$$Y_2^T \sigma_2^{-1} Y_2 = Y_1^T \sigma_1^{-1} Y_1.$$ (4.29)

As before, use is made of the inverse transfer from point 2 to point 1, i.e. $Y_1 = T^{-1} Y_2$ by substituting into (4.29) to give,

$$Y_2^T \sigma_2^{-1} Y_2 = (T^{-1}Y_2)^T \sigma_1^{-1} (T^{-1}Y_2)$$

$$Y_2^T \underbrace{\sigma_2^{-1}} Y_2 = Y_2^T \underbrace{(T^{-1})^T \sigma_1^{-1} T^{-1}} Y_2$$ (4.30)

$$\sigma_2 \equiv T \sigma_1 T^T.$$

This method is of interest since it has the advantage of maintaining the 2×2 matrix formalism.

4.4 EMITTANCE AND ACCEPTANCE

In practice accelerator beams do not comprise a single particle or even a small number of particles, but large numbers, routinely of the order of 10^{10} to 10^{13}. Each particle will have its own motion invariant, but all the ellipses in the same beam will have the same form. Most particles will have small invariants and they will not oscillate far from the equilibrium orbit. At progressively larger oscillation amplitudes there will be fewer and fewer particles. If the beam profile is measured, for example by a flying wire monitor, this will give the projection of the phase-space population onto a transverse coordinate axis (see Figure 4.1). This population will normally have a near-gaussian distribution with tails that stretch to the vacuum chamber walls. The standard deviation of this distribution, σ_y, is a convenient parameter for indicating the size of the beam. Vacuum chamber apertures are often quoted in beam sigmas for example. Sigma is also used to define the *geometrical beam emittance* as,

$$\mathcal{E}_y = \pi \sigma_y^2 / \beta_y.$$ (4.31A)

The emittance is the phase-space area of an ellipse in the (y, y') phase plane with its maximum excursion equal to σ_y. Just as σ_y is representative of the beam size, $\mathcal{E}_y$ is

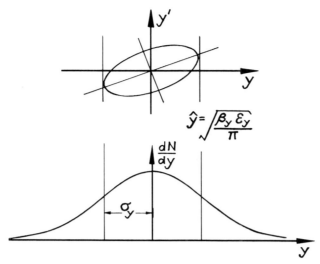

Figure 4.1. Beam profile and emittance definition.

representative of the phase-space area occupied by the beam. The emittance ellipse will have the same form as the single particle invariant ellipses.

Certain simple geometric properties of an ellipse can be used to find the maximum beam excursion, divergence etc. providing that the emittance is known, along with the α_y, β_y and γ_y values at the points of interest in the lattice.

The starting point of this chapter was equation (3.1), which assumed that there would be no acceleration and in this case the geometric emittance (4.31A) is invariant. If acceleration is included, the phase space element $\delta y \delta y'$ is no longer constant and it is the element $\delta y \delta p_y$ that is invariant (see Appendix A). To work in the (y, p_y) phase plane, y' on the vertical axes in Figures 4.1 and 4.2 should be replaced by p_y, where p_y can be written in the normalised form,

$$p_y = \frac{mv_y}{m_0 c} = \beta\gamma y'.$$

The phase-space area (4.31A) then becomes, in the new variables (y, p_y),

$$\varepsilon_n = \beta\gamma\pi\sigma_y^2/\beta_y. \qquad (4.31B)$$

This is called the *normalised, invariant emittance* to distinguish it from the geometric emittance (4.31A), which is not a true invariant.

It can now be seen that acceleration leads to a reduction of the geometric emittance, the so-called *adiabatic damping*. When accelerating from energy γ_1 to γ_2, the normalised emittance (4.31B) stays constant while the geometric emittance is 'damped' by $(\beta\gamma)_1/(\beta\gamma)_2$ and the amplitudes of the particle oscillations by the square root of this.

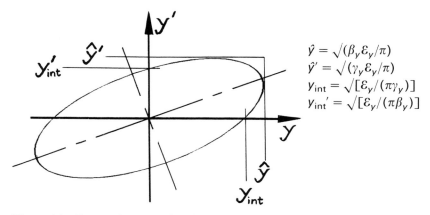

Figure 4.2. Geometric properties of an ellipse.

The *acceptance* of a machine, or beam line, is given by equations (4.31) with σ_y replaced by the aperture a_y, which defines the largest phase-space ellipse that will pass through the lattice.

Unfortunately, not everyone agrees with the emittance definitions (4.31), which are principally supported by the designers of electron machines. Some proton machine designers feel that $2\sigma_y$ is more representative of the beam halfwidth, which makes their definition of emittance four times larger. Readers should also beware that the factor π may or may not be included, which makes four possible definitions. It is recommended that emittances be quoted with π visible, e.g. $10^{-5}\pi$ [m].

4.5 DISTINCTIONS BETWEEN CIRCULAR MACHINES AND TRANSFER LINES

4.5.1 Circular machines

The periodicity of a circular machine imposes the same periodicity on the parameters α, β and γ and determines them uniquely. If the coordinates of a particle are sampled after successive turns the points will fill out the perimeter of an ellipse in phase space. Only one set of α, β and γ values will fit that ellipse; those which are derived from the matched single-turn transfer matrix. It is the periodicity of the machine which ensures that this ellipse is constant turn after turn and for this reason it is called the *matched ellipse*. Suppose now that a beam of particles is injected with a distribution in phase space that does not have the form of the matched ellipse (see Figure 4.3), but defines some new ellipse characterised by different parameters $\alpha^\star$, $\beta^\star$ and $\gamma^\star$. The circular machine will not faithfully return this ellipse after each turn. Instead the

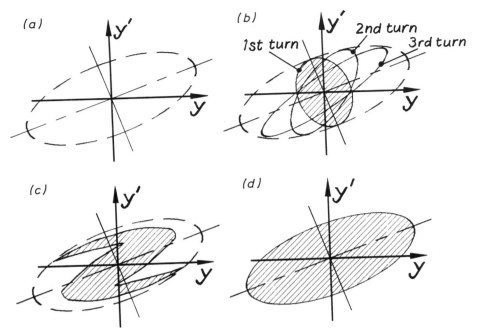

Figure 4.3. Matched, mismatched and filamenting ellipses. (a) Matched ellipse, (b) mismatched beam, (c) filamenting beam, (d) fully filamented beam.

ellipse will tumble round and round filling out a larger ellipse of the matched form, since that is what the individual particles are doing.

In a truly linear system, the original ellipse, defined by the particle distribution, would tumble over and over indefinitely inside the envelope of the matched ellipse and it would maintain its elliptical form and area as if the circular machine were an infinitely long transfer line. In a real lattice, however, there are always small non-linearities which cause an amplitude frequency dependence, which in turn distorts the ellipse as illustrated in Figure 4.3. Liouville's theorem requires that the phase-space density is conserved and in a strict mathematical sense this is so. As the figure becomes more wound up the arms become narrower so that the area does indeed stay constant. It does not take long, however, before the beam is apparently uniformly distributed around the matched ellipse and for all practical purposes the beam emittance has been increased. This is called *dilution of phase space by filamentation* (coarse-graining), which is present to a greater or lesser extent at the injection into all synchrotrons[5].

Since α and β depend on the whole structure, a change at any point in the lattice will change the α and β functions around the whole ring (matched insertions excepted, see Section 4.7).

4.5.2 Transfer lines*

A transfer line is free of the constraints of periodicity. The beam passes only once and the shape of the phase-space ellipse at the entry to the line determines the shape at the exit. Exactly the same transfer line, injected first with one emittance ellipse and then with another, has to be accredited with different α and β values to describe the two cases. Thus α and β depend on the input beam as well as the lattice. Measuring emittance in a transfer line is complicated by this possible mismatch between the beam's parameters and the designer's values for the line, whereas in a circular machine one beam width measurement is sufficient since β is unambiguously known. For a single particle, there will be an infinite number of sets of α and β values to describe its motion through the line. This can be seen from the generalised transfer matrix (4.13) whose four elements (three of which are free) are determined by five parameters α_1, α_2, β_1, β_2 and μ, whereas the machine matrix (4.14) is exactly determined. Thus a transfer line needs an external criterion, such as the matched ellipse coming from a circular machine, to determine its lattice parameters. The underlying transfer matrices, however, depend solely on the lattice. Finally, unlike the circular machine, any change in the lattice will only alter the downstream α and β functions.

4.6 MOTION WITH A MOMENTUM DEVIATION[3,6]

4.6.1 Principal trajectories

Before proceeding to the derivation of an expression for the local dispersion function $D(s)$, which will be the parameterised equivalent of (3.48) from Chapter 3, it is useful to make a small diversion and to investigate the properties of *principal trajectories*†, which were briefly mentioned in Section 4.1.1.

It is customary to abbreviate the terms in (4.11) and (4.12) as,

$$y(s) = \mathscr{C}(s)y(s_0) + \mathscr{S}(s)y'(s_0) \tag{4.32}$$

$$y'(s) = \mathscr{C}'(s)y(s_0) + \mathscr{S}'(s)y'(s_0), \tag{4.33}$$

where

      $\mathscr{C}(s)$ represents the 'cosine-like' trajectory, since $\mathscr{C}(s_0) = 1$ and $\mathscr{C}'(s_0) = 0$,
      $\mathscr{S}(s)$ represents the 'sine-like' trajectory, since $\mathscr{S}(s_0) = 0$ and $\mathscr{S}'(s_0) = 1$,
      s_0 is an arbitrary origin, and
      $'$ denotes differentiation with respect to s.

---

* Linacs fall between the cases of circular machines and transfer lines since periodicity is normally imposed on their focusing structure.
† Reference 6 refers to 'principal trajectories' as 'fundamental solutions'.

In agreement with (4.32) and (4.33), the transfer matrix is written very concisely as,

$$T(s|s_0) = \begin{pmatrix} \mathscr{C} & \mathscr{S} \\ \mathscr{C}' & \mathscr{S}' \end{pmatrix}.$$ (4.34)

Some interesting and useful relationships now become evident. For example, it is possible to express $\mathscr{C}$ and $\mathscr{C}'$ in terms of $\mathscr{S}$ and $\mathscr{S}'$, or vice versa. Consider the identities,

$$\mathscr{C}^2 \frac{d}{ds}\left[\frac{\mathscr{S}}{\mathscr{C}}\right] = \mathscr{C}\mathscr{S}' - \mathscr{S}\mathscr{C}'$$ (4.35)

$$-\mathscr{S}^2 \frac{d}{ds}\left[\frac{\mathscr{C}}{\mathscr{S}}\right] = \mathscr{C}\mathscr{S}' - \mathscr{S}\mathscr{C}'$$ (4.36)

which must be unity, since $|T| = \mathscr{C}\mathscr{S}' - \mathscr{S}\mathscr{C}' = 1$. Thus (4.35) and (4.36) can be integrated to give,

$$\mathscr{S}(s) = \mathscr{C}(s)\int_0^s \frac{1}{\mathscr{C}^2(\sigma)}\,d\sigma, \qquad \mathscr{S}'(s) = \mathscr{C}'(s)\int_0^s \frac{1}{\mathscr{C}^2(\sigma)}\,d\sigma + \frac{1}{\mathscr{C}(s)}$$ (4.37)

$$\mathscr{C}(s) = -\mathscr{S}(s)\int_0^s \frac{1}{\mathscr{S}^2(\sigma)}\,d\sigma, \qquad \mathscr{C}'(s) = -\mathscr{S}'(s)\int_0^s \frac{1}{\mathscr{S}^2(\sigma)}\,d\sigma - \frac{1}{\mathscr{S}(s)}.$$ (4.38)

4.6.2 Use of principal trajectories to express the dispersion function for a circular machine

As was explained in Chapter 3 Section 3.4.2, the solution to the equation of motion (3.37) with a momentum deviation comprises two parts: the complementary function and the particular integral. The former can be written with the aid of (4.32) and (4.33), so that,

$$y(s) = A\mathscr{C}(s) + B\mathscr{S}(s) + \text{PI}(s),$$ (4.39)

where A and B are integration constants, which vary according to the choice made for the particular integral $\text{PI}(s)$.

For the moment it is legitimate to construct any function which works for the particular integral. For example, a possible solution using the principal trajectories of the previous section, is

$$\text{PI}(s) = \mathscr{S}(s)\int_{s_0}^s F(\sigma)\mathscr{C}(\sigma)\,d\sigma - \mathscr{C}(s)\int_{s_0}^s F(\sigma)\mathscr{S}(\sigma)\,d\sigma$$ (4.40)

where $F(s)$ is a driving term, which for a momentum deviation has the form,

$$F(s) = \frac{1}{\rho_0(s)}\frac{\Delta p}{p_0}$$

as seen in (3.37) and s_0 is an arbitrary origin. The substitution of (4.40) into (3.37) verifies that this is a valid solution. The form of (4.40) is a linear sum of $\mathscr{S}(s)$ and $\mathscr{C}(s)$ trajectories with $\mathrm{PI}(s_0) = 0$ and $\mathrm{PI}'(s_0) = 0$. Other valid solutions would be $\mathrm{PI}(s)$ plus any other linear combination of $\mathscr{C}(s)$ and $\mathscr{S}(s)$. In order to find the dispersion function for a circular machine, it is necessary to choose from all these possible solutions the one which is a closed orbit, i.e. the one for which $y_D(s_0) = y_D(s_0 + C)$ and $y'(s_0) = y'(s_0 + C)$. This special solution is denoted by the subscript 'D' as in Section 3.4.2. The application of these boundary conditions to (4.39) gives,

$$A\mathscr{C}(s_0) + B\mathscr{S}(s_0) + \mathrm{PI}(s_0) = A\mathscr{C}(s_0 + C) + B\mathscr{S}(s_0 + C) + \mathrm{PI}(s_0 + C) \tag{4.41}$$

$$A\mathscr{C}'(s_0) + B\mathscr{S}'(s_0) + \mathrm{PI}'(s_0) = A\mathscr{C}'(s_0 + C) + B\mathscr{S}'(s_0 + C) + \mathrm{PI}'(s_0 + C). \tag{4.42}$$

The definitions of $\mathscr{C}$ and $\mathscr{S}$ from (4.32) and (4.33) and the definition of PI from (4.40) impose the following conditions:

$$\left.\begin{array}{lll} \mathscr{C}(s_0) = 1 & \mathscr{C}'(s_0) = 0 & \mathrm{PI}(s_0) = 0 \\ \mathscr{S}(s_0) = 0 & \mathscr{S}'(s_0) = 1 & \mathrm{PI}'(s_0) = 0. \end{array}\right\} \tag{4.43}$$

The solution for the constant A from the above is therefore,

$$A = \frac{\mathscr{S}(s_0 + C)\,\mathrm{PI}'(s_0 + C) + [1 - \mathscr{S}'(s_0 + C)]\,\mathrm{PI}(s_0 + C)}{[1 - \mathscr{C}(s_0 + C)][1 - \mathscr{S}'(s_0 + C)] - \mathscr{C}'(s_0 + C)\mathscr{S}(s_0 + C)}. \tag{4.44}$$

The denominator of (4.44) only contains terms from the transfer matrix (4.34), which is expressed parametrically in (4.14) if, for a full turn, μ_0 is replaced by $2\pi Q$. Thus, with a little manipulation,

$$\text{Denom.} = 2 - \operatorname{Tr} \boldsymbol{T} = 4 \sin^2(\pi Q). \tag{4.45}$$

The numerator can be expressed with the help of (4.40) to give,

$$\text{Num.} = [\mathscr{C}(s_0 + C)\mathscr{S}'(s_0 + C) - \mathscr{C}'(s_0 + C)\mathscr{S}(s_0 + C)] \int_{s_0}^{s_0 + C} F(\sigma)\mathscr{S}(\sigma)\,\mathrm{d}\sigma$$

$$- \mathscr{C}(s_0 + C) \int_{s_0}^{s_0 + C} F(\sigma)\mathscr{S}(\sigma)\,\mathrm{d}\sigma + \mathscr{S}(s_0 + C) \int_{s_0}^{s_0 + C} F(\sigma)\mathscr{C}(\sigma)\,\mathrm{d}\sigma.$$

The first term is the determinant of $\boldsymbol{T}$ and is unity and the remainder of the expression can be re-expressed by substitution from the matrix (4.14) with $\mu_0 = 2\pi Q$ for a full turn, as for the denominator.

$$\text{Num.} = \beta^{\frac{1}{2}}(s_0)2 \sin \pi Q \int_{s_0}^{s_0 + C} \beta^{\frac{1}{2}}(\sigma)F(\sigma) \cos[\mu(\sigma) - \mu(s_0) - \pi Q]\,\mathrm{d}\sigma \tag{4.46}$$

At the origin s_0, the boundary conditions (4.43) are such that,

$$y_D(s_0) = A. \tag{4.47}$$

Furthermore, the origin s_0 was arbitrary, so that in the expressions (4.45) and (4.46), which yield A, s_0 can be replaced by the general coordinate s. Thus the special closed trajectory comes directly from (4.45) to (4.47) as

$$y_D(s) = \frac{\beta^{\frac{1}{2}}(s)}{2 \sin \pi Q} \int_s^{s+C} \beta^{\frac{1}{2}}(\sigma) F(\sigma) \cos[\mu(\sigma) - \mu(s) - \pi Q] \, d\sigma. \tag{4.48}$$

Equation (4.48) gives the closed orbit for any driving term $F(s)$ and is recorded in this general form since it will be of further use in the next chapter for calculating closed orbits with field errors.

Finally, the dispersion function $D(s)$ is found from (4.48) by inserting $F(\sigma) = [1/\rho_0(s)]\Delta p/p_0$, which gives

$$D(s) = \frac{y_D(s)}{\Delta p/p_0} = \frac{\beta^{\frac{1}{2}}(s)}{2 \sin \pi Q} \int_s^{s+C} \frac{\beta^{\frac{1}{2}}(\sigma)}{\rho_0(\sigma)} \cos[\mu(\sigma) - \mu(s) - \pi Q] \, d\sigma. \tag{4.49}$$

4.6.3   Dispersion invariant

Equation (4.49) describes a free betatron oscillation in regions where $1/\rho_0(s)$ is zero, i.e. in regions with no bending. In places where $1/\rho_0(s)$ is finite, the oscillation suffers a phase shift. The amplitude is of course continuous everywhere. Since the dispersion trajectory $D(s)$ is nothing but a free betatron oscillation in regions with no bending, it will obey the motion invariant (4.23). Hence, for *regions with no bending* there is a *dispersion* or *lattice invariant*, H and

$$H = \gamma D(s)^2 + 2\alpha D(s)D'(s) + \beta D'(s)^2 \ (= \text{constant}). \tag{4.50}*$$

In the derivation of the motion invariant (4.23) the sine and cosine of the betatron phase advance were expressed in terms of the trajectory in (4.22). These equations can also be applied to bending-free regions to give the phase shift $\Delta\mu$ between two points as,

$$\left.\begin{aligned} \cos \Delta\mu &= \frac{D_1 D_2 + (\alpha_1 D_1 + \beta_1 D_1')(\alpha_2 D_2 + \beta_2 D_2')}{(\beta_1\beta_2)^{1/2} H} \\ \sin \Delta\mu &= \frac{D_2(\alpha_1 D_1 + \beta_1 D_1') - D_1(\alpha_2 D_2 + \beta_2 D_2')}{(\beta_1\beta_2)^{1/2} H}. \end{aligned}\right\} \tag{4.51}$$

Equations (4.51) will be needed in Section 4.7.3 to determine the phase advance across a matching section. Finally the dispersion can be found anywhere within a bending-free region by applying the dispersion vector to the standard transfer matrix for the

---

* For synchrotron light sources, the function $H(s)$ is evaluated throughout the lattice. Its value in the bending magnets is of importance for the quantum excitation effect that causes growth of the horizontal beam emittance (see Section 10.8.2).

lattice, i.e.

$$\begin{pmatrix} D \\ D' \end{pmatrix}_2 = \begin{pmatrix} t_{11} & t_{12} \\ t_{21} & t_{22} \end{pmatrix} \begin{pmatrix} D \\ D' \end{pmatrix}_1 . \tag{4.52}$$

4.7   A SIMPLE APPROACH TO LATTICE DESIGN

It was natural to consider the early accelerators as single units. Cyclotrons and betatrons had 360° magnets and in the early weak focusing synchrotrons, the introduction of straight sections was thought to be a rather daring innovation. The invention of strong focusing replaced the 'single-unit' concept by the 'multi-segment' machine. A great deal of ingenuity has since gone into designing lattice modules with a small number of elements for specific functions. A modern machine designer will automatically think in terms of these modules and will assemble a lattice from such units as *matched cells, dispersion suppressors, insertions* and *achromats*. In large machines, the different modules will be well defined and the various functions of the lattice easily understood, but often the lack of space in small machines puts many requirements onto the same elements and the functions of the different parts of the lattice are no longer so obvious.

In this section some well-known lattice modules, which might typically be found in a large collider (see Figure 4.4), will be analysed. The aim is to provide approximate

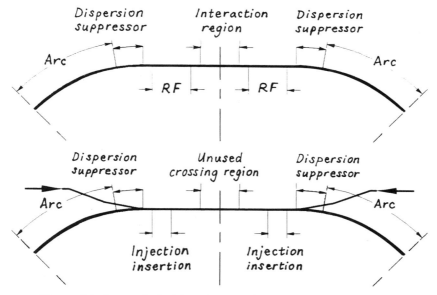

Figure 4.4. Sectors in fictitious large collider.

analytic solutions, which will act as design guides. A computer program for lattice design can then be used to refine the solutions; a task to which it should be well suited. There are many such codes that will adjust the lattice elements to fit any desired parameter configuration by means of minimisation algorithms, but without a guiding design, it is possible to spend much unnecessary effort modifying input values and weighting factors only to obtain a numerical result, which cannot be guaranteed as optimum nor will it give any deep insight.

4.7.1   FODO matched cells in arcs and straight sections

The ubiquitous FODO cell (see Figure 4.5) is used in nearly all long transfer lines, high-energy accelerators and large storage rings. In Figure 4.4 for example, the FODO lattice would extend through the arcs. Dispersion suppressors would be embedded in the FODO lattice, but would not actually interrupt the focusing structure, which would continue right up to the insertion for the interaction region. The injection insertions might have some special features, but the majority of the lattice would be basic FODO cells. One class of storage rings, the synchrotron-light sources, often use special modules called achromats but even in this application the FODO cell is not completely disregarded.

The reader will remember that it was the symmetric FODO cell that was used in Section 3.2.2 to illustrate the 'necktie' stability diagram. The same simple structure will be used here, except that dipoles will be included so as to obtain two simple relationships for the dispersion function. In this analysis, it will be assumed that the quadrupoles can be replaced by thin lenses, that the bending radius in the dipoles is large compared with the individual focal lengths of the lenses and that there are no intermediate drift spaces. There is little point in considering anything beyond this simple situation, since the value of this analysis lies in the physical insight it gives and the simplicity of the expressions, which can be quickly applied for 'back of

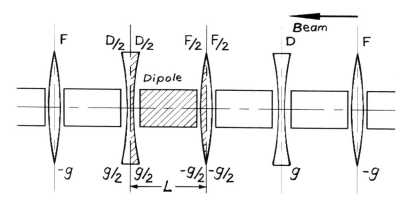

Figure 4.5. Symmetric FODO cell.

envelope' estimates. In order to take account of complications such as long lenses and asymmetries in a specific design, a lattice computer program will in any case be needed. By symmetry, the extrema of the betatron functions are at the centres of the lenses and it is sufficient to consider only a half cell from the mid-point of the F-quadrupole to the mid-point of the D. With these simplifying assumptions, the transfer matrices for a half cell are,

$$
T_{\frac{1}{2}\text{cell}} = \begin{pmatrix} 1 & 0 & 0 \\ \frac{1}{2}g & 1 & 0 \\ 0 & 0 & 1 \end{pmatrix} \begin{pmatrix} \cos(L/\rho_0) & \rho_0 \sin(L/\rho_0) & \rho_0[1 - \cos(L/\rho_0)] \\ -\rho_0^{-1} \sin(L/\rho_0) & \cos(L/\rho_0) & \sin(L/\rho_0) \\ 0 & 0 & 1 \end{pmatrix} \begin{pmatrix} 1 & 0 & 0 \\ -\frac{1}{2}g & 1 & 0 \\ 0 & 0 & 1 \end{pmatrix}.
$$

The central matrix corresponds to a sector dipole with no gradient, as described in Section 3.3.2. This matrix is now simplified to the same order as used in the thin-lens quadrupoles. The cosine terms are replaced by unity and the sine terms by their argument. The first dispersion term, however, is retained to second order, as it would otherwise be zero. In addition, the element in the first column and second row is put to zero. Firstly, this term is small since $\rho \gg 1/g$ and secondly, by setting this element to zero the 2×2 sub-matrix for the betatron motion maintains its unit determinant rigorously through to the final result (4.53). The dipole therefore looks like a drift space with dispersion. The choice of a sector dipole was a convenient starting point, but at the level of approximation applied a rectangular magnet would have served equally well.

$$
T_{\frac{1}{2}\text{cell}} = \begin{pmatrix} 1 & 0 & 0 \\ \frac{1}{2}g & 1 & 0 \\ 0 & 0 & 1 \end{pmatrix} \begin{pmatrix} 1 & L & \frac{1}{2}L^2/\rho_0 \\ -L/\rho_0^2 & 1 & L/\rho_0 \\ 0 & 0 & 1 \end{pmatrix} \begin{pmatrix} 1 & 0 & 0 \\ -\frac{1}{2}g & 1 & 0 \\ 0 & 0 & 1 \end{pmatrix}.
$$

The overall matrix for the half cell is therefore

$$
T_{\frac{1}{2}\text{cell}} = \left(\begin{array}{cc|c} 1 - \frac{1}{2}gL & L & \frac{1}{2}L^2/\rho_0 \\ -\frac{1}{4}g^2L & 1 + \frac{1}{2}gL & (L/\rho_0)(1 + \frac{1}{4}gL) \\ \hline 0 & 0 & 1 \end{array} \right). \tag{4.53}
$$

Since the entry and exit of the half cell were chosen to coincide with the extrema of the betatron functions, these extrema can now be evaluated in terms of the parameters of the cell by directly comparing the matrix elements of the 2×2 matrix for the betatron motion contained in (4.53) with the general transfer matrix (4.13) with $\alpha = 0$. Equation (4.13) is rewritten in terms of the extrema as,

$$
T_{\frac{1}{2}\text{cell}} = \begin{pmatrix} (\hat{\beta}/\check{\beta})^{\frac{1}{2}} \cos(\mu_0/2) & (\hat{\beta}\check{\beta})^{\frac{1}{2}} \sin(\mu_0/2) \\ -(\hat{\beta}\check{\beta})^{-\frac{1}{2}} \sin(\mu_0/2) & (\check{\beta}/\hat{\beta})^{\frac{1}{2}} \cos(\mu_0/2) \end{pmatrix}. \tag{4.54}
$$

where $\check{\beta}$ is taken at the entry and $\hat{\beta}$ at the exit. The ratios of the terms in the leading diagonals of (4.53) and (4.54) and the same ratios in the secondary diagonals yield two equalities linking the betatron extrema and the lens strengths,

$$\hat{\beta}/\check{\beta} = (1 - \tfrac{1}{2}gL)/(1 + \tfrac{1}{2}gL) \qquad \text{and} \qquad (\hat{\beta}\check{\beta})^{-1} = \tfrac{1}{2}g^2. \tag{4.55}$$

The product of the terms in the secondary diagonals yields a third equality, which introduces the phase advance,

$$\sin(\mu_0/2) = \tfrac{1}{2}gL. \tag{4.56}$$

Solution of the first two equalities (4.55) gives relations for $\hat{\beta}$ and $\check{\beta}$ into which the third equality (4.56) can be substituted to give (with a little manipulation) the well-known expressions,

$$\hat{\beta} = 2L \frac{1 + \sin(\mu_0/2)}{\sin \mu_0} \qquad \text{and} \qquad \check{\beta} = 2L \frac{1 - \sin(\mu_0/2)}{\sin \mu_0}. \tag{4.57}$$

The symmetry of the cell also ensures that the extrema of the dispersion will be at the centres of the quadrupoles and the values can be found by transferring the dispersion vector ($D = \hat{D}$, $D' = 0$, $\Delta p/p = 1$) through the matrix (4.53) and imposing the cyclic boundary condition. Matrix (4.53) starts at the mid-F-quadrupole position, so the dispersion will be a maximum at entry and be a minimum at the exit at the centre of the D-quadrupole and yields the two equations,

$$\check{D} = (1 - \tfrac{1}{2}gL)\hat{D} + \tfrac{1}{2}L^2/\rho_0$$

$$0 = -\tfrac{1}{2}g^2 L\hat{D} + (L/\rho_0)(1 + \tfrac{1}{4}gL),$$

which can be solved with the help of (4.56) to give,

$$\hat{D} = \frac{(2L)^2}{\rho_0} \frac{[1 + \tfrac{1}{2}\sin(\mu_0/2)]}{4\sin^2(\mu_0/2)} \qquad \text{and} \qquad \check{D} = \frac{(2L)^2}{\rho_0} \frac{[1 - \tfrac{1}{2}\sin(\mu_0/2)]}{4\sin^2(\mu_0/2)}. \tag{4.58}$$

With a little manipulation the beam parameters can now be found at any position in the cell. For example, an expression that will be needed later is the betatron amplitude function at the mid-point between quadrupoles, which is quickly found by transposing the values from the centre of a lens using the matrix (4.26).

$$\beta_{\mathrm{m}} = \frac{2L}{\sin \mu_0} [1 - \tfrac{1}{2}\sin^2(\mu_0/2)] \tag{4.59}$$

Figure 4.6 contains typical plots of the beam parameters through a half-FODO cell and the universal curves for the betatron function extrema normalised by the cell half-length and the dimensionless parameter $\rho_0 D/L^2$ versus the phase advance. The turning point in the $\hat{\beta}/L$ curve indicates the minimum aperture required for a

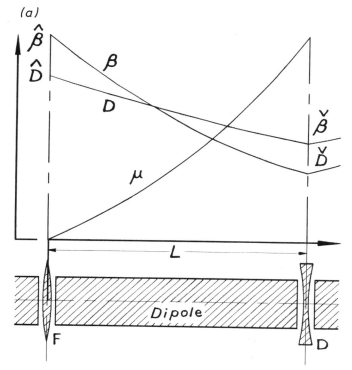

Figure 4.6. Universal curves for matched, symmetric FODO cells. (*a*) Typical beam parameters. (*Continued.*)

beam in a cell of half-length L. The exact position of this minimum can be found by differentiation of $\hat{\beta}$ in (4.57), i.e.

$$0 = \frac{d\hat{\beta}}{d\mu_0} = \frac{-L}{\sin\mu_0}\left[\frac{2\cos\mu_0}{\sin\mu_0}[1 + \sin(\mu_0/2)] - \cos(\mu_0/2)\right]$$

which defines $\mu_0 \simeq 76°$.

The dispersion function D in Figure 4.6(*a*) is close to being a straight line, which suggests that for a machine comprising only FODO cells and no drift spaces the momentum compaction α defined in (3.53) can be estimated as,

$$\alpha = \frac{1}{C_0}\oint\frac{D(s)}{\rho_0}\,ds \simeq \frac{\pi}{C_0}(\hat{D} + \check{D}) \tag{4.60}$$

and since it is often the case that the FODO cells occupy virtually all positions where ρ is finite, (4.60) will be the dominant contribution to the momentum compaction function in many practical machines.

For a machine comprising only N_c FODO cells,

$$2\pi = N_c(2L/\rho_0) \quad \text{and} \quad Q = N_c\mu_0/(2\pi). \tag{4.61}$$

71

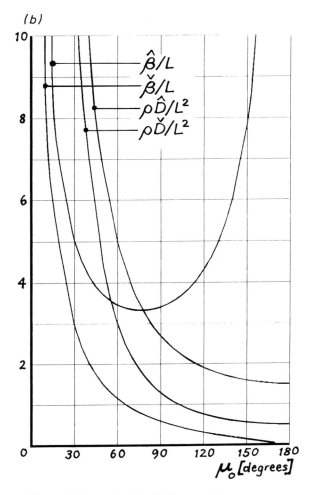

Figure 4.6 (*continued*). (*b*) Universal curves.

The substitution of (4.58) for $\hat{D}$ and $\check{D}$ into (4.60) and the use of (4.61) to re-express the result yields,

$$\alpha = \frac{1}{Q_x^2}\left[\frac{\mu_0/2}{\sin(\mu_0/2)}\right]^2. \tag{4.62}$$

This illustrates the statement made at the end of Section 2.6 that Q_x^{-2} is a useful approximation for α in alternating gradient lattices. For this particular case of FODO cells, the error is less than 5% up to a phase advance of 60° and is still only 11% at 90°.

The *natural chromaticity** of a FODO lattice can also be found by differentiation

* The *natural chromaticity* is the tune change ΔQ, caused by the dipole and quadrupole elements in a lattice, normalised by the corresponding momentum deviation $\Delta p/p_0$. Chromaticity is discussed in Chapter 6.

of (4.61) to give,

$$\frac{dQ}{d\Delta p/p_0} = \frac{N_c}{2\pi} \frac{d\mu_0}{d\Delta p/p_0}. \qquad (4.63)$$

The normalised quadrupole strength depends on momentum according to

$$g = g_0(1 - \Delta p/p_0), \qquad (4.64)$$

and (4.56) relates this strength to the cell's phase advance so that,

$$\tfrac{1}{2}g_0 L = \sin(\mu_0/2) \qquad (\Delta p/p = 0)$$

$$\tfrac{1}{2}g_0 L(1 - \Delta p/p_0) = \sin(\mu/2) \qquad (\Delta p/p \neq 0).$$

Differentiation of the second equation yields,

$$-g_0 L \, d\Delta p/p_0 = \cos(\mu_0/2) \, d\mu$$

so that

$$\frac{dQ}{d\Delta p/p_0} = \frac{N_c}{2\pi} \frac{d\mu_0}{d\Delta p/p_0} = -\frac{N_c}{\pi} \tan(\mu_0/2). \qquad (4.65)$$

As can be seen from (4.65) the natural chromaticity increases rapidly for phase advances above 90°. The usual choices are 45°, 60° and 90°, of which 60° has the best all-round characteristics. It is close to the minimum aperture at $\mu_0 = 76°$ and it is ideal for chromaticity correction schemes (see Chapter 6).

4.7.2   Dispersion suppressors embedded in a FODO lattice[7-10]

The presence of the dipoles in a FODO cell makes little difference to the focusing properties in the first approximation, so that the quadrupoles can continue uninterrupted from the arcs into the rf straight sections (see Figure 4.4). The dispersion function, however, will sense the missing dipoles and become mismatched. The most convenient solution therefore would be to suppress the dispersion at the exit to the arc, so that it remains zero until the next bend. This is also a requirement for the rf cavities in electron machines, as well as a great simplification for the interaction insertion and can be solved in a rather elegant way by acting only on the dipoles, which leaves the focusing unperturbed.

   The dispersion function $D(s)$ is a 'particular integral' solution of the inhomogeneous Hill equation (3.37), discussed in Section 3.4.2. On this occasion, however, it is not the cyclic solution that is required, but one which starts with $D(0) = D'(0) = 0$ and then builds up over N FODO cells, in which there is a rather special dipole distribution, until it matches the values of the cyclic solution for the regular, dipole-filled cells in the arc. These N cells will be the dispersion suppressor (see Figure 4.7). The particular integral (4.40), proposed in Section 4.6.2, will be precisely the

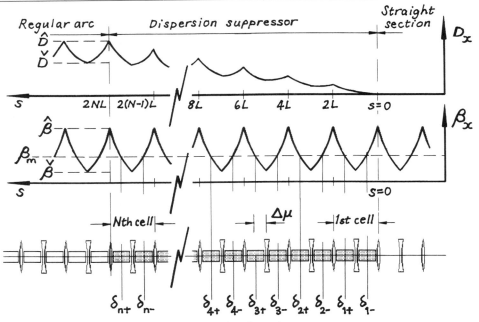

Figure 4.7. Schematic layout of a dispersion suppressor.

solution required, once it is rewritten for $\Delta p/p_0 = 1$. Since the following analysis can be applied to either plane, all suffixes will be omitted for brevity.

$$D(2NL) = \mathscr{S}(2NL) \int_0^{2NL} \frac{\mathscr{C}(s)}{\rho(s)}\, ds - \mathscr{C}(2NL) \int_0^{2NL} \frac{\mathscr{S}(s)}{\rho(s)}\, ds \qquad (4.66)$$

$$D'(2NL) = \mathscr{S}'(2NL) \int_0^{2NL} \frac{\mathscr{C}(s)}{\rho(s)}\, ds - \mathscr{C}'(2NL) \int_0^{2NL} \frac{\mathscr{S}(s)}{\rho(s)}\, ds. \qquad (4.67)$$

The 2×2 focusing matrix for this insertion will be of the form of (4.20),

$$T_{N\,\text{cells}} = \begin{pmatrix} \cos(N\mu_0) & \tilde{\beta}\sin(N\mu_0) \\ -\tilde{\beta}^{-1}\sin(N\mu_0) & \cos(N\mu_0) \end{pmatrix} = \begin{pmatrix} \mathscr{C}(2NL), & \mathscr{S}(2NL) \\ \mathscr{C}'(2NL), & \mathscr{S}'(2NL) \end{pmatrix}, \qquad (4.68)$$

where $\tilde{\beta}$ represents $\hat{\beta}$ when the N cells of the dispersion suppressor are between mid-F positions as shown in Figure 4.7, and $\check{\beta}$ when the N cells are between mid-D positions. Equation (4.68) gives the multipliers of the integrals in (4.66) and (4.67). The evaluation of the integrals appears more complicated but it is sufficient to use the mid-point values of $\mathscr{C}(s)$ and $\mathscr{S}(s)$ in each half cell, since these are linear functions in the dipoles (to first order). These mid-point values are found by the use of the general transfer matrix (4.13),

$$T(\text{mid-dipole}\,|\,s = 0) = \begin{pmatrix} (\beta_{\text{m}}/\tilde{\beta})^{\frac{1}{2}}\cos\mu_n & (\beta_{\text{m}}\tilde{\beta})^{\frac{1}{2}}\sin\mu_n \\ \cdots & \cdots \end{pmatrix}. \qquad (4.69)$$

From Figure 4.7, it can be seen that the phase advance μ_n from the entry to the mid-point of the nth dipole is related to the phase of the D-quadrupole at the centre of the nth cell by,

$$\mu_n = (n - \tfrac{1}{2})\mu_0 \pm \Delta\mu, \tag{4.70}$$

where $\Delta\mu$ is the phase shift from the quadrupole to the centre of the dipole. The value of β_m is the same in all dipoles. Hence the expressions (4.66) and (4.67) for $D(2NL)$ and $D'(2NL)$ can be re-written with the help of (4.68) to (4.70) as sums over the mid-points of the dipoles, to give,

$$D(2NL) = (\tilde{\beta}\beta_m)^{\frac{1}{2}}\left[\sin(N\mu_0)\sum_{n=1}^{N}\{\delta_{n-}\cos[(n-\tfrac{1}{2})\mu_0 - \Delta\mu] + \delta_{n+}\cos[(n-\tfrac{1}{2})\mu_0 + \Delta\mu]\}\right.$$

$$\left. - \cos(N\mu_0)\sum_{n=1}^{N}\{\delta_{n-}\sin[(n-\tfrac{1}{2})\mu_0 - \Delta\mu] + \delta_{n+}\sin[(n-\tfrac{1}{2})\mu_0 + \Delta\mu]\}\right]$$

and likewise,

$$D'(2NL) = (\beta_m/\tilde{\beta})^{\frac{1}{2}}\left[\cos(N\mu_0)\sum_{n=1}^{N}\{\delta_{n-}\cos[(n-\tfrac{1}{2})\mu_0 - \Delta\mu] + \delta_{n+}\cos[(n-\tfrac{1}{2})\mu_0 + \Delta\mu]\}\right.$$

$$\left. + \sin(N\mu_0)\sum_{n=1}^{N}\{\delta_{n-}\sin[(n-\tfrac{1}{2})\mu_0 - \Delta\mu] + \delta_{n+}\sin[(n-\tfrac{1}{2})\mu_0 + \Delta\mu]\}\right],$$

where δ_{n-} and δ_{n+} are the angular kicks ($=L/\rho$) in the dipoles upstream and downstream of the centre quadrupole in the nth cell. These equations simplify rather easily to give,

$$D(2NL) = (\tilde{\beta}\beta_m)^{\frac{1}{2}}\sum_{n=1}^{N}\{\delta_{n-}\sin[(N-n+\tfrac{1}{2})\mu_0 + \Delta\mu] + \delta_{n+}\sin[(N-n+\tfrac{1}{2})\mu_0 - \Delta\mu]\}$$

$$\tag{4.71}$$

$$D'(2NL) = (\beta_m/\tilde{\beta})^{\frac{1}{2}}\sum_{n=1}^{N}\{\delta_{n-}\cos[(N-n+\tfrac{1}{2})\mu_0 + \Delta\mu] + \delta_{n+}\cos[(N-n+\tfrac{1}{2})\mu_0 + \Delta\mu]\}.$$

$$\tag{4.72}$$

Equations (4.71) and (4.72) provide considerable flexibility for controlling the values of D and D' at any later point in the FODO arc and could be the basis of a general computer program. However, practical solutions can be found with far simpler, but more restrictive, equations.

In order to match the dispersion function to the arc only two degrees of freedom will be introduced:

- a gap of an integer number of cells with no dipoles, which is placed at the end of the suppressor before the normal arc (see Figure 4.8), and
- a common strength δ for all the dipoles in the suppressor.

There is of course a cost advantage if solutions can be found that require no new

75

magnets or power supplies. Thus a gap with no dipoles and/or a convenient fractional value of δ, which allows the reduction in bending to be achieved by the partial removal of dipoles from each cell, are particularly useful solutions. In the latter case, the positions of the remaining dipoles can be varied to fine tune the match.

The general equations (4.71) and (4.72) can now be simplified and re-written with a common strength δ for all dipoles and the summation is terminated after i cells of the total of N cells in order to account for a gap at the end of the suppressor, i.e.

$$D(2NL) = 2(\tilde{\beta}\beta_{\mathrm{m}})^{\frac{1}{2}} \delta \cos \Delta\mu \sum_{n=1}^{i} \sin(N - n + \tfrac{1}{2})\mu_0 \tag{4.73}$$

$$D'(2NL) = 2(\beta_{\mathrm{m}}/\tilde{\beta})^{\frac{1}{2}} \delta \cos \Delta\mu \sum_{n=1}^{i} \cos(N - n + \tfrac{1}{2})\mu_0. \tag{4.74}$$

Before proceeding to specific solutions, it is useful to note the following trigonometrical relationships[11] for the summation of these series,

$$\frac{\sin}{\cos}(\alpha) + \frac{\sin}{\cos}(\alpha + \varepsilon) + \frac{\sin}{\cos}(\alpha + 2\varepsilon) + \cdots + \frac{\sin}{\cos}[\alpha + (N-1)\varepsilon]$$

$$= \frac{\sin}{\cos}[\alpha + (N-1)\varepsilon/2] \sin(N\varepsilon/2)/\sin(\varepsilon/2). \tag{4.75}$$

A slight re-arrangement of the standard forms in (4.75) yields the sums of the N terms in both of the series (4.73) and (4.74) when there is no missing-magnet gap, i.e. $i = N$,

$$\sum_{n=1}^{N} \sin[(N - n + \tfrac{1}{2})\mu_0] = \frac{\sin^2(N\mu_0/2)}{\sin(\mu_0/2)} \quad \text{and} \quad \sum_{n=1}^{N} \cos[(N - n + \tfrac{1}{2})\mu_0] = \frac{\sin(N\mu_0)}{2\sin(\mu_0/2)} \tag{4.76}$$

and another manipulation yields the sums of the first i terms in the same series, which corresponds to a gap in the last $(N - i)$ cells, i.e.

$$\text{for } i < N, \sum_{n=1}^{i} \frac{\sin}{\cos}[(N - n + \tfrac{1}{2})\mu_0] = \frac{\sin}{\cos}[(N - i/2)\mu_0] \sin(i\mu_0/2)/\sin(\mu_0/2). \tag{4.77}$$

As a trial solution, equate the dispersion given by (4.73) and (4.74) for a group of i identical F-F cells with strength δ at the start of a series of N cells, to the dispersion at the entry to a regular arc cell (see Figure 4.8). The remaining cells in the suppressor have zero field. With the help of (4.77) this gives,

$$D(2NL) = 2(\hat{\beta}\beta_{\mathrm{m}})^{\frac{1}{2}} \delta \cos \Delta\mu \frac{\sin[(N - i/2)\mu_0] \sin(i\mu_0/2)}{\sin(\mu_0/2)} = \hat{D} \tag{4.78}$$

$$D'(2NL) = 2(\beta_{\mathrm{m}}/\hat{\beta})^{\frac{1}{2}} \delta \cos \Delta\mu \frac{\cos[(N - i/2)\mu_0] \sin(i\mu_0/2)}{\sin(\mu_0/2)} = D' = 0. \tag{4.79}$$

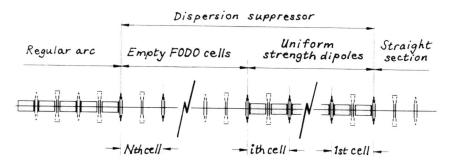

Figure 4.8. Layout of the missing-magnet dispersion suppressor.

Equation (4.79) requires either $\sin(i\mu_0/2) = 0$ or $\cos[(N - i/2)\mu_0] = 0$. The former, when substituted into (4.78), makes δ indeterminate, so this solution is of no use. The latter condition is the one required and can be written as,

$$\mu_0(N - i/2) = (2j - 1)\pi/2 \qquad \text{where } j = 1, 2, 3, \ldots. \qquad (4.80)$$

The phase condition (4.80) has the physical meaning of requiring the phase shift from the mid-point of the group of i cells to the end of the suppressor to be an odd number of quarter betatron wavelengths. This scheme will therefore operate by kicking the beam and arranging for the subsequent oscillation of the dispersion function to reach its maximum at the end of the suppressor and to be equal to the matched value in the arc cells.

It now remains to evaluate (4.78) in order to find the strength of the dipole kick δ but, before this can be done, it is necessary to find the phase shift $\Delta\mu$ from the mid-point in a dipole to the mid-point in the adjacent D-quadrupole. This can be found from the t_{12} term of the corresponding transfer matrix for a half D-quadrupole and a half dipole. The term is first evaluated with the beam parameters using (4.69) and then with the lattice parameters using the thin lens approximation. This yields the equality,

$$(\beta_m \check{\beta})^{\frac{1}{2}} \sin \Delta\mu = L/2$$

so that,

$$\cos \Delta\mu = [1 - (L/2)^2/(\beta_m \check{\beta})]^{\frac{1}{2}}. \qquad (4.81)$$

Equation (4.78) can now be expanded by the substitution of (4.58) for $\hat{D}$, (4.57) for $\hat{\beta}$, (4.59) for β_m and finally (4.81) for $\cos \Delta\mu$. Once expanded in terms of $\sin(\mu_0/2)$ and $\cos(\mu_0/2)$ the expression simplifies to give

$$\frac{1}{2}\frac{L}{\rho_0} = (-1)^{j+1}\delta \sin(i\mu_0/2). \qquad (4.82)$$

It is quickly verified that the two conditions set by (4.80) and (4.82) are satisfied by

Table 4.1. *Examples of missing-magnet
dispersion suppressors*

N	i	j	μ_0	δ
2	1	1	60°	(L/ρ_0)
3	2	1	45°	$(L/\rho_0)/\sqrt{2}$
4	1	2	77°	$-(L/\rho_0)/1.25$
...	...	...	...	...

Table 4.2. *Examples of half-field
dispersion suppressors*

N	j	μ_0	δ
2	1	90°	$\frac{1}{2}(L/\rho_0)$
3	1	60°	$\frac{1}{2}(L/\rho_0)$
4	1	45°	$\frac{1}{2}(L/\rho_0)$
5	2	108°	$\frac{1}{2}(L/\rho_0)$
...	...	...	...

the examples given in Table 4.1, which are known as *missing-magnet dispersion suppressors*. The most convenient configuration is that in which the remaining dipoles retain their full value.

There is also a special class of solutions of (4.80), which occur when $N = i$. This corresponds to there being no empty magnet gap and these suppressors rely on having δ different from the normal value in the arcs. In this case (4.80) reduces to,

$$N\mu_0 = (2j - 1)\pi \qquad \text{where } j = 1, 2, 3, \ldots, \tag{4.83}$$

and (4.82), from which the dipole strength is derived, becomes,

$$\delta = \frac{1}{2}\frac{L}{\rho_0}. \tag{4.84}$$

This extremely simple result has given rise to the name *reduced-field dispersion suppressors*. Table 4.2 lists some specific solutions.

There is a simple descriptive explanation for the mechanism of the half-field suppressors. The maximum dispersion given by (4.58) is inversely proportional to the bending radius. In the half-field cells the matched dispersion would therefore have exactly half the amplitude of the full strength cells. When the dispersion function is launched with $D = D' = 0$ at the entry to the suppressor, it starts to oscillate about the matched value for the half-field cell. The oscillation builds up in half a wavelength to twice the half-field dispersion and thus matches the full-field cell value.

The $60°$ missing-magnet suppressor with $N = 2, i = 1$ and $j = 1$, is a useful solution since it uses full-strength magnets and it is more compact than the half-field version. However, in electron–positron colliders there is an advantage in using the half-field suppressor, because the reduced field in the last magnet projects less synchrotron radiation into the physics experiment in the straight section.

Finally, if the reader returns to (4.78) and (4.79) and rewrites these equations for a dispersion suppressor with N identical D-D cells, then exactly the same results will be obtained.

4.7.3 Matching

It is frequently necessary to match beam parameters over a short distance using quadrupole lenses, for example at each end of a transfer line or for a low-β insertion in a collider. Some analytic solutions do exist and can be of considerable help in guiding numerical matching programs. The early work contains both thin- and thick-lens investigations[12,13], but thin-lens solutions are adequate when followed by numerical matching, e.g. Ref. 14. The examples presented here are variable-geometry multiplets, which use the lens positions as well as the strengths to obtain a match. This type of solution is rather inflexible since it depends on geometry, so if it is known that many widely different optics will be required, as for example in a switch yard, a fixed-geometry solution must be adopted in which only the strengths of the lenses (of which more will be needed) are varied.

(i) *Variable-geometry doublet for α, β matching in a dispersion-free region*[15]

Figure 4.9 shows the layout of a quadrupole doublet. Gradients and positions are taken as variables in order to match α and β in both planes. The insertion is assumed bending-free and dispersion-free.

The transfer matrices of the whole insertion can be represented by,

$$T_h = \begin{pmatrix} t_{h,11} & t_{h,12} \\ t_{h,21} & t_{h,22} \end{pmatrix} \quad \text{and} \quad T_v = \begin{pmatrix} t_{v,11} & t_{v,12} \\ t_{v,21} & t_{v,22} \end{pmatrix}. \tag{4.85}$$

The elements in (4.85) can be evaluated in two ways:

- in terms of the boundary conditions for the beam parameters (α, β, D, $\Delta\mu$) by using the generalised transfer matrix (4.13),
- in terms of the machine elements in the insertion.

The former is simple substitution and the latter will be simplified by using the approximations for thin lenses as shown below,

$$T_h = \begin{pmatrix} 1 & l_5 \\ 0 & 1 \end{pmatrix} \begin{pmatrix} 1 & 0 \\ g_4 & 1 \end{pmatrix} \begin{pmatrix} 1 & l_3 \\ 0 & 1 \end{pmatrix} \begin{pmatrix} 1 & 0 \\ g_2 & 1 \end{pmatrix} \begin{pmatrix} 1 & l_1 \\ 0 & 1 \end{pmatrix}. \tag{4.86}$$

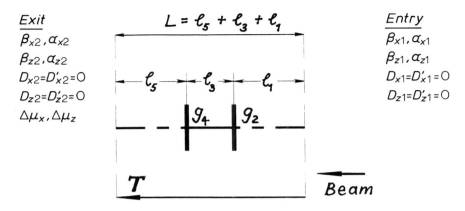

Figure 4.9. Variable-geometry doublet.

The multiplication of these matrices is straightforward and the result is given below in (4.87) with the upper signs ($+$) for the horizontal plane and the lower signs ($-$) for the vertical plane. This matrix embodies all the physical parameters of the doublet.

$$
T_{h,v} = \begin{pmatrix} 1 \pm l_3 g_2 \pm l_5 g_2 & \vdots & l_1 + l_3 + l_5 \pm l_1 l_3 g_2 \pm l_1 l_5 g_2 \\ \pm l_5 g_4 + l_3 l_5 g_2 g_4 & \vdots & \pm l_1 l_5 g_4 \pm l_3 l_5 g_4 + l_1 l_3 l_5 g_2 g_4 \\ \cdots\cdots\cdots\cdots\cdots & + & \cdots\cdots\cdots\cdots\cdots\cdots\cdots\cdots\cdots \\ \pm g_2 \pm g_4 + l_3 g_2 g_4 & \vdots & 1 \pm l_1 g_2 \pm l_1 g_4 \pm l_3 g_4 + l_1 l_3 g_2 g_4 \end{pmatrix}.
\tag{4.87}
$$

By comparison of the terms in (4.87) and (4.85), it will be possible to relate a prescribed set of beam parameters to the physical elements of the insertion. This is the basic idea that is applied in all of the following examples. The algebra becomes rather arduous for the last triplet and quadruplet but it is still tractable by hand. One point to bear in mind is the need to make full use of the information from both planes. This will be done by using a sum and difference notation[15], which will also simplify the algebra.

The matrices $T_{h,v}$ in (4.87) are first re-written as the sum and difference matrices $T_h = T^+ + T^-$ and $T_v = T^+ - T^-$ and their elements are specially grouped so as to make the inter-relationships between them more evident.

$$
T^+ = \begin{pmatrix} [1 + l_5(l_3 g_2 g_4)] & l_1[1 + l_5(l_3 g_2 g_4)] + l_3 + l_5 \\ (l_3 g_2 g_4) & 1 + l_1(l_3 g_2 g_4) \end{pmatrix}
\tag{4.88}
$$

$$
T^- = \begin{pmatrix} [l_3 g_2 + l_5(g_2 + g_4)] & l_1[l_3 g_2 + l_5(g_2 + g_4)] + l_3 l_5 g_4 \\ (g_2 + g_4) & l_1(g_2 + g_4) + l_3 g_4 \end{pmatrix}
\tag{4.89}
$$

where

$$
t_{ij}{}^+ = \tfrac{1}{2}(t_{h,ij} + t_{v,ij})
\tag{4.90}
$$

and

$$t_{ij}^- = \tfrac{1}{2}(t_{h,ij} - t_{v,ij}).\tag{4.91}$$

Inspection of the elements in T^+ and T^- shows directly that,

$$l_1 = (t_{22}^+ - 1)/t_{21}^+\tag{4.92}$$

$$l_5 = (t_{11}^+ - 1)/t_{21}^+\tag{4.93}$$

$$L = l_1 + l_3 + l_5 = (t_{11}^- + t_{22}^-)/t_{21}^-\tag{4.94}$$

and hence

$$l_3 = L - l_1 - l_5.\tag{4.95}$$

A little manipulation of elements t_{21}^- and t_{22}^+ yields a quadratic equation for the gradients, from which the two solutions are found,

$$g_{2,4} = \tfrac{1}{2}\{t_{21}^- \pm \sqrt{[(t_{21}^-)^2 - 4t_{21}^+/l_3]}\} \qquad (g_2 \text{ takes } + \text{ sign}).\tag{4.96}$$

The insertion parameters expressed in (4.92) to (4.96) will be non-physical in certain cases. The distances must of course be positive and the gradients in (4.96) must be real, i.e.

$$(t_{21}^-)^2 \geq 4t_{21}^+/l_3.$$

However, the above is not sufficient to ensure a consistent solution, since the doublet has only five free variables (l_1, g_2, l_3, g_4, l_5), while the beam is described by eight matrix elements, which have six degrees of freedom (the condition for unit matrices reduces the degrees of freedom by two). Thus it is not possible to find a solution for all sets of beam parameters. Consistent solutions will only exist when the transfer matrix elements are related in a suitable way. This extra boundary condition can be found from (4.88) and (4.89) by expressing t_{12}^-, which has not been used so far, in terms of the other matrix elements.

$$t_{12}^- = l_1 t_{11}^- + l_5 t_{22}^- - l_1 l_5 t_{21}^-$$

$$(t_{21}^+)^2 t_{12}^- - (t_{22}^+ - 1)t_{11}^- t_{21}^+ - (t_{11}^+ - 1)t_{22}^- t_{21}^+ + (t_{22}^+ - 1)(t_{11}^+ - 1)t_{21}^- = 0\tag{4.97}$$

In the present formulation it will be assumed that the phase advances can be varied at will in order to satisfy (or nearly satisfy) this condition. An example is shown in Figure 4.10 for a symmetric low-β insertion starting at the central low-β point and matching onto a FODO lattice terminated by a half-strength F-lens (i.e. $\alpha_x = \alpha_z = 0$). The results of the search over a large area of the (μ_x, μ_z) parameter space are given. In the solution set, the line dividing the positive and negative mismatch factors indicates the presence of valid solutions, 'i' indicates imaginary gradients and '$-$' indicates negative lengths. The 'mismatch factor' is found by evaluating the boundary condition (4.97), which must be zero for a perfect solution. The second search over

81

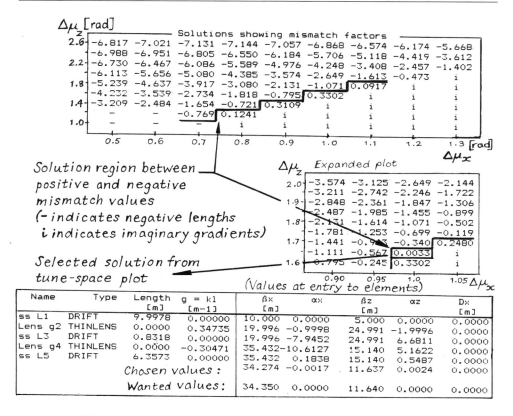

Figure 4.10. Search for a low-β solution with a thin-lens doublet. (Matching: low-β, $\beta_x = 10$ m, $\beta_z = 5$ m, $\alpha_{x,z} = 0$ to half-strength F-quadrupole terminating FODO cell with $\beta_x = 34.35$ m, $\beta_z = 11.64$ m, $\alpha_{x,z} = 0$.)

a reduced region finds a good solution. A standard lattice program with a minimisation routine will quickly optimise the parameters of the thin-lens solution into the equivalent thick-lens one. Reference 15 describes an application of doublet matching.

(ii) *Variable-geometry triplet (I) for α, β matching in a dispersion-free region*

The quadrupole triplet shown in Figure 4.11 is similar to the doublet analysed in the previous section, inasmuch as it has the same number of free parameters. In this case, it is better to treat the insertion in two stages and to first evaluate the transfer matrix P of the central section and then the overall matrix Q. The details of the derivation will be left to the reader as an exercise, but it is useful to note the following relationships for constructing the sum and difference terms for the matrix Q.

Main term	Sum term	Difference term
$q_{ij} \longrightarrow$	q_{ij}^{+}	q_{ij}^{-}
$g_n q_{ij} \longrightarrow$	$g_n q_{ij}^{-}$	$g_n q_{ij}^{+}$
$g_n g_m q_{ij} \longrightarrow$	$g_n g_m q_{ij}^{+}$	$g_n g_m q_{ij}^{-}$

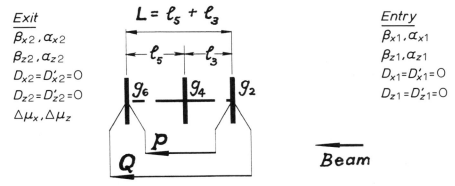

Exit
β_{x2}, α_{x2}
β_{z2}, α_{z2}
$D_{x2} = D'_{x2} = 0$
$D_{z2} = D'_{z2} = 0$
$\Delta\mu_x, \Delta\mu_z$

$L = \ell_5 + \ell_3$

$g_6 \quad g_4 \quad g_2$

P
Q

Entry
β_{x1}, α_{x1}
β_{z1}, α_{z1}
$D_{x1} = D'_{x1} = 0$
$D_{z1} = D'_{z1} = 0$

Beam

Figure 4.11. Variable-geometry triplet I. (The parameters are numbered from g_2 in order that the results can be directly used in the next section where drift spaces will be added.)

The final equations for determining the parameters of the insertion are:

$$\left.\begin{array}{ll} g_2 = (q_{11}^{+} - 1)/q_{12}^{-}, & g_6 = (q_{22}^{+} - 1)/q_{12}^{-} \\ g_4 = q_{21}^{-} - g_2 - g_6 - g_2 g_6 q_{12}^{-}, & \\ l_3 = (q_{22}^{-} - g_6 q_{12}^{+})/g_4, & l_5 = (q_{11}^{-} - g_2 q_{12}^{+})/g_4. \end{array}\right\} \quad (4.98)$$

As in the case of the doublet, a further boundary condition must be fulfilled for a solution to be valid. This condition is found by re-expressing the q_{21}^{+} term with the other matrix elements to give,

$$(q_{12}^{-})^2 q_{21}^{+} - (q_{22}^{+} - 1)(q_{11}^{-} q_{12}^{-} - q_{12}^{+} q_{11}^{+} + q_{12}^{+})$$
$$- (q_{11}^{+} - 1)(q_{22}^{-} q_{12}^{-} - q_{12}^{+} q_{22}^{+} + q_{12}^{+}) - (q_{11}^{+} - 1)(q_{22}^{+} - 1)q_{12}^{+} = 0. \quad (4.99)$$

Solutions are again found by searching the (μ_x, μ_z) tune space and evaluating the boundary condition (4.99) at each point as a mismatch factor for guiding the search. This insertion is useful for creating liaisons between different sections of a dispersion-free lattice.

(iii) *Variable-geometry triplet (II) for α, β, D matching in a bending-free region*[16]

The quadrupole triplet, shown in Figure 4.12, has two additional drift spaces and hence seven variables, which in principle is sufficient to satisfy all matrices. This extra flexibility makes it possible to also match the dispersion in a bending-free region, i.e. no dipoles are allowed inside the insertion.

The results from the previous section can be used directly. The overall transfer matrix is given by $T = [l_7]Q[l_1]$. This final 'layer' comprises only a pair of drift spaces, but they greatly complicate the algebra for solving the triplet. In this case, the sum and difference matrices have similar forms and can be written as,

$$T^\pm = \begin{pmatrix} q_{11}{}^\pm + l_7 q_{21}{}^\pm & l_1 q_{11}{}^\pm + q_{12}{}^\pm + l_1 l_7 q_{21}{}^\pm + l_7 q_{22}{}^\pm \\ q_{21}{}^\pm & l_1 q_{21}{}^\pm + q_{22}{}^\pm \end{pmatrix}. \tag{4.100}$$

An intermediate inversion of each term yields,

$$\left.\begin{aligned}
q_{21}{}^\pm &= t_{21}{}^\pm \\
q_{11}{}^\pm &= t_{11}{}^\pm - l_7 t_{21}{}^\pm \\
q_{22}{}^\pm &= t_{22}{}^\pm - l_1 t_{21}{}^\pm \\
q_{12}{}^\pm &= t_{12}{}^\pm - l_1 t_{11}{}^\pm - l_7 t_{22}{}^\pm - l_1 l_7 t_{21}{}^\pm
\end{aligned}\right\} . \tag{4.101}$$

Thus all the elements of Q are known in terms of the elements of T, l_1 and l_7. Since the matrix Q was fully analysed in the previous section, all the remaining parameters of the insertion are known in terms of the elements in Q. It only remains to determine l_1 and l_7. One of these lengths, say l_1, will be used as a variable when searching for solutions and the other, l_7, can be found by equating $q_{12}{}^-$ from (4.101) to $q_{12}{}^-$ from the boundary condition (4.99). This leads to a very long expression, which is quadratic

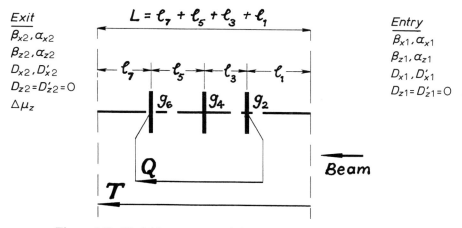

Figure 4.12. Variable-geometry triplet II.

in l_7, but simple to solve with a short computer program.

$$l_7{}^2(ebt_{21}{}^- - edt_{21}{}^+) + l_7(abt_{21}{}^+ + eat_{21}{}^- - ebt_{11}{}^- - ect_{21}{}^+ + def + fb^2)$$
$$+ (a^2t_{21}{}^+ - eat_{11}{}^- + ecf + fab) = 0$$

where:
$$a = t_{12}{}^- - l_1t_{11}{}^-, \qquad b = l_1t_{21}{}^- - t_{22}{}^-$$
$$c = t_{21}{}^+ - l_1t_{11}{}^+, \qquad d = l_1t_{21}{}^+ - t_{22}{}^+$$
$$e = t_{22}{}^+ - l_1t_{21}{}^+ - 1, \qquad f = t_{11}{}^+ - 1.$$

$$(4.102)$$

The substitution of the set of equations (4.101) into the solutions (4.98) obtained in the previous section gives the final expressions for this triplet,

$$g_2 = (f - l_7t_{21}{}^+)/(a + l_7b), \qquad g_6 = e/(a + l_7b)$$
$$g_4 = t_{21}{}^- - g_2 - g_6 - g_2g_6(a + l_7b),$$
$$l_3 = [-b - g_6(c + l_7d)]/g_4, \qquad l_5 = [t_{11}{}^- - l_7t_{21}{}^- - g_2(c + l_7d)]/g_4.$$

$$(4.103)$$

A search is then made for suitable solutions by introducing a range of values for l_1 and $\Delta\mu_z$. Since there are now seven variables, it was possible to satisfy the boundary condition (4.98) and all solutions will be exact. Horizontal dispersion matching is introduced by the restriction this imposes on the horizontal phase advance through the insertion. This has already been described in Section 4.6.3 and is expressed by (4.51). This triplet usually yields an abundance of solutions and it is quite feasible to apply a further restriction on the vertical phase advance in order to match the vertical dispersion as well. An example of a symmetric low-β insertion in which the horizontal dispersion is forced to zero at the crossing point (with a large divergence) is given in Figure 4.13. In this example, the matching is made from the exit of the F-quadrupole of a FODO cell to the low-β point.

Reference 16 also contains the equivalent thick-lens solution based on the techniques developed in Refs 12 and 13.

(iv) *Variable geometry quadruplet for α, β, D matching in a bending-free region*

A quadruplet can also be solved by hand by using the results already obtained for the triplet (I) and by analysing the addition of a drift space l_1 and a quadrupole g_0, see Figure 4.14.

In this case, values will be assumed for the gradient g_0 and the search for a convenient solution will be made by varying this parameter. The length l_1 is found in an analogous way to that used for l_7 in the last example.

$$l_1{}^2(fbd + aj - ad^2) + l_1(2adt_{12}{}^- - ah - ej - fbt_{12}{}^- - fdt_{22}{}^-)$$
$$+ (eh + ft_{22}{}^-t_{12}{}^- - a[t_{12}{}^-]^2) = 0$$

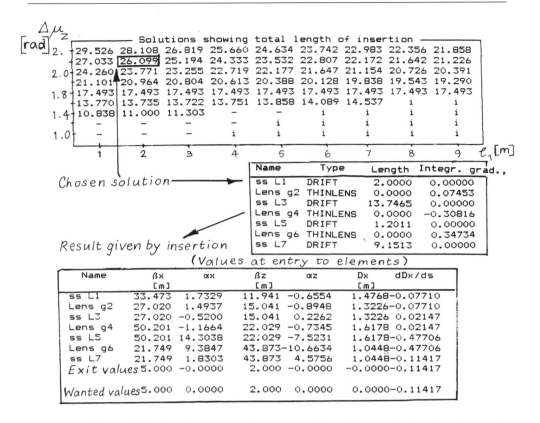

$\frac{\Delta\mu_z}{[\text{rad}]}$

Solutions showing total length of insertion

2.	29.526	28.108	26.819	25.660	24.634	23.742	22.983	22.356	21.858
	27.033	26.099	25.194	24.333	23.532	22.807	22.172	21.642	21.226
2.0	24.260	23.771	23.255	22.719	22.177	21.647	21.154	20.726	20.391
	21.101	20.964	20.804	20.613	20.388	20.128	19.838	19.543	19.290
1.8	17.493	17.493	17.493	17.493	17.493	17.493	17.493	17.493	17.493
	13.770	13.735	13.722	13.751	13.858	14.089	14.537	i	i
1.4	10.838	11.000	11.303	−	−	i	i	i	i
	−	−	−	−	i	i	i	i	i
1.0	−	−	−	i	i	i	i	i	i

$\ell_1\,[m]$

Chosen solution ⟶

Name	Type	Length	Integr. grad.,
ss L1	DRIFT	2.0000	0.00000
Lens g2	THINLENS	0.0000	0.07453
ss L3	DRIFT	13.7465	0.00000
Lens g4	THINLENS	0.0000	−0.30816
ss L5	DRIFT	1.2011	0.00000
Lens g6	THINLENS	0.0000	0.34734
ss L7	DRIFT	9.1513	0.00000

Result given by insertion

(*Values at entry to elements*)

Name	βx [m]	αx	βz [m]	αz	Dx [m]	dDx/ds
ss L1	33.473	1.7329	11.941	−0.6554	1.4768	−0.07710
Lens g2	27.020	1.4937	15.041	−0.8948	1.3226	−0.07710
ss L3	27.020	−0.5200	15.041	0.2262	1.3226	0.02147
Lens g4	50.201	−1.1664	22.029	−0.7345	1.6178	0.02147
ss L5	50.201	14.3038	22.029	−7.5231	1.6178	−0.47706
Lens g6	21.749	9.3847	43.873	−10.6634	1.0448	−0.47706
ss L7	21.749	1.8303	43.873	4.5756	1.0448	−0.11417
Exit values	5.000	−0.0000	2.000	−0.0000	−0.0000	−0.11417
Wanted values	5.000	0.0000	2.000	0.0000	0.0000	−0.11417

Figure 4.13. Search for a low-β solution with a thin-lens triplet. (Matching: exit to FODO cell $\beta_x = 33.47$ m, $\beta_z = 11.94$ m, $\alpha_x = 1.73$, $\alpha_z = -0.655$ to low-β, $\beta_x = 5$ m, $\beta_z = 2$ m, $\alpha_{x,z} = 0$.)

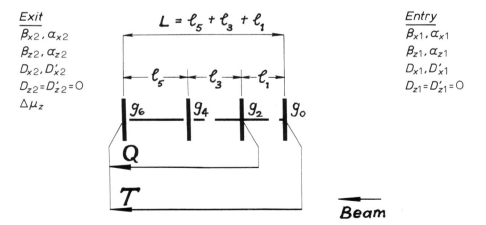

Exit
β_{x2}, α_{x2}
β_{z2}, α_{z2}
D_{x2}, D'_{x2}
$D_{z2} = D'_{z2} = 0$
$\Delta\mu_z$

$L = \ell_5 + \ell_3 + \ell_1$

ℓ_5 ℓ_3 ℓ_1

g_6 g_4 g_2 g_0

Q

T

Beam

Entry
β_{x1}, α_{x1}
β_{z1}, α_{z1}
D_{x1}, D'_{x1}
$D_{z1} = D'_{z1} = 0$

Figure 4.14. Variable-geometry quadruplet.

where,

$$
\left.
\begin{aligned}
a &= t_{21}^{+} - g_0 t_{22}^{-}, & b &= t_{21}^{-} - g_0 t_{22}^{+} \\
c &= t_{11}^{+} - g_0 t_{12}^{-}, & d &= t_{11}^{-} - g_0 t_{12}^{+} \\
e &= t_{22}^{+} - 1, & f &= c - 1 \\
h &= d t_{12}^{-} - f t_{12}^{+}, & j &= d^2 - c^2 + c.
\end{aligned}
\right\}
\tag{4.104}
$$

One l_1 has been found from the quadratic expression (4.104), the remaining parameters can be determined from,

$$
\left.
\begin{aligned}
g_2 &= f/(t_{12}^{-} - l_1 d), & g_6 &= (e - l_1 a)/(t_{12}^{-} - l_1 d) \\
g_4 &= b - g_2 - g_6 - g_2 g_6 (t_{12}^{-} - l_1 d), & & \\
l_3 &= [t_{22}^{-} - l_1 b - g_6(t_{12}^{+} - l_1 c)]/g_4, & l_5 &= [d - g_2(t_{12}^{+} - l_1 c)]/g_4.
\end{aligned}
\right\}
\tag{4.105}
$$

This quadruplet is very flexible and is ideal for making the liaison in a short distance between a regular lattice and an element such as a kicker with special requirements.

The full derivations by hand for the last two examples are rather long and clearly they would be well suited to a computer program using symbolic logic. Reference 17 describes such solutions for a quadruplet with entry and exit drift spaces.

4.7.4   Injection and extraction

(i) *Single-turn injection and extraction*

Figure 4.15 shows the schematic layout of the elements for *single-turn injection and extraction*. The design details will vary with respect to apertures, clearance, types of septa and number of kickers, because of the very different energies and intensities, but all schemes will be governed by the same basic optics. For the present purpose, the beam will be followed out of the machine, independent of whether the insertion is

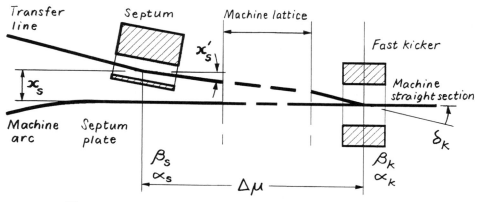

Figure 4.15. Single-turn injection or extraction insertion.

for injection or extraction. The application of the boundary conditions at the kicker* ($x_k = 0$ and $x_k' = \delta_k$) in the trajectory equations (4.1)' and (4.21) determines the subsequent position and angle of the beam at the septum†.

$$x_s = (\delta_k \beta_k^{1/2})\beta_s^{1/2} \sin \Delta\mu \qquad (4.106)$$

$$x_s' = (\delta_k \beta_k^{1/2})\beta_s^{-1/2}(\cos \Delta\mu - \alpha_s \sin \Delta\mu) \qquad (4.107)$$

This trajectory from the kicker must match the design orbit in the septum, so that

$$\delta_k = \frac{x_s}{(\beta_k \beta_s)^{1/2}} \frac{1}{\sin \Delta\mu} \qquad (4.108)$$

$$x_s' = x_s \beta_s^{-1}(\cot \Delta\mu - \alpha_s). \qquad (4.109)$$

The offset x_s must not be less than the halfwidth of the circulating beam with an allowance for closed orbit distortions and alignment errors plus the halfwidth of the extracted/injected beam and the septum thickness. The kicker is usually limited by technology and/or economics, so it is advantageous to make $\Delta\mu$ close to $\pi/2$ and β_k large, so that δ_k is reduced. The minimum strength of the septum depends on how much more angular deflection is needed to take the beam clear of the next machine magnet, but this is helped by making sure α is large and negative. A large value for β_s reduces δ_k, but has the adverse effects of increasing the beam size and hence x_s and reducing x_s'.

Single-turn schemes normally inject, or extract, directly from central orbit. However, provision is usually made for closed-orbit bumps (see Chapter 5) in the vicinities of the kicker and septum in order to adjust the closed orbit locally and to minimise the strength of the injection elements by bringing the central orbit closer to the septum. When the injected beam is bunched and is transferred from rf bucket to rf bucket between two machines, the injection is known as *box-car stacking*.

(ii) *Multi-turn injection*

Multi-turn injection is similar to single-turn injection, but is combined with a variable orbit bump, or a shrinking orbit, which makes room for the successive turns. This is known as *stacking in betatron phase space* and was already tried in weak focusing synchrotrons[18].

Stacking can be carried out in either of the betatron phase spaces, or in the longitudinal phase space, or, in principle, in any combination of these. *Stacking in*

---

* The kicker is a pulsed dipole magnet with rise and fall times that are short on the timescale of the bunch spacing.

† The septum is a specially designed dipole magnet with a thin partition (i.e. septum), which encloses the deflecting field. The stray field outside the septum, where the main circulating beam passes, is made as nearly zero as possible. The kicker trajectory must have sufficient amplitude to pass inside the septum. Electrostatic septa are weaker, but are often used since they can be made extremely thin.

the longitudinal phase space was proposed in the MURA papers[19]. The MURA idea led to its use in the ISR[20] where operational stacks of tens of amperes comprised several hundred pulses. In this case, some adjustment to the tune was needed on the injection orbit as the intensity built up, but it was still possible to keep a constant matching at the end of the transfer line (see also Section 12.2.4).

(iii) *H⁻ injection*

H⁻ injection is also a form of multi-turn injection and is one of a number of schemes which increase the phase-space density in defiance of Liouville's theorem (see Appendix A). This technique was pioneered at Novosibirsk[21] and later at ANL[22] and has since become a preferred method of injection into high-intensity machines. The first stage is to produce a beam of negative hydrogen ions, which are transported up to the injection point and put on an orbit tangential to the main ring orbit. At this point, Liouville's theorem is cheated by fully stripping the ions and converting them to protons in the middle of the phase space occupied by the circulating proton beam. In principle, this can be done with any particle for which a negative-ion source is available. The stripping can be done by a gas jet[21], but is more usually done by a thin foil. The foil is typically made of polyparaxylene[22], carbon[23,24], or aluminium oxide[25,26].

A variant of this technique is to accelerate the negative ions to top energy and then to use the same procedure for extraction. Another variant, suggested by O'Neill[27], is to use particle decay instead of stripping. This technique was used to fill the CERN g-2 ring.

(iv) *Injection with radiation damping*

Electron machines have always made good use of the damping due to synchrotron radiation (see Chapter 10) and injection is an example of this. For example in LEP[28], a fast bump is used to bring the already damped circulating beam close to a septum at the moment of injection. The acceptance ellipse of the main ring, once displaced in this way, overlaps the injection line aperture (see Figure 4.16). As can be seen from Figure 4.16, it is advantageous to match the incoming beam to a rather tall thin ellipse. Directly after injection the bump is removed and during the next few turns the injected bunch is damped down into the circulating bunch. Once this has been achieved, the process can be repeated. The kickers are fast enough to select specific rf buckets for the transfers.

4.8   GENERAL REMARKS

The numerical view of the linear lattice developed in Chapter 3, with the use of 3×3 transfer matrices for each element, has now been complemented by an analytical

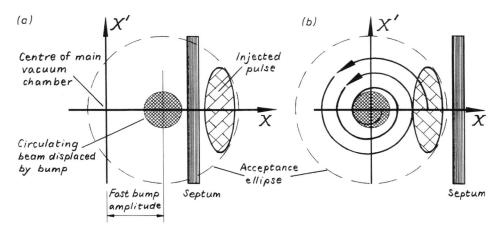

Figure 4.16. Phase-space ellipses for injection with damping [normalised coordinates (X, X')]. (a) At injection, (b) after injection.

description. One advantage of this description is its power to guide the numerical analysis and a number of ways of doing this have been discussed. It is also the foundation of most of the theoretical treatises concerning the transverse motion and for this reason some care has been taken to introduce useful concepts such as normalised phase, principal trajectories, the σ-matrix for the transmission of beam parameters etc.

One important type of lattice cell that has not yet been mentioned is the *achromat*. This is the essential 'building block' for designing a high-brightness lattice for a synchrotron light source. Since this is intimately related to the effects of radiation on the particle motion, this topic has been left until Chapter 10.

In the section on injection and extraction, the main omissions are *resonance injection*[29] and *resonance extraction*[30], which is also known as *slow extraction*. Detailed descriptions have been left out since a full analytical treatment of non-linear resonances is beyond the scope of this book. For a more complete review of injection and extraction see Ref. 31.

CHAPTER 5

Imperfections and resonances

This chapter treats in some detail the actions of imperfections in the linear lattice, including the useful applications for local-orbit bumps and tune control. Higher-order imperfections are described, but in far less detail, since the full analysis requires an advanced perturbative treatment of the Hamiltonian. Linear coupling is described using a simplified theory and the results of the full Hamiltonian analysis are quoted where needed. Much of the first part of the chapter can be found in Ref. 1 or is a logical extension of that reference.

5.1   CLOSED-ORBIT DISTORTION FROM DIPOLE KICKS

Equation (3.37) for the particle motion with a momentum deviation can be directly adapted to the motion with field errors by replacing the driving term $\rho^{-1}(s)\,\Delta p/p$, on the righthand side, with the analogous expression $\Delta B(s)/(B_0\rho_0)$. Some care is needed over the signs but a little thought gives,

$$\frac{d^2}{ds^2}\,y + K(s)y = \mp\frac{\Delta B(s)}{B_0\rho_0}, \tag{5.1}$$

where the upper sign $(-)$ refers to the horizontal plane and the lower sign $(+)$ refers to the vertical plane. This double sign will reappear throughout this section.

The general solution for a closed orbit in equations of the type (3.37) and (5.1) was derived in some detail in Section 4.6.2. Thus the solution for (5.1) can be written down directly from (4.48).

$$y(s) = \mp\frac{\beta_y^{\frac{1}{2}}(s)}{2\sin\pi Q_y}\int_s^{s+C}\beta_y^{\frac{1}{2}}(\sigma)\,\frac{\Delta B(\sigma)}{B_0\rho_0}\cos[\mu_y(\sigma)-\mu_y(s)-\pi Q_y]\,d\sigma. \tag{5.2}$$

Equation (5.2) gives the *closed-orbit distortion* with respect to the ideal orbit without errors, i.e. $y(s) = 0$. The suffixes will be dropped in the later equations where there is no risk of confusion.

Inspection of (5.2) reveals that the closed orbit will become unstable as the tune approaches an integer, due to the 'sin πQ' term in the denominator. This is easily visualised. If the tune is integer, the oscillation caused by a field error is always returned with the same phase and its amplitude will grow indefinitely. This condition is known as the *integer resonance*.

It can also be seen that, for a given error, the distortion is proportional to $\sqrt{\beta}$, so that there is an advantage in placing beam position monitors at peaks in the betatron amplitude function. Likewise the effect of a dipole kick is proportional to its local $\sqrt{\beta}$, so that orbit correctors also benefit from being placed at positions with large β-values.

A very frequent problem is to write a short computer program, which will predict the closed-orbit distortions from known field errors using a data base of machine parameters. The form of the integral in (5.2) is inconvenient for this task, since it either requires the data to be re-ordered with the origin at the observer's position s or to test when $\mu(\sigma) < \mu(s)$ for adding $2\pi Q$ to the argument of the cosine term. This can be avoided by using the following transformation of the integral in (5.2),

$$\int_s^{s+C} \cdots \, d\sigma = \int_s^C \cdots \, d\sigma + \int_C^{s+C} \cdots \, d\sigma.$$

By virtue of the machine's periodicity the terms in the last integral can be re-written as,

$$\beta(\sigma) = \beta(\sigma - C)$$

$$\Delta B(\sigma) = \Delta B(\sigma - C)$$

$$\cos[\mu(\sigma) - \mu(s) - \pi Q] = \cos[\mu(C) + \mu(\sigma - C) - \mu(s) - \pi Q]$$

$$= \cos[\mu(\sigma - C) - \mu(s) + \pi Q].$$

Thus the limits on the last integral can be shifted by $-C$ and $(\sigma - C)$ simply becomes σ to give,

$$\int_s^{s+C} \cdots \, d\sigma = \int_s^C \cdots \cos[\mu(\sigma) - \mu(s) - \pi Q] \, d\sigma + \int_0^s \cdots \cos[\mu(\sigma) - \mu(s) + \pi Q] \, d\sigma.$$

Since cosine is an even function, two parts of the integral can be combined into the more compact form of (5.3), which is then independent of the origin of the data base and thus more convenient for computations.

$$y(s) = \mp \frac{\beta^{\frac{1}{2}}(s)}{2 \sin \pi Q} \int_0^C \beta^{\frac{1}{2}}(\sigma) \frac{\Delta B(\sigma)}{B_0 \rho_0} \cos[|\mu(\sigma) - \mu(s)| - \pi Q] \, d\sigma. \qquad (5.3)$$

For computations, it is also more convenient to replace the integral sign in (5.3) by a summation over short-field errors. This would be used, for example, when calculating the effects of steering magnets.

$$y(s) = \frac{\beta^{\frac{1}{2}}(s)}{2 \sin \pi Q} \sum_n \beta_n^{\frac{1}{2}} \delta_n \cos[|\mu_n - \mu(s)| - \pi Q]$$

where

$$\delta_n = \mp (l\Delta B)_n / (B_0 \rho_0).$$

(5.4)

'Short' means that β, μ and ΔB can be regarded as constant over the length l of the field error. The term δ_n is then equal to a sharp angular kick $\Delta y_n'$ given to the beam at μ_n. This formulation has the advantages of being brief and of removing the sign difference between the planes. It will be used later when calculating orbit and bumps.

When visualising the closed orbit it is useful to re-express (5.2) with the normalised amplitude $Y(s) = y(s)/\sqrt{[\beta(s)]}$, which was introduced in Section 4.1 and to replace the independent variable s by the betatron phase μ, so that,

$$Y(\mu) = \mp \frac{Q}{2 \sin \pi Q} \int_\mu^{\mu + 2\pi} \beta^{3/2}(\psi) \frac{\Delta B(\psi)}{B_0 \rho_0} \cos[\psi - \mu - \pi Q] \, d\psi.$$

(5.5)

In (5.5) μ is the phase of the observer and ψ the phase of the error. This formulation removes the β-modulation, so that the orbit becomes a superposition of sections of sine waves. This will render the closed orbit more easily analysed visually and will be used for many of the diagrams. In the following sections, the normalised and the unnormalised representations will be used freely, including the use of the normalised phase, which was also introduced in Section 4.1.

5.1.1 Qualitative description of a closed orbit

From the above equations, it is difficult to appreciate exactly how the closed orbit is formed and what characteristics it will have. The two descriptions given below will hopefully add some physical insight.

(i) First approach

Consider the distortion due to a single field error δ using the normalised amplitude representation of (5.5) with the argument of the cosine term written in the form of (5.3).

$$Y(\mu) = \underbrace{\frac{1}{2 \sin \pi Q} \beta^{\frac{1}{2}}(\psi)\delta}_{\text{constant amplitude}} \quad \underbrace{\cos[|\psi - \mu| - \pi Q]}_{\text{cosine oscillation}}$$

(5.6)

This is a sinusoidal oscillation, which is launched symmetrically in the two directions round the ring from the point at which the error occurs (ψ). At the error, the phase angle is $\pm \pi Q$ and the phase shift is $2\pi q$, where q is the fractional part of Q. The

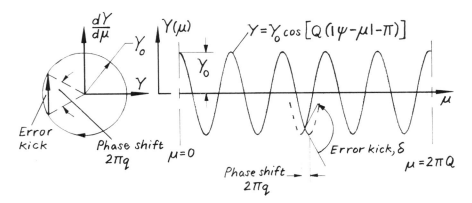

Figure 5.1. First approach to constructing the closed orbit from a single field error [normalised coordinates (μ, Y)].

amplitude at the error is,

$$Y(\psi = \mu) = \tfrac{1}{2}\beta^{\frac{1}{2}}(\psi)\delta \cot(\pi Q) \qquad (5.7)$$

and the orbit will have, by symmetry, a maximum or a minimum exactly opposite the error, i.e. πQ away in betatron phase (see Figure 5.1 in which ψ has been chosen equal to πQ to illustrate the last point). Many dipole kicks can be treated by the superposition of the individual closed orbits for each one. Thus the final orbit will be sinusoidal in regions with no field errors, but at each error there will be a jump in the derivative corresponding to the kick at that point.

(ii) *Second approach*

In this case, the construction of a closed orbit is based on plotting the particular integral and complementary function solutions (see Section 3.4.2). First plot the trajectory of a particle, which enters on axis, is then deviated by a kick and as a consequence continues to oscillate around the machine. This is a particular-integral solution of the inhomogeneous equation (5.1). If a trajectory for the complementary function is added to this particular integral, but with the constants chosen such that the sum of the two solutions returns the input conditions after one turn, then the situation in Figure 5.2 is obtained. The addition of these two orbits, which are individually not closed round the machine, makes the desired closed orbit. Inspection reveals that the second component orbit is in fact given by the first with $s \to -s$.

Figures 5.1 and 5.2 show the characteristic peak or cusp due to the phase discontinuity at a field error. Conversely the existence of a peak or cusp in a closed orbit is indicative of a field error. Normally, this is not directly visible due to the betatron modulation and the spacing of the beam position monitors around the ring. However, by translating the monitor readings into normalised values and fitting sine

94

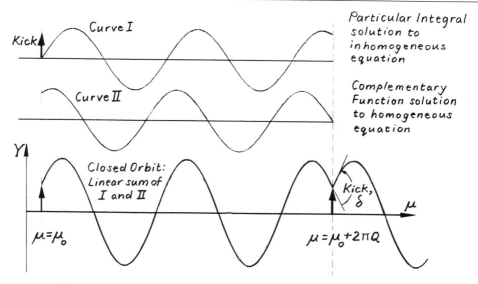

Figure 5.2. Second approach to constructing the closed orbit from a single field error [normalised coordinates (μ, Y)].

curves to contiguous points, discontinuities can be detected and their positions and magnitudes estimated. This technique provides a useful on-line diagnostic tool for large errors (see Section 11.1.4(iii)).

5.1.2 Closed-orbit bumps

(i) *Half-wavelength bumps*

It can be seen directly from equation (4.1) that a betatron oscillation launched by a dipole kick δ_1 can be exactly compensated half a wavelength later or earlier (or at any integer number of half wavelengths, once the excursion is again zero), by a similar kick δ_2. Care must be taken to adjust this kick to the local betatron amplitude, but this is straight-forward. The excursion and slope of the oscillation are given by (4.1) and are,

$$y = A\beta^{\frac{1}{2}} \cos(\mu + B) \quad \text{and} \quad y' = -A\beta^{-\frac{1}{2}}[\alpha \cos(\mu + B) + \sin(\mu + B)].$$

If the first kick $\delta_1(=y_1')$ is put at $\mu_1 = -\pi/2$, then $B = 0$ and $A = \delta_1\beta_1^{\frac{1}{2}}$. At $\mu = \pi/2$, just half a wavelength later, the excursion will again be zero. At this point the oscillation can be killed by a second kick δ_2, which is equal and opposite to the trajectory slope at this point, so that $\delta_2 = -y_2'$, which gives the conditions,

$$\left.\begin{array}{ll} \mu_2 - \mu_1 = \pi & \text{(imposed condition)} \\ \delta_1\beta_1^{\frac{1}{2}} = \delta_2\beta_2^{\frac{1}{2}} & \text{(derived conditions).} \end{array}\right\} \tag{5.8}$$

95

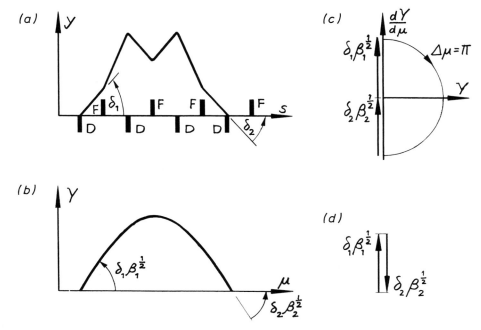

Figure 5.3. 2-magnet, half-wavelength, closed-orbit bump. (a) Real-space (y, s), (b) normalised coordinates (μ, Y), (c) normalised phase-space (Y, Y'), (d) vector diagram.

By inspection it is seen that (5.8) can be generalised for extended bumps of n half wavelengths so that,

$$\mu_2 - \mu_1 = n\pi \qquad \text{and} \qquad \delta_1 \beta_1^{\frac{1}{2}} = (-1)^{n+1} \delta_2 \beta_2^{\frac{1}{2}}. \qquad (5.8)'$$

Figure 5.3 illustrates the simple half-wavelength bump.

As an exercise, the above derivation will be repeated using the closed orbit distortion equation (5.4). The suffixes 1 and 2 denote parameters at the position of the first and second dipole kicks and s the position of the observer (see Figure 5.4).

$$y_s = \frac{\beta_s^{\frac{1}{2}}}{2 \sin(\pi Q)} [\beta_1^{\frac{1}{2}} \delta_1 \cos(|\Delta\mu_{1,s}| - \pi Q) + \beta_2^{\frac{1}{2}} \delta_2 \cos(|\Delta\mu_{2,s}| - \pi Q)] \qquad (5.9)$$

where $\Delta\mu_{1,s} = (\mu_1 - \mu_s)$ and $\Delta\mu_{2,s} = (\mu_2 - \mu_s)$.

The condition $\beta_1^{\frac{1}{2}} \delta_1 = \beta_2^{\frac{1}{2}} \delta_2$ is imposed on the magnet kicks so that (5.9) can be reduced to,

$$y_s = \frac{\beta_s^{\frac{1}{2}} \beta_1^{\frac{1}{2}} \delta_1}{\sin(\pi Q)} \cos \tfrac{1}{2}(|\Delta\mu_{1,s}| + |\Delta\mu_{2,s}| - 2\pi Q) \cos \tfrac{1}{2}(|\mu_{1,s}| - |\Delta\mu_{2,s}|). \qquad (5.10)$$

If $\mu_s < \mu_1$ or $\mu_s > \mu_2$, then the observer is outside the pair of adjacent magnets and

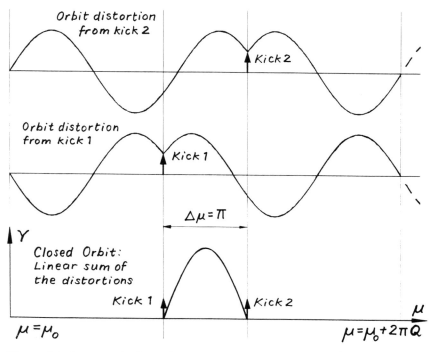

Figure 5.4. 2-magnet, half-wavelength, closed-orbit bump shown as the summation of the closed-orbit distortions from two kicks [normalised coordinates (μ, Y)].

the last cosine term in (5.9) becomes,

$$\cos \tfrac{1}{2}(|\Delta\mu_{1,s}| - |\Delta\mu_{2,s}|) = \cos \tfrac{1}{2}(\mu_2 - \mu_1), \tag{5.11}$$

whereas, if $\mu_1 < \mu_s < \mu_2$, then the observer is between the magnets and the last cosine takes a different form,

$$\cos \tfrac{1}{2}(|\Delta\mu_{1,s}| - |\Delta\mu_{2,s}|) = \cos[\mu_s - \tfrac{1}{2}(\mu_1 + \mu_2)]. \tag{5.12}$$

When the condition $(\mu_2 - \mu_1) = \pi$ is imposed on the magnet separation, (5.11) is zero and hence y_s is zero for all points outside the adjacent magnets, but (5.12) is a half sine wave with its maximum at the mid-point $\mu_s = \tfrac{1}{2}(\mu_1 + \mu_2)$. Thus equation (5.9) represents a half-wavelength bump with the conditions already derived and given in (5.8).

(ii) *3-magnet bumps*

It is rare that magnets can be placed with a phase separation of exactly π and even when this is possible it makes the lattice inflexible for future developments. It is therefore useful to know how to correct the residual error with a third dipole. This

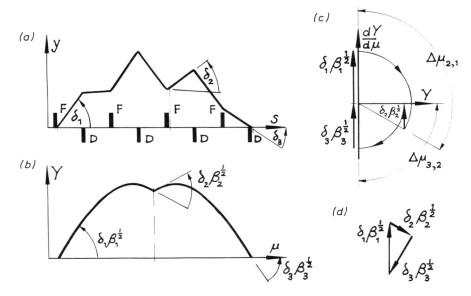

Figure 5.5. 3-magnet, closed-orbit bump. (*a*) real-space (*s, y*), (*b*) normalised coordinates (μ, Y), (*c*) normalised phase-space (Y, Y'), (*d*) vector diagram.

can be demonstrated by using the generalised transfer matrix (4.13) to track the beam trajectory through three dipole kicks.

Track δ_1 to δ_2:
$$\begin{cases} y_{2f} = (\beta_1\beta_2)^{\frac{1}{2}} \sin(\Delta\mu_{2,1})\delta_1 \\ y_{2f}' = (\beta_1/\beta_2)^{\frac{1}{2}}[\cos(\Delta\mu_{2,1}) - \alpha_2 \sin(\Delta\mu_{2,1})]\delta_1 \end{cases}$$

Back-track δ_3 to δ_2:
$$\begin{cases} y_{2b} = (\beta_3\beta_2)^{\frac{1}{2}} \sin(-\Delta\mu_{2,3})\delta_3 \qquad (\text{n.b. } y_{3f}' = -\delta_3) \\ y_{2b}' = (\beta_3/\beta_2)^{\frac{1}{2}}[\cos(-\Delta\mu_{2,3}) - (-\alpha_2) \sin(-\Delta\mu_{2,3})]\delta_3 \end{cases}$$

The forward- and back-tracked amplitudes at δ_2 must be equal and the difference in the derivatives must be matched by the dipole kick δ_2, i.e. $y_{2f} = y_{2b}$ and $\delta_2 = -y_{2b}' - y_{2f}'$ (see Figure 5.5). A little manipulation yields,

$$\frac{\delta_1\beta_1^{\frac{1}{2}}}{\sin(\Delta\mu_{3,2})} = \frac{\delta_2\beta_2^{\frac{1}{2}}}{\sin(\Delta\mu_{1,3})} = \frac{\delta_3\beta_3^{\frac{1}{2}}}{\sin(\Delta\mu_{2,1})}. \tag{5.13}$$

An array of 3-magnet bumps is frequently used in closed orbit correction schemes and for making aperture scans.

(iii) *4-magnet bumps*

It is often necessary to control both amplitude and slope at a given position in a ring and to do this for a number of different optics configurations. This occurs most frequently for injection and extraction schemes and the minimum solution is a

4-magnet bump. Two upstream dipoles are used to match to the required amplitude and slope at the specified position and then two downstream magnets are similarly matched. As before, this is done by applying the generalised transfer matrix (4.13).

5.1.3   Harmonic response of the closed orbit

The original differential equation for the closed orbit (5.1) can be rewritten in normalised amplitude and phase coordinates using the relationships (4.3) to (4.6) to give,

$$\frac{d^2}{ds^2}(Y\beta^{\frac{1}{2}}) + K(s)Y\beta^{\frac{1}{2}} = \mp\frac{\Delta B(s)}{B_0\rho_0}$$

$$\frac{1}{Q^2\beta^{3/2}}\frac{d^2}{d\Phi^2}Y + Y\frac{d^2}{ds^2}\beta^{\frac{1}{2}} + K(s)Y\beta^{\frac{1}{2}} = \mp\frac{\Delta B(s)}{B_0\rho_0}.$$

This can be simplified by using (4.2) to remove the second order differential of $\beta^{\frac{1}{2}}$ so that,

$$\frac{d^2}{d\Phi^2}Y + Q^2Y = \mp Q^2\beta^{3/2}\frac{\Delta B(\Phi)}{B_0\rho_0} \tag{5.14}$$

which has the form of a simple harmonic oscillator with a forcing term on the righthand side. The forcing term due to the dipole errors can be decomposed into a Fourier series,

$$f(\Phi) = \mp\beta^{3/2}\frac{\Delta B(\Phi)}{B_0\rho_0} = \sum_n f_n\, e^{jn\Phi} \tag{5.15}$$

where,

$$f_n = \frac{1}{2\pi}\int_0^{2\pi} f(\Phi)\, e^{jn\Phi}\, d\Phi = \frac{\mp 1}{2\pi Q}\int_0^C \beta^{\frac{1}{2}}\frac{\Delta B(s)}{B_0\rho_0}\, e^{jns}\, ds. \tag{5.16}$$

Note that the definition of $f(\Phi)$ has been adapted to remove the sign asymmetry. Equation (5.14) can now be rewritten with its driving term expressed using the Fourier series (5.15) to give,

$$\frac{d^2}{d\Phi^2}Y + Q^2\beta^2K(\Phi)Y = Q^2\sum_n f_n\, e^{jn\Phi}. \tag{5.17}$$

It is natural to construct a trial solution for (5.17), which also has the form of a Fourier series,

$$Y(\Phi) = \sum_n F_n\, e^{jn\Phi} \tag{5.18}$$

and to make the substitution back into (5.17) to give,

$$(-n^2 + Q^2)F_n = Q^2 f_n, \tag{5.19}$$

which yields the final equation for the closed orbit from the combination of (5.18) and (5.19),

$$Y(\Phi) = \sum_n f_n \frac{Q^2}{(Q^2 - n^2)} e^{jn\Phi}.$$ (5.20)

Equation (5.20) is a very interesting formulation of the closed orbit since it makes the resonant response of the machine apparent. Those harmonics in the distribution of the field errors which are close to Q have an amplified effect and those for which $|n| > \sqrt{2}Q$ have a diminished effect. This provides the basis for a closed orbit correction scheme (see Section 11.1.3(ii)). It also makes it possible to predict how a closed orbit will evolve when the tune is changed and in Section 8.5 the converse is used to find a tune shift due to an intense stack.

5.2   GRADIENT DEVIATIONS

5.2.1   Stability and tune shifts

Suppose that a gradient deviation, that can be represented by a thin lens as defined in Section 3.1.2, is introduced into a machine. The single-turn transfer matrix in the unperturbed machine has already been derived in (4.14), so the matrix for the perturbed machine will now become,

$$T(s + C|s) = \begin{pmatrix} \cos 2\pi Q_0 + \alpha \sin 2\pi Q_0 & \beta \sin 2\pi Q_0 \\ -\gamma \sin 2\pi Q_0 & \cos 2\pi Q_0 - \alpha \sin 2\pi Q_0 \end{pmatrix} \begin{pmatrix} 1 & 0 \\ kl & 1 \end{pmatrix}$$

$$T(s + C|s) = \begin{pmatrix} \cos 2\pi Q_0 + \alpha \sin 2\pi Q_0 + \beta kl \sin 2\pi Q_0 & \beta \sin 2\pi Q_0 \\ -\gamma \sin 2\pi Q_0 + kl(\cos 2\pi Q_0 - \alpha \sin 2\pi Q_0) & \cos 2\pi Q_0 - \alpha \sin 2\pi Q_0 \end{pmatrix}.$$ (5.21)

In Section 3.2.1 the stability of the particle motion in a periodic structure was investigated and a condition for stability was found (3.20), which required the modulus of half the trace of the transfer matrix for one period to be less than or equal to unity. This condition can be applied to (5.21) to give,

$$|\cos 2\pi Q_0 + \tfrac{1}{2}\beta kl \sin 2\pi Q_0| = |\cos 2\pi(Q_0 + \Delta Q)| \leq 1$$ (5.22)

where $(Q_0 + \Delta Q)$ is the tune value for the perturbed machine. If $\cos 2\pi Q_0$ is already close to unity, that is the tune Q_0 is close to an integer or half-integer number, then it is clear from (5.22) that a small gradient deviation could drive the machine unstable. This condition is known as the *half-integer resonance*.

   If it is assumed that the tune shift ΔQ is small then,

$$\cos 2\pi(Q_0 + \Delta Q) \simeq \cos 2\pi Q_0 - 2\pi \, \Delta Q \sin 2\pi Q_0$$

which, when compared to (5.22), gives an expression for the tune shift in terms of the added gradient, i.e.

$$\Delta Q = -\frac{1}{4\pi} \beta kl. \tag{5.23}$$

This tune shift arises from a small modulation of the betatron function all around the machine. So long as this modulation remains small the effects of several small gradient deviations can be added and (5.23) can be extended to

$$\Delta Q = \mp \frac{1}{4\pi} \sum_n (\beta kl)_n \quad \text{or} \quad \mp \frac{1}{4\pi} \int_0^C \beta(s)k(s)\,\mathrm{d}s \tag{5.24}$$

where the upper sign $(-)$ applies to the horizontal plane and the lower sign $(+)$ applies to the vertical plane. Equation (5.24) could be used, for example, to predict the effect of a series of tuning quadruples. If the probable tune shift due to random errors, for example arising from manufacturing tolerances, were to be needed, (5.24) can be reformulated as,

$$\langle (\Delta Q)^2 \rangle^{\frac{1}{2}} = \frac{1}{4\pi} \left[\sum_n (\beta kl)_n{}^2 \right]^{\frac{1}{2}} \langle (\Delta k/k)^2 \rangle^{\frac{1}{2}}. \tag{5.24'}$$

Close to half-integer values of the tune Q_0, there are regions where (5.22) may *not* be satisfied. These regions are known as *half-integer stopbands*. The existence of stopbands can be illustrated using the same example as above of a single error kl. When Q_0 is close to a half integer, the argument $2\pi Q_0$ in (5.22) can be written as $(\pi n + 2\pi \Delta Q_0)$ where $\Delta Q_0 = (Q_0 - n/2)$ and $n = 1, 2, 3, \ldots$. On the assumption that βkl is small an expansion of (5.22) to second order then yields,

$$\tfrac{1}{2} \operatorname{Tr} \boldsymbol{T} \simeq (-1)^n [1 - \tfrac{1}{2}(2\pi \Delta Q_0)^2 + \tfrac{1}{2}\beta kl(2\pi \Delta Q_0)].$$

The limits of the stopband are found by setting the square bracket to unity, which gives,

$$\Delta Q_0 = 0 \quad \text{and} \quad \Delta Q_0 = \frac{1}{2\pi} \beta kl.$$

Between these limits, the ΔQ_0 term will dominate the $(\Delta Q_0)^2$ term making the beam unstable. The stopbands occur at each half-integer tune, which includes the integer values. It is worth noting that the stopband width has twice the magnitude of the detuning as found in (5.23). The probable stopband width for random errors will therefore be twice the probable value for the detuning given in (5.24)'.

5.2.2   Betatron amplitude modulation

In a similar manner to the above, it is possible to evaluate the modulation of the betatron function around the machine. Consider an observer at position s and a small gradient error at σ of length $\mathrm{d}\sigma$, which can be represented by the thin lens

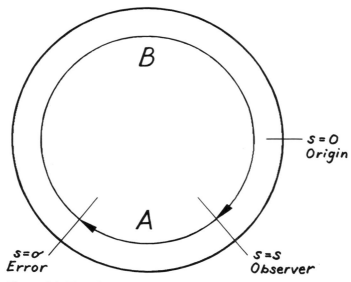

Figure 5.6. Transfer matrices from observer to error and from error to observer.

formulation as before. Let the unperturbed machine matrix from s to σ be A and from σ to s be B (see Figure 5.6).

The matrix for a single turn in the perturbed machine, as seen by the observer at s, will therefore be,

$$T(s + C|s) = \begin{pmatrix} b_{11} & b_{12} \\ b_{21} & b_{22} \end{pmatrix} \begin{pmatrix} 1 & 0 \\ k\,\mathrm{d}\sigma & 1 \end{pmatrix} \begin{pmatrix} a_{11} & a_{12} \\ a_{21} & a_{22} \end{pmatrix}. \tag{5.25}$$

It is only necessary to evaluate the t_{12} term in (5.25), which is

$$t_{12} = b_{11}a_{12} + b_{12}(k\,\mathrm{d}\sigma\,a_{12} + a_{22}) = t_{0\,12} + \mathrm{d}\sigma\,a_{12}b_{12},$$

where $t_{0\,12}$ is from the unperturbed matrix found by putting $k\,\mathrm{d}\sigma = 0$. Thus the variation in the t_{12} term due to the perturbation is,

$$\Delta t_{12} = k\,\mathrm{d}\sigma\,a_{12}b_{12}. \tag{5.26}$$

Equation (5.26) can be expressed in terms of the Courant and Snyder beam parameters by comparison of the individual terms with the t_{12} term in the generalised transfer matrix (4.13), which gives

$$\Delta(\beta_s \sin 2\pi Q_0) = k\,\mathrm{d}\sigma\,\beta_s\beta_\sigma \sin(\mu_\sigma - \mu_s)\sin(\mu_s - \mu_\sigma)$$

$$\Delta(\beta_s \sin 2\pi Q_0) = k\,\mathrm{d}\sigma\,\beta_s\beta_\sigma \sin(\mu_\sigma - \mu_s)\sin[2\pi Q_0 - (\mu_\sigma - \mu_s)].$$

This can be generalised to include many errors by integrating around the ring with the position of the error σ running from the observer's position s to $s + C$. This yields,

$$\Delta(\beta_s \sin 2\pi Q_0) = \beta_s \int_s^{s+C} k(\sigma)\beta_\sigma \sin(\mu_\sigma - \mu_s) \sin[2\pi Q_0 - (\mu_\sigma - \mu_s)] \, d\sigma. \quad (5.27)$$

The lefthand and righthand sides of (5.27) can be expanded to give,

$$\Delta\beta_s \sin 2\pi Q_0 + 2\pi\beta_s \Delta Q \cos 2\pi Q_0$$

$$= -\beta_s \int_s^{s+C} k(\sigma)\beta_\sigma \tfrac{1}{2}\{\cos 2\pi Q_0 - \cos 2[(\mu_\sigma - \mu_s) - \pi Q_0]\} \, d\sigma.$$

The second term on the lefthand side cancels with the first term on the righthand side by virtue of the tune shift relationship (5.24). This leaves the final expression for the betatron amplitude modulation around the ring, generalised for the two planes, as,

$$\Delta\beta_y(s) = \mp \frac{\beta_y(s)}{2 \sin \pi Q_{0y}} \int_s^{s+C} k(\sigma)\beta_y(\sigma) \cos 2[(\mu_y(\sigma) - \mu_y(s) - \pi Q_{0y}] \, d\sigma. \quad (5.28)$$

The formulation in (5.28) now looks very similar to the expression for the closed orbit distortion in real-space coordinates due to distributed field errors as written in (5.2). The principal difference is that the betatron modulation advances with the frequency $2Q$ rather than Q.

5.3 WEAK LINEAR COUPLING

Coupling is the periodic exchange of energy between two oscillators and is a widespread phenomenon in physics. This action can take place via the amplitudes, velocities or accelerations of the oscillators. The traditional example of amplitude coupling is a pair of pendulums coupled by a weightless spring. This example also nicely illustrates the concept of normal modes (see Figure 5.7), which can be used in linear combination to express any subsequent motion of the system.

The first storage ring projects in the 1960s and early 1970s stimulated an interest in betatron coupling. Initially the concern was over the coupling from random quadrupole tilts, which introduce skew quadrupole field errors around the ring, but later attention turned to the use of solenoids for physics detectors, which apply a longitudinal field to the beam. References 2 and 3 are early examples of the analysis of skew quadrupole errors in electron storage rings. There are two comprehensive references in the subsequent literature, Ripken[4], in which the Courant, Livingston and Snyder theory is extended to 4-dimensional phase space and Guignard[5], in which the theory of all sum and difference resonances in 3-dimensional magnetic fields is developed by a perturbative treatment of the Hamiltonian for the single particle motion in a synchrotron. The method and essential results of this last reference were in fact anticipated by Morton[6]. In the present chapter, a simplified theory for betatron

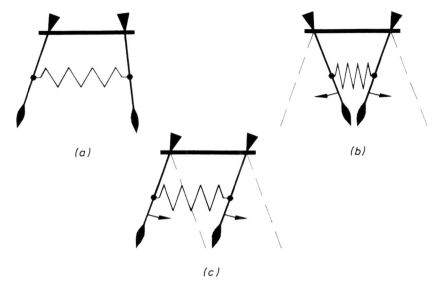

Figure 5.7. Amplitude-coupled pendulums. (*a*) General motion, (*b*) anti-symmetric normal mode, (*c*) symmetric normal mode.

coupling in skew quadrupole and longitudinal fields will be presented[7]. The results from the detailed theory in Ref. 5 will simply be quoted where needed.

5.3.1   Coupling in uniform skew quadrupole and longitudinal fields: basic equations

Since the coupling action builds up over many turns, it is likely to be dependent on global and average parameters rather than local lattice functions. On this premise, the *sinusoidal or smooth approximation* will be used for the motion in the horizontal and vertical planes.

The section title 'linear coupling' means more precisely the *second-order, zero-harmonic, difference resonance*, $Q_x - Q_z = 0$. This resonance is driven by:

- *skew quadrupole fields* usually due to random quadrupole tilts
- *longitudinal fields* usually due to physics detectors.

The skew quadrupoles cause amplitude coupling, while the longitudinal fields are an example of velocity coupling*. Since the zero-order harmonic of the fields is the driving harmonic, the simple model of uniformly distributed fields will be used. This will still demonstrate all the principal features of coupling and by taking the zero

---

* The equivalent of acceleration coupling in inductance-coupled a.c. circuits does not exist in accelerators.

$$\mathrm{d}^2 q_1/\mathrm{d}t^2 + (R_1/L_1)\,\mathrm{d}q_1/\mathrm{d}t + [1/(L_1 C_1)]q_1 = -M_{12}\,\mathrm{d}^2 q_2/\mathrm{d}t^2.$$

harmonic of a practical distribution, it will be possible to do useful back-of-envelope calculations. Only when designing a compensation scheme is the full theory necessary and in order to cover this eventuality the relevant references and formula for the driving term will be quoted.

The sinusoidal approximation to the particle motion in either plane with driving terms will have the form,

$$\frac{d^2}{dt^2} y + (Q\Omega)^2 y = \text{forcing terms (force/mass).}$$ (5.29)

It is convenient to change the independent variable from time to distance and at the same time the driving term is evaluated in general terms to give,

$$\frac{d^2}{ds^2} y + (Q\Omega/v_s)^2 y = \frac{e}{m_0\gamma v_s^2}(E + v \times B).$$

The cyclotron motion relationship (2.6) and the expression for the longitudinal momentum $p = m_0\gamma v_s$ reduce this equation to,

$$\frac{d^2}{ds^2} y + (Q/R)^2 y = \frac{-1}{B_0\rho_0 v_s}(E + v \times B).$$ (5.30)

It is now necessary to evaluate the driving terms for the righthand side of (5.30), which cause the coupling:

For a longitudinal field,
$$\begin{cases} (v \times B)_x = v_z B_s i_x = v_s B_s \dfrac{d}{ds} z\, i_x \\[2mm] (v \times B)_z = -v_x B_s i_z = -v_s B_s \dfrac{d}{ds} x\, i_z. \end{cases}$$ (5.31)

For a skew quadrupole,
$$\begin{cases} (v \times B)_x = -v_s\left[\dfrac{\partial B_z}{\partial z}\right]_0 z\, i_x \\[2mm] (v \times B)_z = v_s\left[\dfrac{\partial B_x}{\partial z}\right]_0 x\, i_z \end{cases}$$ (5.32)

where i_x, i_s, i_z are the unit vectors along the x, s, z axes.

Before the substitution of (5.31) and (5.32) into (5.30) is made, it is useful to make some simplifying definitions to render the subsequent equations less cumbersome to handle, i.e.

$$k_s = \frac{1}{B_0\rho_0}\left[\frac{\partial B_x}{\partial x}\right]_0 = \frac{-1}{B_0\rho_0}\left[\frac{\partial B_z}{\partial z}\right]_0$$ (5.33)

$$b = \frac{1}{B_0\rho_0} B_s$$ (5.34)

$$\kappa_x = (Q_x/R) \quad \text{and} \quad \kappa_z = (Q_z/R).$$ (5.35)

so that,

$$
\left.
\begin{array}{l}
x'' + \kappa_x{}^2 x = -bz' - k_s z \\[4pt]
z'' + \kappa_z{}^2 z = bx' - k_s x
\end{array}
\right\}
\tag{5.36}
$$

where ' indicates differentiation with respect to distance along the trajectory. For the solution of (5.36) consider trial functions for the x and z motions with the forms of the uncoupled motions modified by envelope terms, i.e.

$$
x = X(s)\, e^{j\kappa_x s} \qquad z = Z(s)\, e^{j\kappa_z s}.
\tag{5.37}
$$

With the restriction of weak coupling, it is possible to neglect the X'' and Z'' terms when substituting into (5.36) since the envelopes will be slowly varying functions with respect to the betatron wavelength. Thus,

$$
\left.
\begin{array}{l}
2j\kappa_x X' = e^{-j\delta s}[-bZ' - (k_s + j\kappa_z b)Z] \\[4pt]
2j\kappa_z Z' = e^{j\delta s}[bX' - (k_s - j\kappa_x b)X]
\end{array}
\right\}
\tag{5.38}
$$

where,

$$
\delta = (\kappa_x - \kappa_z) = (Q_x - Q_z)/R.
\tag{5.39}
$$

In (5.38) the further mutual substitution between the equations for X' and Z' on the righthand side shows that these terms are second order. Once b^2 and bk are neglected (5.38) reduces to,

$$
\left.
\begin{array}{l}
X' = j\,\dfrac{e^{-j\delta s}}{2\kappa_x}\,(k_s + j\kappa_z b)Z \\[12pt]
Z' = j\,\dfrac{e^{j\delta s}}{2\kappa_z}\,(k_s - j\kappa_x b)X.
\end{array}
\right\}
\tag{5.40}
$$

The linear coupling resonance will only be of consequence when $Q_x \simeq Q_z$ which makes it possible to consider the tunes as equal, except when their difference appears in the equations and to put

$$
\kappa \simeq \kappa_x \simeq \kappa_z \qquad \text{(close to resonance)}
\tag{5.41}
$$

and

$$
\left.
\begin{array}{l}
X' = j\,\dfrac{e^{-j\delta s}}{2\kappa}\,(k_s + j\kappa b)Z \\[12pt]
Z' = j\,\dfrac{e^{j\delta s}}{2\kappa}\,(k_s - j\kappa b)X.
\end{array}
\right\}
\tag{5.42}
$$

Equations (5.42) are in a standard form for coupled mode analysis, i.e. $X' = c_{12}Z$ and $Z' = -c_{12}{}^\star X$, where $\star$ indicates the complex conjugate. The minus sign indicates positive energy flow in both oscillations.

The general solutions can be verified with a little algebra and are,

$$x = e^{j\kappa_x s}\left[A\, e^{(j/2)(\eta-\delta)s} - B\frac{(\eta-\delta)\kappa}{k_s - j\kappa b}\, e^{-(j/2)(\eta+\delta)s}\right]\left.\begin{array}{c} \\ \\ \\ \\ \end{array}\right\}$$
$$z = e^{j\kappa_z s}\left[B\, e^{-(j/2)(\eta-\delta)s} + A\frac{(\eta-\delta)\kappa}{k_s + j\kappa b}\, e^{(j/2)(\eta+\delta)s}\right]$$

(5.43)

where,

$$\eta = \sqrt{[\delta^2 + (k_s/\kappa)^2 + b^2]}$$

(5.44)

and A and B are complex constants.

At first sight these equations do not convey a great deal. More physical insight is obtained by taking the squares of the moduli of the amplitudes X and Z, which form the envelopes of the oscillations.

$$XX^\star = |X|^2 = |A|^2 + \frac{|B|^2(\eta-\delta)^2}{(k_s/\kappa)^2 + b^2} + \frac{2|AB^\star|(\eta-\delta)}{\sqrt{[(k_s/\kappa)^2 + b^2]}}\cos(\eta s - \phi)\left.\begin{array}{c} \\ \\ \\ \end{array}\right\}$$
$$ZZ^\star = |Z|^2 = |B|^2 + \frac{|A|^2(\eta-\delta)^2}{(k_s/\kappa)^2 + b^2} - \frac{2|AB^\star|(\eta-\delta)}{\sqrt{[(k_s/\kappa)^2 + b^2]}}\cos(\eta s - \phi)$$

(5.45)

where

$$\phi = \tan^{-1}(bk_s/\kappa).$$

(5.46)

Thus there is a sinusoidal exchange of energy with the interchange wavelength,

$$\lambda = \frac{2\pi}{\eta} = \frac{2\pi}{+\sqrt{[\delta^2 + (k_s/\kappa)^2 + b^2]}}$$

(5.47)

and the sum of the squares of the amplitudes is conserved,

$$E_{max} = |X|^2 + |Z|^2 = \frac{2\eta}{(\eta+\delta)}(|A|^2 + |B|^2).$$

(5.48)

It is worth considering the special case of a beam being kicked in one plane. The ensuing oscillation will be coherent and the amplitude interchange will be visible to a pick-up in either plane. Figure 5.8 illustrates this situation for a kick in the horizontal plane. Let $s = 0$ at the kick, then $|X|^2$ is a maximum (E_{max}) and $|Z|$ is zero. Equation (5.43) gives with these boundary conditions,

$$B = -\frac{A(\eta-\delta)\kappa}{k_s + j\kappa b}$$

and (5.45) gives the interchange amplitude,

$$E_T = \frac{4|AB^\star|(\eta-\delta)}{\sqrt{[(k_s/\kappa)^2 + b^2]}}.$$

After a little manipulation with use made of (5.44) for the definition of η, the following

107

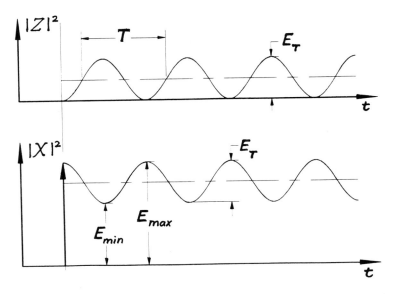

Figure 5.8. The envelope functions of the coherent oscillations following a kick in the horizontal plane.

expression for the modulation of the signal can be deduced,

$$R_{\mathrm{m}} = \frac{E_{\mathrm{min}}}{E_{\mathrm{max}}} = \frac{\delta^2}{\eta^2}. \tag{5.49}$$

The above equations contain the essential features of linear coupling and the situation in Figure 5.8 is the basis of a method for measuring the excitation of this resonance. It is, however, convenient and usual to define coupling coefficients at this stage, which effectively replace the definitions of k and b, i.e.

For skew gradients

$$C_{\mathrm{q}} = \left[\frac{R^2}{Q} \right]\left[\frac{1}{B_0 \rho_0} \right]\left[\frac{\partial B_x}{\partial x} \right]_0, \tag{5.50}$$

For axial fields

$$C_{\mathrm{b}} = \left[\frac{1}{B_0 \rho_0} \right] B_{\mathrm{s}}. \tag{5.51}$$

These can be combined into a global coupling coefficient,

$$C = +\sqrt{(C_{\mathrm{q}}^2 + C_{\mathrm{b}}^2)}. \tag{5.52}$$

The period T of the energy interchange [from (5.47)] and the modulation of the beam envelope [from (5.49)] can now be written in terms of more easily observed quantities, i.e.

Interchange period,

$$T = \frac{1}{f_{\mathrm{rev}}\sqrt{[\Delta^2 + |C|^2]}} \tag{5.53}$$

Modulation,
$$R_{\mathrm{m}} = \frac{E_{\min}}{E_{\max}} = \frac{\Delta^2}{\Delta^2 + |C|^2} \qquad (5.54)$$

where,

$$\Delta = Q_x - Q_z \qquad (5.55)$$

and f_{rev} is the revolution frequency.

The separation of the uncoupled tunes Δ cannot be measured directly in the presence of coupling. However, Δ can be eliminated from the above equations giving independent expressions for both $|C|$ and $|\Delta|$,

$$|C| = \frac{1}{f_{\mathrm{rev}} T} \sqrt{(1 - R_{\mathrm{m}})} \qquad (5.56)$$

$$|\Delta| = \frac{\sqrt{R_{\mathrm{m}}}}{f_{\mathrm{rev}} T}. \qquad (5.57)$$

This provides a very precise and elegant method[8] for measuring $|C|$. Despite the very simplified model used above, it turns out that the above results are quite general. The difference between this and the complete theory is how C_{q} and C_{b} are computed from the error distribution.

For simple estimations it is sufficient to average the fields around the machine so that,

$$C_{\mathrm{q}} = \frac{1}{2\pi Q} \frac{1}{B_0 \rho_0} \oint \left[\frac{\partial B_x}{\partial x} \right]_0 \mathrm{d}s \qquad (5.58)$$

$$C_{\mathrm{b}} = \frac{1}{2\pi} \frac{1}{B_0 \rho_0} \oint B_s \, \mathrm{d}s. \qquad (5.59)$$

The exact computation of the coupling coefficient is given later in Section 5.3.4.

5.3.2 Normal modes

The normal modes are of interest because they give some physical insight into what is happening, which will make it possible to propose a method for measuring the separate coefficients for the amplitude coupling C_{q} and the velocity coupling C_{b}.

A quick indication of the normal modes can be found by re-writing (5.43) to give,

$$\left. \begin{array}{l} x = A\, \mathrm{e}^{\mathrm{j}(\kappa + \eta/2)s} - \dfrac{B(\eta - \delta)}{\sqrt{[(k_s/\kappa)^2 + b^2]}}\, \mathrm{e}^{\mathrm{j}(\kappa - \eta/2)s + \mathrm{j}\phi} \\[4mm] z = B\, \mathrm{e}^{\mathrm{j}(\kappa - \eta/2)s} + \dfrac{A(\eta - \delta)}{\sqrt{[(k_s/\kappa)^2 + b^2]}}\, \mathrm{e}^{\mathrm{j}(\kappa + \eta/2)s - \mathrm{j}\phi} \end{array} \right\} \qquad (5.60)$$

where

$$\kappa_x = (\kappa + \delta/2), \qquad \kappa_z = (\kappa - \delta/2), \qquad \kappa = (Q_x + Q_z)/R \qquad (5.61)$$

$$\phi = \tan^{-1}(\kappa b/k_s) = \tan^{-1}(C_b/C_q). \qquad (5.62)$$

The normal modes can be found by putting either of the two arbitrary constants A or B to zero. This gives,

$$\frac{x}{z} = \left[\frac{\Delta \pm \sqrt{(\Delta^2 + |C|^2)}}{C} \right] e^{j\phi}. \qquad (5.63)$$

Equation (5.63) shows the characteristics of a normal mode that x and z bear a constant relationship, i.e. the mode retains its 'shape' at all times. The interpretation of (5.63) is that the normal modes lie on inclined planes and the complex phase shift indicates that they are elliptically polarised (see Figures 5.9 and 5.10). This information

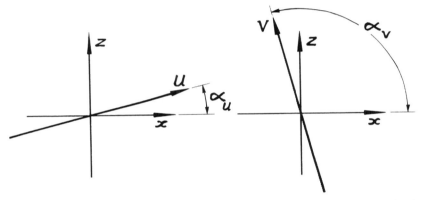

Figure 5.9. Inclined normal modes due to skew gradient coupling (U is the near-horizontal mode, V is the near-vertical mode).

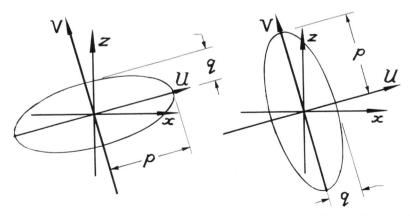

Figure 5.10. Inclined, elliptically-polarised normal modes due to skew gradient and longitudinal field coupling.

suggests an alternative way of solving the basic equations of motion (5.36), which in turn will suggest a way of measuring C_q and C_b individually. First the motion equations (5.36) are re-written using (5.61).

$$\left.\begin{array}{l} x'' + (\kappa^2 + \delta\kappa)x = -bz' - kz \\ z'' + (\kappa^2 - \delta\kappa)z = \quad bx' - kx \end{array}\right\} \tag{5.64}$$

where $\kappa_{x,z} = (\kappa \pm \delta/2)^2 = \kappa^2 \pm \delta k + \delta^2/4$ and $\delta^2/4$ is neglected.

It is possible to decouple these equations partially by making a transformation to an inclined coordinate system (U, V)

$$\left.\begin{array}{l} x = U \cos \alpha - V \sin \alpha \\ z = U \sin \alpha + V \cos \alpha \end{array}\right\}. \tag{5.65}$$

The substitution of (5.65) into (5.64) yields two equations from which first V'' can be eliminated and then U'' to give,

$$U''\left(\tan \alpha + \frac{1}{\tan \alpha}\right) + U\left[(\kappa^2 - \delta k)\tan \alpha + \frac{(\kappa^2 + \delta k)}{\tan \alpha} + 2k\right]$$

$$= -V'b\left(\tan \alpha + \frac{1}{\tan \alpha}\right) + V\left[k_s\left(\tan \alpha - \frac{1}{\tan \alpha}\right) + 2\delta\kappa\right] \tag{5.66}$$

$$V''\left(\tan \alpha + \frac{1}{\tan \alpha}\right) + V\left[(\kappa^2 + \delta k)\tan \alpha + \frac{(\kappa^2 - \delta k)}{\tan \alpha} - 2k\right]$$

$$= U'b\left(\tan \alpha + \frac{1}{\tan \alpha}\right) + U\left[k_s\left(\tan \alpha - \frac{1}{\tan \alpha}\right) + 2\delta\kappa\right]. \tag{5.67}$$

By choosing α such that the last term in (5.66) and (5.67) is zero, the equations become uncoupled for k_s. This value of α is given by,

$$\tan \alpha = \frac{-\delta \pm \sqrt{[\delta^2 + (k_s/\kappa)^2]}}{k_s/\kappa} = \frac{-\Delta \pm \sqrt{(\Delta^2 + C_q^2)}}{C_q}. \tag{5.68}$$

The angle of inclination α of the axes can be identified in (5.63) by putting $C_b = 0$. The two solutions for α can be called α_u and α_v and it is quickly verified that $(\alpha_v - \alpha_u) = \pi/2$. Equation (5.68) yields the quadrupole coupling coefficient directly

$$C_q = \Delta \tan 2\alpha. \tag{5.69}$$

At the privileged angle of α in the (U, V) system,

$$\left.\begin{array}{l} U'' + (\kappa^2 + \gamma\kappa)U = -bV' \\ V'' + (\kappa^2 - \gamma\kappa)V = bU' \end{array}\right\} \tag{5.70}$$

where,

$$\gamma\kappa = \delta\kappa \cos 2\alpha + k_s \sin 2\alpha. \tag{5.71}$$

In the rotated system (U, V) only the axial field coupling is apparent, except for the

$\pm\gamma/2$ shifts to get the uncoupled frequencies in the (U, V) system. Thus from (5.71) with the use of (5.33), (5.34), (5.35), (5.51), (5.55) and (5.69) and some manipulation,

$$\Delta_{uv} = (\Delta \cos 2\alpha + C_q \sin 2\alpha) = \Delta/\cos 2\alpha. \qquad (5.72)$$

The same equations that were used in the (x, z) system can now be applied in the (U, V) system to get,

$$T_{uv} = \frac{1}{f_{rev}\sqrt{[\Delta_{uv}{}^2 + C_b{}^2]}} \qquad (5.73)$$

and

$$R_{uv} = \frac{\Delta_{uv}{}^2}{\Delta_{uv}{}^2 + C_b{}^2}. \qquad (5.74)$$

By the substitution of (5.69) and (5.72) into (5.74), it can be shown that $T_{uv} = T$ of the original xz-system, which is intuitively expected. Thus the period of amplitude exchange is constant and only R_m and Δ change between the two systems. As was done before, Δ_{uv} can be eliminated between the two equations (5.73) and (5.74) to give,

$$C_b = \frac{1}{Tf_{rev}}\sqrt{(1 - R_{uv})}. \qquad (5.75)$$

This suggests that if a rotating kicker and pickup were used to search for a minimum in R or $|C|$, then at the angle corresponding to the minimum, C_b would be given by (5.75) and C_q by (5.69) using the value of Δ from the first measurement with $\alpha_u = 0$[8,9].

5.3.3   Observations with coupling

(i) *Tune measurements*

In (5.60) there are two frequencies, which are the frequencies of the normal modes,

$$\kappa_{u,v} = \frac{1}{R}[Q \pm \tfrac{1}{2}\sqrt{(\Delta^2 + C_q{}^2 + C_b{}^2)}]. \qquad (5.76)$$

If $C_q = C_b = 0$, then $\kappa_u = Q/R$ and $\kappa_v = Q_z/R$, i.e. the normal unperturbed betatron motion. When $C_q \neq 0$ and/or $C_b \neq 0$, the effect on the normal mode frequencies (and hence on the observed tune of the machine) depends on Δ. If Δ is large (i.e. the working point is far from the resonance), there is little impact on the normal modes. The effect increases as Δ is reduced, until at $\Delta = 0$ the frequency split is at its maximum and is entirely due to the coupling coefficients. In addition to the frequency split between the normal modes, it was also shown that the modes become inclined due to C_q and elliptically polarised due to C_b. Again when Δ is large the effect is small, but as Δ is reduced the effect increases until at $\Delta = 0$ the modes are inclined

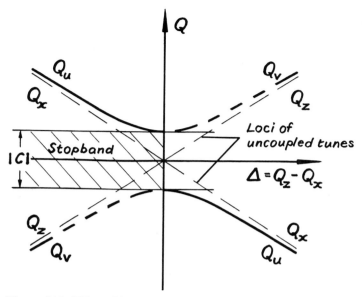

Figure 5.11. Effect of linear coupling on tune measurements.

at 45° [see (5.68) with $\Delta = 0$]. At this point the machine is working exactly on resonance with $Q_x = Q_z$ and is said to be fully coupled.

If one were to carry out the above experimentally, that is increasing Q_x and reducing Q_z in the vicinity of the coupling resonance, while also trying to measure the tunes with horizontally and vertically mounted kickers and pickups, then the results would be similar to Figure 5.11.

The pickups will 'see' the frequencies of the U- and V-modes. While these modes are 'nearly horizontal' and 'nearly vertical' the measurements will be reliable. As the modes rotate the horizontal plane pickup will start to 'see' the V-mode as well as the U-mode and vice versa. Operational difficulties for the tune measurement will appear at around $|C/\Delta| = 1$. Once $|C/\Delta| > 1$, tune measurements will become unreliable and the tune readings will start to jump back and forth from the U-mode to the V-mode value, spanning the stopband shown in Figure 5.11. At $\Delta = 0$ the pickups respond equally well to both modes, since they are now at 45° inclination. The difference in the mode frequencies $(Q_u - Q_v)$ equals $|C|$, which provides another way of measuring $|C|$. As $|\Delta|$ increases on the far side of the resonance normality will be restored.

(ii) *Other methods for the measurement of coupling*

As an extension to the method of measuring $|C|$ by observing the oscillations following a kick, it is theoretically possible to deduce all parameters from this one waveform[9], although this has never been demonstrated as a practical method to the knowledge of the authors. A rather different method, which can be used on 'quiet' beams (i.e. the physics beams in colliders) uses the coupling transfer function, which is obtained by correlating a horizontal transverse perturbation imposed on the beam to the resulting vertical coherent motion[10].

5.3.4 Results from a more exact analysis

A more rigorous analysis for coupling in an alternating-gradient lattice can be found in Ref. 5. This analysis confirms the formulae for T, S, κ_x and κ_z, but more importantly it shows how to evaluate C in terms of the field errors. It therefore replaces the simple averaging formulae (5.58) and (5.59). This is essential for the more specialised tasks such as designing a compensation scheme. In the formalism of Ref. 5, C is complex and is given by,

$$
C(\Theta_0) = \frac{1}{2\pi R} \int_{\Theta_0}^{\Theta_0 + 2\pi} \sqrt{(\beta_x \beta_z)} \left[K(\Theta) + \frac{M(\Theta)}{2} R \left\{ \frac{\alpha_x}{\beta_x} - \frac{\alpha_z}{\beta_z} \right\} \right.
$$

$$
\left. - j \frac{M(\Theta)}{2} R \left\{ \frac{1}{\beta_x} + \frac{1}{\beta_z} \right\} \right] e^{j[\{\mu_x - \mu_z\} - \{\mu_x(\Theta_0) - \mu_z(\Theta_0)\} - \Delta(\Theta - \Theta_0)]} \, d\Theta \qquad (5.77)
$$

where, Θ_0 is the origin at which C is evaluated (observed)

Θ is s/R the mean azimuthal angle and

$$
K_s(\Theta) = \frac{1}{2} \frac{R^2}{|B\rho|} \left[\frac{\partial B_x}{\partial x} - \frac{\partial B_z}{\partial z} \right] \qquad (5.78)
$$

$$
M(\Theta) = \frac{1}{|B\rho|} B_{\text{axial}}. \qquad (5.79)
$$

The $|C(\Theta)|$ is constant for all Θ and is exactly equivalent to $|C|$ from (5.52). Furthermore the real part of C is equivalent to C_q and the imaginary part is equivalent to C_b.

The most important feature of (5.77) is the exponential term which makes it possible for the skew gradient of (5.78) and the longitudinal fields of (5.79) to both contribute to the real and imaginary parts of C. Thus there is a mixing between the C_q and C_b terms of the simple theory. If the phasing of this exponential term is handled correctly, then skew quadrupoles can be used to compensate solenoid detector magnets. This is excluded in the simple theory and before the situation was fully understood, it was often said that such compensation would be impossible. In a qualitative way the action of this phase term can be seen by considering a particle oscillating in an inclined plane. By moving a short distance through the lattice one can see that the oscillation will become elliptical if the phase advances are unequal in the two planes. This is taken into account by the $(\mu_x - \mu_z)$ term in (5.77), which will typically oscillate ± 0.5 radian in a regular FODO lattice.

Finally, $K_s(\Theta)$ has a more general form than $k(s)$ in order to take account of the end fields of solenoids, which can either be circularly symmetric for round apertures or 2-dimensional for slots. When calculating $C(\Theta_0)$ for a solenoid in a collider, it will most probably be inside a low-beta insertion. In this case, the term $\frac{1}{2}M(\alpha_x/\beta_x - \alpha_z/\beta_z)$ will be important and since β_x and β_z will be varying quickly the choice of the length of the solenoid and the type of end plate can lead to situations with self-compensation.

Theoretically it is equally possible to base a compensation scheme on either skew quadrupoles or solenoids. However, the substitution of practical values into (5.77) shows that for the same value of C skew quadrupoles are far less space and power consuming.

The theory given in Ref. 5 is *not* limited to weak coupling. However, in order to fully compensate at the level of the 4×4 matrices the $Q_x - Q_z = 0$ and the $Q_x + Q_z = P$ resonances must be included[11]. This implies the need for four families of skew quadrupoles unless there are some preferential phases between the errors and the correctors.

5.3.5 Vertical dispersion and compensation schemes

In addition to coupling, skew quadrupoles will excite vertical dispersion when located in a region of finite horizontal dispersion. At the position $x = D_x \Delta p/p_0$ of an off-momentum particle, the skew gradient will give a vertical kick $x k_s l$, which is directly proportional to the momentum error. The substitution of this momentum-dependent kick into the equation for closed-orbit distortion (5.4) yields an expression for the vertical dispersion function D_z,

$$D_z(s) = \frac{\beta^{\frac{1}{2}}(s)}{2 \sin \pi Q_z} \sum_n (\beta_x^{\frac{1}{2}} D_x k_s l)_n \cos[|\mu_n - \mu(s)| - \pi Q_z]. \tag{5.80}$$

The combined effect of vertical and horizontal dispersion is to cause the median plane of the beam to tilt across the aperture.

This effect can be avoided by placing the skew quadrupoles in regions with zero horizontal dispersion. In this case two lenses that are separated by as nearly as possible $\pi/2$, in the phase term in (5.77), will be able to compensate the real and imaginary coupling vectors. This strategy is very common, although in practice there will often be several lenses distributed uniformly around the machine and it is usual to have four families rather than two, in order that the nearest sum resonance $Q_x + Q_z = P$ can also be compensated[11].

However, it may not be possible to locate the quadrupoles in dispersion-free regions and indeed just the opposite is needed when the vertical dispersion is also to be compensated[12]. There are two main approaches to planning combined coupling/vertical-dispersion compensation schemes.

(i) *Harmonic method*

The vertical dispersion given by (5.80) can be expressed as a harmonic series in an exactly analogous way to the closed orbit in Section 5.1.3, so that,

$$D_z(s) = \beta^{\frac{1}{2}}(s) \sum_{n=1}^{N} [f_{c,n} \cos(n\Phi) + f_{s,n} \sin(n\Phi)] \qquad (5.81)$$

where the harmonic amplitudes are given by,

$$\left.\begin{matrix} f_{c,n} \\ f_{s,n} \end{matrix}\right\} = \frac{1}{\pi Q_z} \left[\frac{Q_z^2}{Q_z^2 - n^2} \right] \sum_{n=1}^{N} (\beta_x^{\frac{1}{2}} D_x k_s l)_n \begin{Bmatrix} \cos(n\Phi_n) \\ \sin(n\Phi_n). \end{Bmatrix} \qquad (5.82)$$

The coupling coefficient (5.77) can be written in a similar way for just skew lenses and the observer at the origin $\Theta_0 = 0$, i.e.

$$\left.\begin{matrix} \mathrm{Re}(C) \\ \mathrm{Im}(C) \end{matrix}\right\} = \frac{1}{2\pi} \sum_{n=1}^{N} (\sqrt{(\beta_x \beta_z)} k_s l)_n \begin{Bmatrix} \cos[(\mu_x - \mu_z) - \Theta\Delta] \\ \sin[(\mu_x - \mu_z) - \Theta\Delta], \end{Bmatrix} \qquad (5.83)$$

where $\Delta = Q_x - Q_z$ in agreement with the nomenclature of this section. Thus with N skew lenses, the equations for $(N-2)$ harmonics of $D_z(f_{c,n}, f_{s,n})$ and two further equations for $\mathrm{Re}(C)$ and $\mathrm{Im}(C)$ can be formulated. By inverting these equations, the current distributions can be found for exciting the chosen dispersion harmonics and the coupling. To get good results the skew lenses should be uniformly distributed in betatron phase. Not all the dispersion harmonics can be controlled but the attenuation factor $Q_z^2/(Q_z^2 - n^2)$ will suppress the higher-order errors. It is better to omit the zero-order dispersion harmonic since this is very similar to the coupling coefficient.

(ii) *Insertion method*

Just as the position and slope of the closed orbit at a specific point can be controlled by a 4-magnet bump, it is equally possible to control the median-plane tilt using skew quadrupoles, for example, at the injection kicker or the crossing point in a collider. By a judicious positioning of the lenses, the bump in vertical dispersion can be made to have either a zero net effect on the coupling or say a contribution to the real coupling coefficient while a second bump would contribute to the orthogonal imaginary coupling coefficient. In an irregular or mismatched lattice, this may require some considerable searching and repeated evaluations of (5.82), but in a regular lattice it is often quite straightforward.

5.4   NON-LINEAR RESONANCES

In the earlier sections the basic formulae for errors in the dipole–quadrupole (or linear) lattice have been derived. It will come as no surprise that higher-order multipole errors will also be present in all machines and will pose a potential threat to the efficient operation. This subject leads rapidly into advanced non-linear mechanics and the state-of-the-art leaves many problems unanswered, such as the seemingly simple question of how to predict when a particle is stable for all time in the presence of any given distribution of multipole errors. In this section only a few of the most basic ideas will be presented with some references for further reading.

The general condition for a resonance to occur is both simple and easy to remember and has the form,

$$nQ_x \pm mQ_z = P \tag{5.84}$$

where

$$n + m = N \tag{5.85}$$

which is known as the order of the resonance. The integers n and m are positive and one may be zero. Thus for a given resonance of order N there are $(N + 1)$ separate resonance conditions of the form (5.84). The integer N defines the principal driving multipole, which will have $2N$-poles and P is the order of the azimuthal harmonic in the distribution of the $2N$th multipole error that is effective in driving the resonance concerned. The Pth harmonic will drive an Nth order resonance if the tunes are close to satisfying the condition (5.84). The positive sign in (5.84) is for resonances known as *sum resonances* and the negative sign is for *difference resonances*. The lowest order sum resonances correspond to dipole errors ($N = 1$) and quadrupole errors ($N = 2$), which are in fact linear and were discussed earlier in this chapter. The lowest-order difference resonance, the so-called *linear coupling resonance*, $Q_x = Q_z$, was discussed in Section 5.3.

When P is a multiple of the machine superperiodicity, the resonance excitation is sensitive to systematic errors in the lattice elements. These resonances are called *systematic* or *structure resonances* and are likely to be strong. The families of resonance lines represented by (5.84) fill the whole tune space, or *tune diagram*, and it is frequently difficult to find a *working point, line or region*, which is sufficiently free of low-order resonances for operation. Furthermore, all resonances have stopbands, which further reduces the 'safe' space. This problem is most critical in colliders, where beam–beam interaction and the exotic optics are sources of resonance excitation. Colliders also require very low loss rates from their beams in order to keep background events in the physics detectors to a minimum. Accelerators have far less of a problem, since the beam storage time is short and losses less critical.

Figure 5.12 shows the resonance pattern up to 8th order between the tune values of 8.5 and 9, which was the working range of the CERN ISR. The figure also includes the ISR working line 8C ('8' refers to the lowest order resonance in the stack and

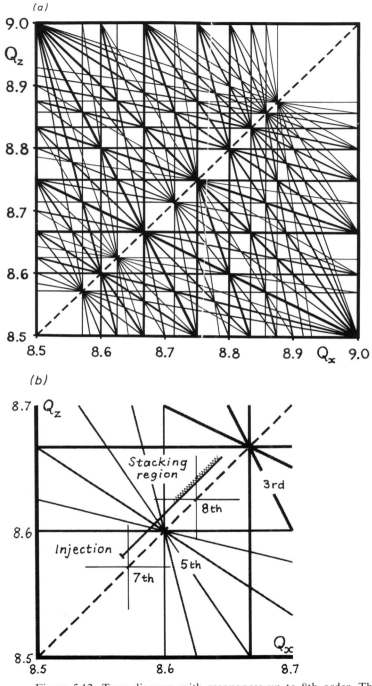

Figure 5.12. Tune diagram with resonances up to 8th order. The ISR working line 8C is shown in (b).

'C' indicates the line had dynamic space-charge compensation, see Chapter 8). The direction of increasing tune along this line corresponds to increasing momentum, so that the chromaticity (see Chapter 6) is positive in both planes for reasons of stability (see Chapter 9). The injection point is separated from the stack and in the intermediate space there are 5th order resonances. Beam pulses were able to cross this region quickly without loss, but had the main stack been put there the background would have been prohibitively high. On this working line, it was possible to collide beams of 30 A for many hours with loss rates of 1 or 2 parts per million per minute. The top of the stack is limited by the 3rd order resonance, which would inflict catastrophic losses if crossed. The dashed line $Q_x = Q_z$ is the linear coupling resonance and in order to maintain the emittance ratio of about 3, this had to be avoided with a minimum tune difference of 0.01. During stacking the space-charge tune shifts moved the line of the order of 0.03 in Q towards the bottom right before the on-line compensation system acted. The smallest controllable tune shift was about 0.001. The stacking and maintenance of this line was therefore quite delicate.

In the CERN SPS proton–antiproton collider, the beams are bunched and the collision is head-on, which increases the beam–beam excitation and makes resonance problems more acute than in the ISR with its coasting beams colliding at a small angle (14°). Resonances are visible up to the order of 23 and the beams are adjusted to avoid all orders up to 14.

The compensation of resonances is difficult and in most cases impossible. In colliders, unwanted collisions can be avoided by separating the beams, so reducing the beam–beam interaction. In general, the resonance excitation increases with a power of the emittance depending on the order of the resonance, so by scraping away the halo of large-amplitude particles an improvement can be made in the loss rate and hence the signal-to-noise factor for a physics detector, which far outweighs the reduction in event rate due to the current loss. The optimisation of the working line in a collider is very dependent on the quality of the measurements made when setting up the line and the non-destructive diagnostics needed to monitor the stack when it is in the machine. More complete treatments of non-linear resonances can be found in Refs 13–15.

More exotic non-linear effects such as *Arnold diffusion* and *stochastic motion* are treated in Ref. 16, which gives a comprehensive coverage of non-linear dynamics in particle accelerators.

CHAPTER 6

Chromaticity

6.1   CHROMATIC EFFECTS

Chromatic effects are caused by the momentum dependence of the focal properties
of lattice elements. The momentum of a particle is closely analogous to the frequency
of light in classical optics and it is for this reason that the name *chromaticity* has
been adopted (Gk. *chroma* colour). Figure 6.1 illustrates the effect of the momentum
dependence of the focal length of a lens, first for a beam in a dispersion-free region,
which is the normal case in classical optics and secondly for a beam with finite
dispersion, which is the normal case in the arcs of an accelerator. The differential
bending in dipoles gives rise to the dispersion function $D(s)$, which is also strictly
a chromatic effect, but since the dispersion function is well understood and
readily calculated by optics programs, it is not usually included under the title of
chromaticity.

Optical systems can be made achromatic by combining lenses made from different
glasses, whose refractive indices behave differently with wavelength, but the focal
strength of a magnetic lens always varies with momentum according to the same
law, so that accelerators have no equivalent of a 'different glass'. Instead, the
accelerator designer uses sextupoles, set in regions with finite dispersion. This can be
rather complex and far from ideal. The complexity arises because errors that occur
in a dispersion-free region, as shown in Figure 6.1(*a*), cannot be corrected locally. It
is not ideal because sextupoles introduce the problem of resonance excitation.

6.2   EVALUATION OF THE CHROMATICITY

6.2.1   Chromaticity and natural chromaticity

Chromatic aberration is first seen as a change in tune and then as a change
in the betatron amplitude. Consequently, in the early days of synchrotrons, the

120

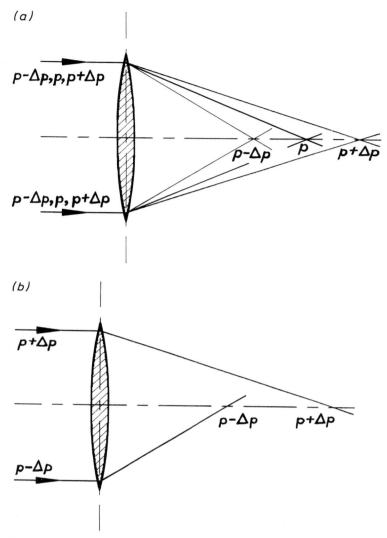

Figure 6.1. Examples of chromatic aberration. (*a*) Input beam without dispersion, (*b*) input beam with dispersion.

*chromaticity** was considered as a single number and was defined as the incremental change in tune normalised by the corresponding fractional change in momentum

$$\frac{\Delta Q}{\Delta p/p_0} = Q'. \tag{6.1}$$

* Chromaticity is sometimes defined as $(\Delta Q/Q)/(\Delta p/p_0)$. The useful approximation of $-Q$ for the natural chromaticity becomes -1, when using this alternative definition.

This number is the averaged result of all the chromatic aberrations around the machine to first order. The contributions to (6.1) arising from only the pure dipole–quadrupole elements in a lattice, give what is often called the *natural chromaticity*.

An approximate evaluation of the natural chromaticity can be made with the tune-shift formula (5.24) derived in Section 5.2.1.

$$\Delta Q = \mp \frac{1}{4\pi} \int_0^C \beta(s)\,\Delta k(s)\,ds \qquad (5.24)$$

where the upper sign $(-)$ applies to the horizontal plane, the lower sign $(+)$ applies to the vertical plane and $\Delta k(s)$ is the gradient error seen by the beam due to its momentum error. In order to evaluate Δk, the definition of $k(s)$ given in (2.16) must be generalised for different momenta and expressed in terms of the momentum error with respect to the design orbit by the use of (2.6) as

$$k(\Delta p/p_0) = \frac{e}{p_0(1 + \Delta p/p_0)}\left[\frac{\partial B_z}{\partial x}\right]_0, \qquad (6.2)$$

which can be expanded with respect to $\Delta p/p_0$ to give,

$$k(\Delta p/p_0) = k_0[1 - \Delta p/p_0 + (\Delta p/p_0)^2 - (\Delta p/p_0)^3 + \cdots]. \qquad (6.3)$$

When limited to the terms linear in $\Delta p/p_0$, Δk becomes,

$$\Delta k = -k_0\,\Delta p/p_0 \qquad (6.4)$$

and the natural chromaticity is therefore given by,

$$Q' = \frac{\Delta Q}{\Delta p/p} = \pm\frac{1}{4\pi}\int_0^C k_0(s)\beta(s)\,ds. \qquad (6.5)$$

This parameter can be large and negative and, in most cases, will be approximately equal to $-Q$. For example, this result is found by rewriting the expression (4.63) from Section 4.7.1 for the natural chromaticity of a machine comprised solely of FODO cells,

$$\frac{\Delta Q}{\Delta p/p_0} = -\frac{N_c}{\pi}\tan(\mu_0/2) = -\frac{2Q}{\mu_0}\tan(\mu_0/2) \simeq -Q \qquad (6.6)$$

and it is quickly verified that the approximation $Q' \simeq -Q$ is within 10% for phase advances up to 60° per cell and within 30% up to 90° per cell.

Equation (6.5) omits contributions from dipoles and their edge-focusing fields. These omissions tend to be more apparent in small rings. A far more accurate expression can be obtained by setting the sextupole contributions to zero in the formulae presented in the next section.

6.2.2 A more careful evaluation of the chromaticity

In addition to the various contributions omitted in (6.5), higher-order field components in the lattice elements also introduce chromatic effects. The net effect of all these contributions on the tune is simply referred to as the *chromaticity*. The limited space between non-linear resonance lines (see Section 5.3) and the need to ensure the beam is stabilised by Landau damping (to be discussed in Chapter 9) make it essential to compensate the chromaticity and to set it precisely to certain values in both planes. This is done by introducing sextupoles into the lattice.

Most lattice programs will calculate the chromaticity, but for small rings it has not been unusual for their results to disagree. The user is therefore advised to make enquiries as to whether the version of the program being used is adapted to his application. Reference 1 compares the results from several programs and Refs 1–3 provide a number of analytical methods for calculating the lattice chromaticity with fields up to sextupole. Reference 3 gives a very transparent derivation of the natural chromaticity. The most complete formulation for the chromatic contribution of a magnet element, expressed in terms of the central-orbit lattice functions and with fields up to sextupole, is to be found in Ref. 2 and is quoted below. The terms in boxes correspond to the simpler results in Refs 1 and 3, which should be sufficient in most cases.

$$
\frac{\Delta Q_x}{\Delta p/p_0} = \boxed{-\frac{1}{4\pi}\left[\int_{s_1}^{s_2}\left[(\rho_0^{-2}-k)\beta_x + k^{(1)}D_x\beta_x + \rho_0^{-1}(2\beta_x D_x k + 2\alpha_x D' - \gamma_x D_x)\right]ds\right.}
$$

$$
-\left[-\beta_x(\rho_0^{-1}+2D_xk)\tan\varepsilon_1 + \rho_0^{-1}(\beta_x D_x' - 2\alpha_x D_x + \rho_0^{-1}\beta_x D_x\tan\varepsilon_1)\tan^2\varepsilon_1\right.
$$

$$
\left. - \tau_1\rho_0^{-1}\beta_x D_x\right]_{\text{entry edge}}
$$

$$
+\left[\beta_x(\rho_0^{-1}+2D_xk)\tan\varepsilon_2 + \rho_0^{-1}(\beta_x D_x' - 2\alpha_x D_x + \rho_0^{-1}\beta_x D_x\tan\varepsilon_2)\tan^2\varepsilon_2\right.
$$

$$
\left.\left. - \tau_2\rho_0^{-1}\beta_x D_x\right]_{\text{exit edge}}\right] \tag{6.7}
$$

$$
\frac{\Delta Q_z}{\Delta p/p_0} = \boxed{-\frac{1}{4\pi}\left[\int_{s_1}^{s_2}\left[k\beta_z - k^{(1)}D_x\beta_z - \rho_0^{-1}D_x(k\beta_z + \gamma_z)\right]ds\right.}
$$

$$
+\left[\beta_z(\rho_0^{-1}+2D_xk)\tan\varepsilon_1 - \rho_0^{-1}(\beta_z D_x' - 2\alpha_z D_x - \rho_0^{-1}\beta_z D_x\tan\varepsilon_1)\tan^2\varepsilon_1\right.
$$

$$
\boxed{-\beta_z\rho_0^{-1}D_x'} - \tau_1\rho_0^{-1}\beta_z D_x\Big]_{\text{entry edge}}
$$

$$
+\left[\beta_z(\rho_0^{-1}+2D_xk)\tan\varepsilon_2 + \rho_0^{-1}(\beta_z D_x' - 2\alpha_z D_x + \rho_0^{-1}\beta_z D_x\tan\varepsilon_2)\tan^2\varepsilon_2\right.
$$

$$
\left.\boxed{+\beta_z\rho_0^{-1}D_x'} - \tau_2\rho_0^{-1}\beta_z D_x\Big]_{\text{exit edge}}\right] \tag{6.8}
$$

where

$$
\textit{Normalised sextupole gradient, } k^{(1)}(s) = -1/(B_0\rho_0)\,(\partial^2 B/\partial x^2) \tag{6.9}
$$

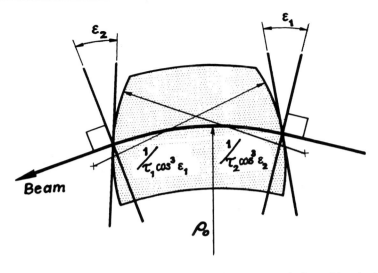

Figure 6.2. Sign convention for slope and curvature of edges of hard-edge magnet model. All parameters are positive as shown.

$\varepsilon_{1,2}$, slopes of entry and exit edges (see Figures 3.10 and 6.2), $\tau_{1,2} \cos^3 \varepsilon_{1,2}$, radii of curvature of entry and exit edges (Figure 6.2) (for exact correspondence with Ref. 2 $\varepsilon \to \theta$, $\rho_0^{-1} \to h$, $k^{(1)} \to r$ and expressions should be normalised by the tune).

6.2.3   Higher orders of the chromaticity

The tendency in colliders has been towards lower β-values at the crossings, which exacerbates chromatic effects, so that higher order terms are needed to adequately describe the chromaticity, i.e.

$$Q' = \Delta Q / \Delta p / p_0; \qquad Q'' = \Delta Q / (\Delta p / p_0)^2; \qquad Q''' = \Delta Q / (\Delta p / \rho_0)^3; \dots .$$

These higher-order components have to be calculated numerically or measured. Octupole lenses are sometimes used to compensate Q'' and in the CERN ISR poleface windings were able to shape the main field up to and including the decapole.

6.3   ADJUSTING THE CHROMATICITY

6.3.1   Compensation of the natural chromaticity of a quadrupole

A sextupole can be regarded as a quadrupole whose gradient varies linearly across its aperture and passes through zero on the central orbit. In a region of finite dispersion, the different momenta in the beam are also spread linearly across the

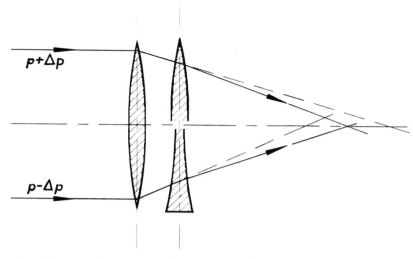

Figure 6.3. An achromatic quadrupole–sextupole doublet.

aperture, so that a sextupole provides the possibility of matching the gradient to the momentum. This effect can be exploited to make the normal lattice quadrupoles achromatic.

The gradient seen by an off-momentum particle in a sextupole with finite dispersion is given by,

$$\Delta k = k^{(1)} D_x \, \Delta p / p_0 \tag{6.10}$$

and the corresponding gradient error in a quadrupole, $-k_0 \, \Delta p / p_0$, has already been derived from (6.3). These effects can be opposed and cancelled as illustrated in Figure 6.3, so creating an achromatic quadrupole–sextupole doublet. The achromatic condition is,

$$k^{(1)} l_s = k_0 l_q / D_x \qquad \text{(for } D_x \neq 0) \tag{6.11}$$

where l_s and l_q are the lengths of the sextupole and quadrupole respectively. This can be achieved in a flexible way by mounting a sextupole next to the quadrupole or, in a more permanent way, by building the necessary sextupole profile directly into the quadrupole.

6.3.2   A simple scheme for chromaticity control

If sextupoles were to be added beside each lattice quadrupole in the arcs, it would be natural, on the basis of the above, to power all those close to F-quadrupoles in series and all those close to a D-quadrupole in series. In this case, the two families are designated SF and SD according to which type of quadrupole they are located beside. By powering these two sextupole families according to (6.11) the lattice in the arcs can be fully compensated and made achromatic. If now the current in one of the families is incremented it will create a finite chromaticity, which can be found with the tune-shift formula (5.24) by summing all the gradient increments given by (6.10). If this is done separately for the two sextupole families in each plane, then two expressions are obtained for the resultant chromaticities,

$$
\left.\begin{aligned}
\Delta Q_x' &= -\frac{l_s}{4\pi}\left[\Delta k_{\mathrm{SF}}^{(1)} \sum_{n=1}^{N_c} (\beta_x D_x)_n + \Delta k_{\mathrm{SD}}^{(1)} \sum_{n=1}^{N_c} (\beta_x D_x)_n\right] \\
\Delta Q_z' &= \frac{l_s}{4\pi}\left[\Delta k_{\mathrm{SF}}^{(1)} \sum_{n=1}^{N_c} (\beta_z D_x)_n + \Delta k_{\mathrm{SD}}^{(1)} \sum_{n=1}^{N_c} (\beta_z D_x)_n\right]
\end{aligned}\right\} \tag{6.12}
$$

where $\Delta k^{(1)}$ indicates the increment in the normalised sextupole gradient w.r.t. the value needed for a compensated lattice. Since β_x will be large for the F-sextupoles near to the F-quadrupoles and similarly β_z will be large for the D-sextupoles the coefficients in (6.12) should differ sufficiently to provide an efficient control of the chromaticities. If this scheme is implemented without the full complement of sextupoles the chromaticities can still be controlled, but local chromatic effects (changes in β) will be present.

For many accelerators this simple scheme is quite adequate, but the introduction of low-β insertions into colliders made it necessary to add more sophistication to account for the large chromatic errors arising in the dispersion-free low-β quadrupoles. This problem led machine designers to formulate a theory for chromatic effects in terms of a first-order perturbation of the lattice functions for off-momentum particles. First developed by Zyngier[4], this theory was elaborated by Montague[5] whose analysis is followed here.

6.4 THE w-VECTOR FORMULATION OF CHROMATIC EFFECTS

6.4.1 Basic theory

Let the suffixes $_0$ and $_1$ denote the central orbit and an off-momentum orbit respectively. The differences in the lattice functions between these two orbits will be due to the chromatic perturbations in the lattice. These two sets of parameters will be used to define new variables $A(s)$, $B(s)$, $\psi(s)$ and $\Delta K(s)$, which embody the chromatic effects,

$$
\left.\begin{aligned}
B &= \frac{(\beta_1 - \beta_0)}{(\beta_0\beta_1)^{1/2}}; & A &= \frac{(\alpha_1\beta_0 - \alpha_0\beta_1)}{(\beta_0\beta_1)^{1/2}} \\
\psi &= \tfrac{1}{2}(\mu_0 + \mu_1); & \Delta K &= (K_1 - K_0)
\end{aligned}\right\} \tag{6.13}
$$

where K is the generalised focusing constant described in Section 3.1. Note that exceptionally in *this* section μ_0 refers to the general betatron phase advance on central orbit and not to the specific value for a cell. The basic relationships (4.2), (4.4) and (4.8) developed in Chapter 4 can be applied to both trajectories individually, i.e.

$$\frac{d\mu_{0,1}}{ds} = \frac{1}{\beta_{0,1}} \tag{6.14}$$

$$\frac{d\beta_{0,1}}{ds} = -2\alpha_{0,1} \tag{6.15}$$

$$\frac{d^2}{ds^2} \beta_{0,1}{}^{\frac{1}{2}} + K_{0,1}(s)\beta_{0,1}{}^{\frac{1}{2}} = \beta_{0,1}{}^{-3/2}. \tag{6.16}$$

The differentiation of (6.16) with substitution from (6.15) yields,

$$\frac{d\alpha_{1,2}}{ds} = K_{0,1}(s) - \frac{(1 + \alpha_{0,1}{}^2)}{\beta_{0,1}}. \tag{6.17}$$

The differentiation of B from (6.13) w.r.t. s with substitutions from (6.14) and (6.15) and the differentiation of A from (6.13) w.r.t. s with substitutions from (6.14), (6.15) and (6.17) yield with a little manipulation the exact equations,

$$\left.\begin{aligned} dB/ds &= -2A \, d\psi/ds \\ dA/ds &= 2B \, d\psi/ds + (\beta_0\beta_1)^{\frac{1}{2}} \, \Delta K. \end{aligned}\right\} \tag{6.18}$$

It follows that in an achromatic region, where $\Delta K = 0$,

$$\left.\begin{aligned} dB/d\psi &= -2A; \qquad dA/d\psi = 2B \\ d^2A/d\psi^2 &+ 4A = 0 \qquad \text{(identical equation for } B) \end{aligned}\right\} \tag{6.19}$$

and that

$$A^2 + B^2 = \text{Constant}. \tag{6.20}$$

It can be seen from (6.19) that in an achromatic region A and B will oscillate at twice the average betatron frequency and that (6.20) is an invariant of the motion. In non-achromatic regions, ΔK becomes finite and both A and B become modified. The chromatic error term, ΔK, in principle contains all effects from dipoles, edge focusing and multipoles. Up to this point the theory is exact and the accuracy of any calculation depends only on the ability of a computer program to calculate the lattice functions on the two trajectories correctly.

In most machines the lattice quadrupoles in the arcs can be compensated as explained in Section 6.3.1 and the relatively small contributions from the dipoles can be included by use of the scheme described in Section 6.3.2, so that the basic lattice can be made achromatic. The remaining errors will then come principally from the quadrupoles located in the dispersion-free regions. In this case, the chromatic gradient error term, ΔK can be simplified to,

$$\Delta K = -(k_1 - k_0) \tag{6.21}$$

since the dipole effects are both small and can be considered as already compensated. In (6.21) k_1 can be developed with respect to the momentum error as,

$$k_1(\Delta p/p_0) = [1 - \Delta p/p_0 + (\Delta p/p_0)^2 - \cdots]$$
$$\times [k_0 + D_x \Delta p/p_0 k^{(1)} + (1/2!)(D_x \Delta p/p_0)^2 k^{(2)} + \cdots]. \tag{6.22}$$

Equation (6.22) is similar to (6.3) except that higher-order gradient terms have been included for a more general description of the field on the off-axis trajectory. The substitution of (6.22) into (6.21) yields for ΔK,

$$\Delta K = -[D_x k^{(1)} - k_0](\Delta p/p_0) - [(1/2!)D_x^2 k^{(2)} - D_x k^{(1)}](\Delta p/p_0)^2 - \cdots. \tag{6.23}$$

For each order in $\Delta p/p_0$ this equation shows a correspondence between a gradient term and the next higher derivative of the gradient combined with the local dispersion function. It should also be noted that no multipole higher than quadrupole can contribute to the chromatic error when the dispersion is zero.

For the present purpose, it is convenient to normalise the chromatic variables A, B and ΔK by the momentum deviation $\Delta p/p_0$ and to re-define them in the limit as $\Delta p/p_0$ tends to zero, i.e.

$$\left. \begin{array}{ll} a = \underset{\Delta p/p_0 \to 0}{\text{limit}} \dfrac{A}{\Delta p/p_0}; & b = \underset{\Delta p/p_0 \to 0}{\text{limit}} \dfrac{B}{\Delta p/p_0} \\[4mm] \Delta k_n = \underset{\Delta p/p_0 \to 0}{\text{limit}} \dfrac{-\Delta K}{\Delta p/p_0}; & \psi \to \mu_0 \end{array} \right\} \tag{6.24}$$

so that the basic equations (6.18) become,

$$\left. \begin{array}{l} db/ds = -2a \, d\mu_0/ds \\[2mm] da/ds = 2b \, d\mu_0 ds - (\beta_0 \beta_1)^{\frac{1}{2}} \Delta k_n \end{array} \right\} \tag{6.25}$$

and when $\Delta k_n = 0$, then (6.19) becomes,

$$\left. d^2 a/d\mu_0^2 + 4a = 0 \quad \text{(identical equation for } b\text{).} \right\}$$

It is also useful to define a complex vector w,

$$w = (b + ja). \tag{6.26}$$

By comparison with the invariant of the motion (6.20), it can be seen that $|w|$ will also be an invariant in an achromatic region.

By defining a, b and Δk_n in the limit as $\Delta p/p_0 \to 0$, the equations (6.25) have been linearised since Δk_n will contain only the quadrupole and sextupole terms that were linear in $\Delta p/p_0$ in (6.23).

The w-vector defined in (6.26) is a useful tool for designing chromaticity compensation schemes. In achromatic regions w has a constant amplitude and rotates at twice the average betatron phase advance. When passing through an uncompensated quadrupole or sextupole its amplitude is modified and then to a lesser degree its phase. Thus *w* provides a measure of the chromatic aberration throughout the lattice.

6.4.2 The *w*-vector in a FODO lattice

As an exercise, it is interesting to plot the *w*-vector in (b, a) space as it advances through a matched FODO cell. Since the lattice functions are periodic in a matched cell the *w*-vector will also be periodic and the plot will be a closed figure. This has been done in Figure 6.4 for a regular FODO cell with a 60° phase advance (the plots are similar in both planes). The first plot shows the natural chromaticity, the second shows the cell compensated by sextupoles and the third shows the effect of incrementing the D-sextupole from the compensation setting. These figures have the common features of circular arcs in the drift spaces and dipoles, which are basically achromatic, and jumps parallel to the *a*-axis in the quadrupoles and sextupoles. The angle subtended by an arc is equal to twice the betatron phase advance in the dipole and drift spaces separating the F and D lenses. Unequal lens strengths or lens separations distort the plots, but the essential features remain unchanged.

6.4.3 Strategy for the chromaticity correction in a collider

In modern colliders it is common to design the lattice with dispersion-free regions in which to locate the low-β insertions. The result is that the low-β quadrupoles,

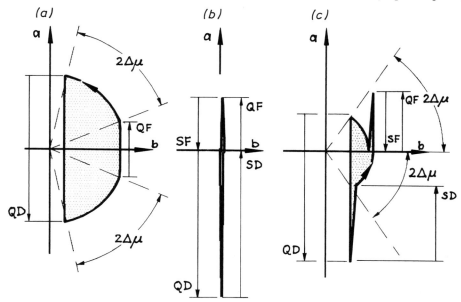

Figure 6.4. Loci of *w* in (b, a) space in a matched FODO lattice for the vertical plane, $k_F = -k_D$ and equal lens separations. (*a*) Uncompensated, (*b*) compensated, (*c*) SD incremented.

which are the strongest in the machine, cannot be compensated locally and must instead be compensated by the sextupoles in the distant arcs. This can only be a partial compensation and the aim is to control two particular aspects:

- the distortion of β in both planes at the crossing point;
- the chromaticities (tune spreads) in both planes.

The strategy is to mount sextupoles adjacent to the normal lattice quadrupoles in the arcs. These are first powered so as to make the arcs achromatic, as described in Section 6.3. The section of the machine from the crossing point to the centre of the adjacent arc is then treated as a transfer line and the w-vector is set to zero at the crossing point and tracked through the dispersion-free section to the arc. In the arc, the w-vector will simply rotate with constant amplitude since the lattice has been made achromatic in this region. Finally the sextupoles are divided into families and are adjusted so as to progressively reduce the amplitude of the w-vector to zero by the time the centre of the arc is reached. In this way, the chromatic effects are confined to a local bump in w and all first order effects are set to zero at the crossing point of the beams.

Figure 6.5 shows schematically the modulus of the w-vector in the vertical plane tracked from the crossing point, through the rf straight section to the arc. The plot illustrates that the first quadrupole of the low-β insertion is the main source of error. For example, in an early LEP design, the first low-β quadrupole excited an error of $|w| \simeq 101$, whereas the uncompensated 60° FODO lattice generated a maximum of only $|w| \simeq 1.8$.

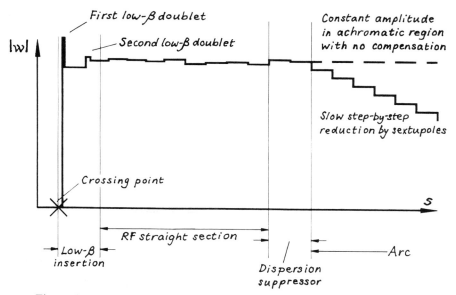

Figure 6.5. Example tracking of the w-vector.

The detailed design of such a scheme should not be attempted by trial and error. The problem can be solved first analytically, as will be shown in the next Sections and then this result can be refined by the use of a program such as HARMON[6].

6.5 ANALYTIC EXPRESSIONS FOR THE CHROMATIC VARIABLES

6.5.1 Approximate expressions for a and b

It is always possible to resort to the computer to find a and b, but it is more informative to have analytical expressions. This is very simple in the thin-lens approximation. It can be seen from (6.25) that by concentrating a chromatic error in a thin lens $\Delta b = 0$, since the input and output β-values will be equal, and only Δa is finite and equal to,

For a quadrupole: $\Delta a = -(\beta_0\beta_1)^{\frac{1}{2}} \Delta k_n \Delta s \simeq \beta_0 k_0 l_q$

For a sextupole: $\Delta a = -(\beta_0\beta_1)^{\frac{1}{2}} \Delta k_n \Delta s \simeq -\beta_0 D_x k^{(1)} l_s.$ (6.27)

In this simplified model with thin lenses, the w-vector will rotate at twice the betatron frequency through achromatic regions and suffer jumps in the imaginary component a at quadrupoles and sextupoles in regions of finite dispersion according to (6.27). This is a familiar concept and is similar to the normalised phase-space plots for closed-orbit bumps, in which the particles rotate at the betatron frequency at constant radius with jumps parallel to the Y'-axis due to the dipole kicks (see Section 5.1.2).

Thus an observer at some point downstream of a thin-lens chromatic error in an otherwise achromatic lattice would see,

For a quadrupole: $\Delta a(\mu) \simeq \beta_0 k_0 l_q \cos(2\mu),$

$\Delta b(\mu) \simeq \beta_0 k_0 l_q \sin(2\mu);$

For a sextupole: $\Delta a(\mu) \simeq -\beta_0 D_x k^{(1)} l_s \cos(2\mu),$ (6.28)

$\Delta b(\mu) \simeq -\beta_0 D_x k^{(1)} l_s \sin(2\mu).$

In fact these approximations are quite good as can be seen by comparison with the more exact expressions in the next section.

6.5.2 Exact expressions for a and b

Autin and Verdier[7] have found more exact expressions for the chromatic variables a and b excited by quadrupoles and sextupoles of finite length. (Note. Reference 7 does not refer to a and b, but to the equivalent forms $\Delta\beta/\beta$ and $\Delta\alpha - \alpha\Delta\beta/\beta$.) Their analysis is too long and specialised to be reproduced here, so that the results are quoted for reference.

The equations equivalent to (6.28) for the variables a and b downstream of a chromatic perturbation with finite length in an achromatic structure are,

$$\Delta a(\mu) = [- \Sigma \cos(2\mu) + \Gamma \sin(2\mu)]/(\Delta p/p_0) \left.\right\}$$
$$\Delta b(\mu) = [\Gamma \cos(2\mu) + \Sigma \sin(2\mu)]/(\Delta p/p_0) \quad (6.29)$$

where

$$\Gamma = 2\left(c_{11} - \frac{\alpha_i}{\beta_i} c_{12}\right) \quad \text{and} \quad \Sigma = \beta_i c_{21} + 2\alpha_i c_{11} + \frac{1 - \alpha_i^2}{\beta_i} c_{12} \quad (6.30)$$

and the subscript i refers to the lattice parameters on central orbit at the entry to the quadrupole or sextupole.

The coefficients $c_{i,j}$ refer to a matrix which describes the chromatic perturbation and has to be calculated for each type of lens. These are derived in Ref. 7, but are quoted here as given in Ref. 8.

$$c_{11} = -\tfrac{1}{2}\sin^2\{|k_0|^{\frac{1}{2}}l_q)(\Delta p/p_0)$$
$$c_{12} = \tfrac{1}{2}[|k_0|^{-\frac{1}{2}}\sin(|k_0|^{\frac{1}{2}}l_q)\cos(|k_0|^{\frac{1}{2}}l_q) - l_q](\Delta p/p_0) \left.\right\} \text{ F-quadrupole} \quad (6.31)$$
$$c_{21} = \tfrac{1}{2}[|k_0|l_q + |k_0|^{\frac{1}{2}}\sin(|k_0|^{\frac{1}{2}}l_q)\cos(|k_0|^{\frac{1}{2}}l_q)](\Delta p/p_0)$$

$$c_{11} = \tfrac{1}{2}\sinh^2(|k_0|^{\frac{1}{2}}l_q)(\Delta p/p_0)$$
$$c_{12} = \tfrac{1}{2}[|k_0|^{-\frac{1}{2}}\sinh(|k_0|^{\frac{1}{2}}l_q)\cosh(|k_0|^{\frac{1}{2}}l_q) - l_q](\Delta p/p_0) \left.\right\} \text{ D-quadrupole} \quad (6.32)$$
$$c_{21} = -\tfrac{1}{2}[|k_0|l_q + |k_0|^{\frac{1}{2}}\sinh(|k_0|^{\frac{1}{2}}l_q)\cosh(|k_0|^{\frac{1}{2}}l_q)](\Delta p/p_0)$$

$$c_{11} = \mp\tfrac{1}{2}k^{(1)}(D_{x,i} + \tfrac{2}{3}D_{x,i}'l_s)l_s^2(\Delta p/p_0)$$
$$c_{12} = \mp\tfrac{1}{2}k^{(1)}(\tfrac{2}{3}D_{x,i} + \tfrac{1}{2}D_{x,i}'l_s)l_s^3(\Delta p/p_0) \left.\right\} \text{ Sextupole (upper sign horizontal)} \quad (6.33)$$
$$c_{21} = \pm\tfrac{1}{2}k^{(1)}(2D_{x,i} + D_{x,i}'l_s)l_s.$$

If l_q and l_s are made small then the principal terms in (6.31), (6.32) and (6.33) become the first term in c_{21} in each case. When these terms are substituted back into (6.30) and then into (6.29) the approximate expressions of (6.28) are refound.

6.5.3 Summation over a series of sextupoles[8]

In preparation for designing a sextupole compensation scheme, it is useful to have an expression for the summation of the chromatic variables a and b from n series-powered sextupoles, which have been regularly set into an arc comprising many identical matched cells. Assume that each sextupole is separated from its neighbour on either side by m cells. This *sextupole family* will be referred to as SF1 (see Figure 6.6). The contribution of the first sextupole SF1$_1$ at the end of n groups of m lattice cells will be given by (6.29) with $\mu = nm\mu_0$, where μ_0 has reverted to its normal definition of the phase advance through a cell. The next sextupole SF1$_2$ will contribute at the same point in a similar way, except that $\mu = (n - 1)m\mu_0$ and so on.

132

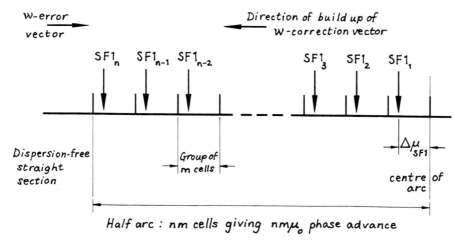

Figure 6.6. Sextupole family SF1 in a regular bending lattice.

Since the basic equations (6.25) are linear these contributions can be summed directly. Use can again be made of the trigonometrical relationship (4.75) applied in Section 4.7.2 to give,

$$\left.\begin{aligned}
\Delta b &= \{\Gamma_1 \cos[(n+1)m\mu_0] + \Sigma_1 \sin[(n+1)m\mu_0]\}\frac{\sin(nm\mu_0)}{\sin(m\mu_0)}\frac{1}{(\Delta p/p_0)}\\[2mm]
\Delta a &= \{-\Sigma_1 \cos[(n+1)m\mu_0] + \Gamma_1 \sin[(n+1)m\mu_0]\}\frac{\sin(nm\mu_0)}{\sin(m\mu_0)}\frac{1}{(\Delta p/p_0)}.
\end{aligned}\right\} \quad (6.34)$$

The coefficients Γ_1 and Σ_1 depend upon the plane. They contain the strength of the SF1 sextupoles and their lattice parameters as defined in (6.30) with substitution from (6.33).

In the following analysis, equal phase advances will be imposed on the two planes and it will be assumed that the members of a given sextupole family are separated in betatron phase by an integer number of π, so that the *w*-vector will rotate an integer number of full turns between sextupoles. The members of the same sextupole family are then in phase and have the maximum effect. Under these conditions, the expressions (6.34) simplify by the use of

$$\lim_{m\mu_0 \to \text{int. } \pi} \frac{\sin(nm\mu_0)}{\sin(m\mu_0)} = \begin{cases} n & \text{when } n \text{ is even} \\ -n & \text{when } n \text{ is odd} \end{cases}$$

to give

$$\left.\begin{aligned}
\Delta b &= \{\text{sign}\}n\{\Gamma_1 \cos[(n+1)m\mu_0] + \Sigma_1 \sin[(n+1)m\mu_0]\}/(\Delta p/p_0)\\
\Delta a &= \{\text{sign}\}n\{-\Sigma_1 \cos[(n+1)m\mu_0] + \Gamma_1 \sin[(n+1)m\mu_0]\}/(\Delta p/p_0)
\end{aligned}\right\} \quad (6.35)$$

133

where

$$\{\text{sign}\} = \begin{cases} 1 \text{ when } n \text{ is odd} \\ -1 \text{ when } n \text{ is even.} \end{cases}$$

The above equations give the combined effect of an n-sextupole family at a point $nm\mu_0$ in betatron phase after the first sextupole. In order to combine the effects of several families their vectors must be referred to a common point. If this is taken as the exit of the nth and last cell then the vector from the SF1 family must be back-tracked by $\Delta\mu_{\text{SF1}}$ (see Figure 6.6).

$$w_{\text{exit}} = (b + ja)\, e^{-j2\Delta\mu_{\text{SF1}}}. \tag{6.36}$$

6.6   PLANNING SEXTUPOLE FAMILIES

For convenience the following will be discussed with reference to a simple FODO lattice, but the principles are quite general. It is assumed that sextupoles have been mounted beside each lattice quadrupole in the arcs of the collider. It is therefore possible to render the lattice in the arcs achromatic and to adjust the tune spreads to whatever value is required by incrementing the SF- and SD-sextupoles with respect to the settings needed for compensating the lattice, as described in Section 6.3.2. In general, the closer the sextupoles are to the quadrupoles the better the compensation of the natural chromaticity and the larger the difference in betatron amplitudes and dispersion between the F- and D-lenses. This difference helps to decouple the horizontal and vertical planes in (6.12) when adjusting the tune spreads.

So far the sextupoles have been grouped into type F and type D in order to adjust the tune spreads and to compensate the natural chromaticity. They must now be further divided into families of F-type and D-type lenses, so that a specified w-vector can be progressively built up in the arc to compensate that coming from the low-β insertion.

6.6.1   Interleaved schemes

Let the F-type and D-type sextupoles be divided into $(N + 1)$ families. The first $(N + 1)$ cells in the arc will contain the first members of each family, the second group of $(N + 1)$ cells the second members and so on up to the nth and last group as illustrated in Figure 6.7. The betatron phase shift between adjacent members of the same family was defined earlier as $m\mu_0$ and was restricted to being an integer number of π. From Figure 6.7 it can be seen that

$$m\mu_0 = (N + 1)\mu_0 = \text{integer no. } \pi. \tag{6.37}$$

If now the SF1 family is incremented by $\Delta k_{\text{SF1}}{}^{(1)}$ then a small step in the w-vector will appear at the exit of the first lens in that family. Since the arc has been made

$$\{|S_F, S_D|S_{F1}, S_{D1}|S_{F2}, S_{D2}|\cdots|S_{FN-1}, S_{DN-1}|S_{FN}, S_{DN}|\}$$

$\longleftarrow$ ——— 1st Group of $(N + 1)$ cells ——— $\longrightarrow$

$$\{|S_F, S_D|S_{F1}, S_{D1}|S_{F2}, S_{D2}|\cdots|S_{FN-1}, S_{DN-1}|S_{FN}, S_{DN}|\}$$

$\longleftarrow$ ——— 2nd Group of $(N + 1)$ cells ——— $\longrightarrow$

$$\{|S_F, S_D|S_{F1}, S_{D1}|S_{F2}, S_{D2}|\cdots|S_{FN-1}, S_{DN-1}|S_{FN}, S_{DN}|\}$$

$\longleftarrow$ ——— nth Group of $(N + 1)$ cells ——— $\longrightarrow$

Figure 6.7. $(N + 1)$ interleaved sextupole families.

achromatic, this vector will rotate with constant amplitude at twice the betatron frequency until it reaches the next member of the SF1 family and since $m\mu_0$ has been made an integer number of π it will arrive with the same phase and will be incremented in amplitude. This will repeat for each member of the family. If $m\mu_0$ were not an integer number of π, a large vector could still be built up but less efficiently.

Each of the sextupole families will be able to build up a w-vector in a similar way. The only difference will be that an observer at the exit to the last cell will see the various w-vectors at different phases in a–b space. The phase shift in ψ between the w-vectors of two families will be equal to twice the betatron phase shift between the two families in the lattice. A minimum of two families will be needed, providing they are not degenerate, to create a given w-vector. Thus for the horizontal and vertical planes, four families will be needed to compensate a distant error. A further two families are then needed to control the chromaticities Q_x' and Q_z' for the overall machine making a minimum of six in all. Finally in order that the sextupoles do not excite third-order resonances, each family must be arranged so that its members are self-compensating, that is they should be in anti-phase for the resonance condition and ideally there should be an even number. These conditions can be summarised as,

$$
\left.
\begin{array}{ll}
(N + 1)\mu_0 = i\pi & \text{(phase condition)} \\
3(N + 1)\mu_0 = (2k + 1)\pi & \text{(resonance condition)}
\end{array}
\right\}
\qquad (6.38)
$$

where i and k are integers.

The two conditions in (6.38) require that $3i = (2k - 1)$, so that i is constrained to be odd. Some possible combinations are listed below:

$$
\begin{array}{lll}
\mu_0 = \pi/2 & \text{for } i = 1, N = 1 & \text{i.e. 4 families;} \\
\mu_0 = \pi/3 & \text{for } i = 1, N = 2 & \text{i.e. 6 families;} \\
\mu_0 = \pi/4 & \text{for } i = 1, N = 3 & \text{i.e. 8 families;} \\
\mu_0 = 3\pi/5 & \text{for } i = 3, N = 4 & \text{i.e. 10 families} \ldots .
\end{array}
$$

135

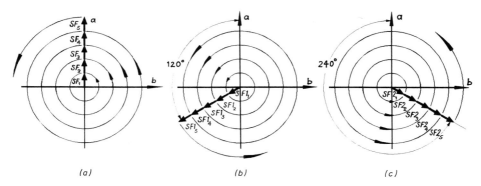

Figure 6.8. Periodic build-up of *w*-vectors in a 60° lattice. (*a*) SF vector, (*b*) SF1 vector, (*c*) SF2 vector.

The most popular phase advance is 60° per cell. This satisfies all of the above conditions and it is the first solution to give the required six families. The solution with 90° per cell is also possible and is, for example, employed as one of the optics for the CERN LEP collider. In this latter case there are only four families, which is strictly insufficient, but the problem can be circumvented by calculating in advance the chromatic error from the dispersion-free region and then adjusting the betatron phase at the entry to the first sextupole, such that its *w*-vector has the right phase for compensation. This solves the immediate problem, but lacks the flexibility to account for later changes in the optics of the low-β insertion. The options with more than six families can be simplified by connecting the excess families in series with the SF and SD and leaving these two enlarged families adjusted to the compensation setting for the natural chromaticity. This has the disadvantage that there are fewer active sextupoles for compensating the error from the low-β insertion, which is likely to be much larger than the natural chromaticity.

Figure 6.8 shows how the individual *w*-vectors would build up for the three F-type families in a lattice of 60° cells. If all three families are first set to the compensation value for the natural chromaticity and then all three are incremented by an equal amount, it can be seen that no net *w*-vector is created [see case (*a*) in Figure 6.9]. If the same is done with the D-type families then the scheme can be operated according to the prescriptions in Section 6.3.2 for controlling the chromaticity. The next stage is to superimpose on the basic chromaticity correction of Figure 6.9(*a*) combinations of (SF1 + SF2) and (SF1 − SF2), see cases (*b*) and (*c*) in Figure 6.9. These are orthogonal vectors which can reach all parts of the (*b*, *a*) space. By including the SD1 + SD2 and the SD1 − SD2 vectors, the two planes can also be made orthogonal. Figure 6.9(*d*) shows the situation in 90° lattices where the two vectors are degenerate.

To compensate an error propagating from another part of the machine, the settings of the sextupoles will be unbalanced and they will affect the chromaticities in both planes of the machine. Thus it will be necessary to adjust the average settings of all

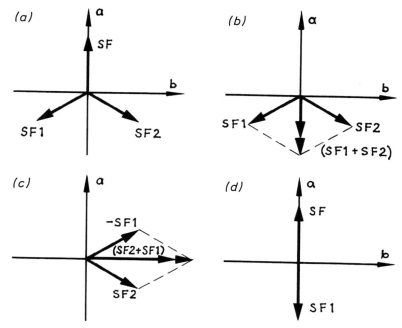

Figure 6.9. Creation of *w*-vectors in 60° and 90° lattices. (*a*) SF + SF1 + SF2, no net vector (60° lattice); (*b*) SF1 + SF2, resultant vector (60° lattice); (*c*) SF2 − SF1, resultant vector (60° lattice); (*d*) SF + SF1, degenerate (90° lattice).

F- and all D-type lenses so as to keep the machine chromaticities constant when making such a correction. On the other hand equal changes to all F- or all D-type lenses will not excite *w*-vectors.

6.6.2   Non-interleaved schemes

The interleaved scheme is based on the principle that the contributions from each sextupole can be added linearly. This is only true for small momentum deviations. One partial solution that has been suggested is to separate the sextupole families. First the lattice is compensated as before and then the families are placed one after the other in the arc, but unless the arc is very long there will clearly be rather few sextupoles in each family.

A hybrid scheme is to first interleave the SF1 and SD1 families in the first half of the arc and then to interleave the SF2 and SD2 families in the second half. The SF and SD sextupoles should occupy all other positions to ensure the lattice is fully compensated. The logic here is that since the betatron amplitudes at the F-type and D-type lenses differ strongly (the F-type act predominantly on the horizontal plane and the D-type on the vertical) they are almost invisible to each other and can therefore be interleaved (see Figure 6.10). In practice, it appears that fully interleaved

$$\xleftarrow{\hspace{1cm}} \text{SF1 and SD1 interleaved} \xrightarrow{\hspace{1cm}}$$

$$\{|S_F, S_D|S_{F1}, S_{D1}|S_F, S_D|\cdots|S_F, S_D|S_{F1}, S_{D1}|S_F, S_D|\}$$

$$\xleftarrow{\hspace{1cm}} \text{First half of arc} \xrightarrow{\hspace{1cm}}$$

$$\xleftarrow{\hspace{1cm}} \text{SF2 and SD2 interleaved} \xrightarrow{\hspace{1cm}}$$

$$\{|S_F, S_D|S_F, S_D|S_{F2}, S_{D2}|\cdots|S_F, S_D|S_F, S_D|S_{F2}, S_{D2}|\}$$

$$\xleftarrow{\hspace{1cm}} \text{Second half of arc} \xrightarrow{\hspace{1cm}}$$

Figure 6.10. Partially interleaved families in a 60° compensated lattice.

schemes are better, since the active families which create the *w*-vectors have more members. This reduces the strength needed in each sextupole and makes the distribution more uniform, which reduces the resonance excitation.

6.7   NON-LINEARITIES

The simple exercise of plotting, with the aid of a lattice computer program, the *w*-vector with increasing momentum deviations between the two reference orbits will reveal at which point non-linearities become important for a particular lattice. The non-linearity will first appear in the phase of *w* rather than its amplitude.

It is not unusual to have two or three orders of magnitude difference between the betatron amplitudes at the crossing point in a low-β insertion and the maximum in the first quadrupole. With such a high magnification, it is not surprising that higher order chromatic and geometric aberrations become visible. In the presence of a strong low-β insertion, it is in fact hard to maintain a constant betatron amplitude across the aperture at the interaction point for more than a few per mil momentum spread. Similarly, it is hard to maintain the higher order tune spreads Q', Q'', Q''' etc. close to zero.

Once the compensation scheme has been designed using the linear theory, a program such as HARMON[6] should be used to refine the results. HARMON minimises the excitation of a wide range of resonances and finds the best sextupole settings to equalise the various non-linearities. There is still no guarantee that the beam will be stable and it is normal to use one of the many tracking programs[9–12] to simulate the beam behaviour with random closed-orbit distortions and other errors. The aperture in which the beam is found to be stable is known as the *dynamic aperture*. A detailed review of the analytical and numerical tools available for calculating the dynamic aperture can be found in Ref. 13. With strong low-β insertions and superconducting magnets, which are liable to have a rather poorer field quality than conventional magnets, it is sometimes difficult to have a dynamic aperture as large as the physical one.

CHAPTER 7

Longitudinal beam dynamics

The Lorentz force on a charged particle in an electromagnetic field is given by*,

$$F = eE + e(v \times B)$$ (7.1)

and the rate at which work is done on the particle is,

$$F \cdot v = eE \cdot v + e(v \times B) \cdot v.$$ (7.2)

The term $e(v \times B) \cdot v$ is always zero, that is magnetic fields can never do any work directly and in this respect magnet systems are passive. The only way to give energy to the particle is via the electric field, which can be expressed as,

$$E = -\nabla \phi - \frac{\partial}{\partial t} A$$ (7.3)

where ϕ is a scalar potential and A is a vector potential satisfying,

$$B = \nabla \times A.$$ (7.4)

In (7.3) the scalar potential term applies to acceleration in d.c. accelerators such as the Cockroft–Walton or Van de Graaff (see Section 1.1), whilst the vector potential applies to all other accelerators.

From (7.3) and (7.4) Faraday's law is obtained in differential form as

$$\nabla \times E = -\frac{\partial}{\partial t} B$$ (7.5)

and in integral form, by the use of Stoke's theorem applied to a closed integration loop C which encloses an area S, as

$$\oint_C E \cdot \mathrm{d}s = \oint_S \nabla \times E \cdot \mathrm{d}S = -\frac{\partial}{\partial t} \oint_S B \cdot \mathrm{d}S.$$ (7.6)

---

* For a multi-charged particle replace e by $q = ne$ throughout.

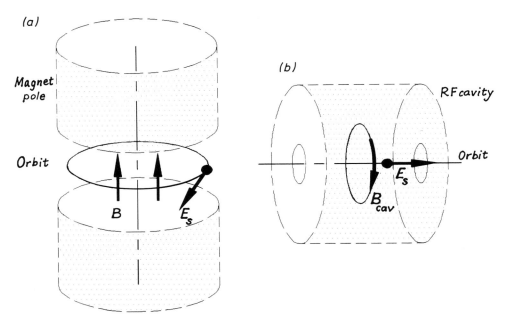

Figure 7.1. Configurations for acceleration.

(a) Betatron configuration, $\oint E_s \cdot ds = -\int_{\text{pole}} \dfrac{\partial B}{\partial t} \cdot dS;$

(b) cavity configuration, $\int_{-\infty}^{\infty} E_s \cdot ds = -\int_{\text{cavity}} \dfrac{\partial B_{\text{cav}}}{\partial t} \cdot dS.$

Acceleration using (7.6) is possible with two basic topologies:

- the *betatron configuration* where the beam encircles the time-varying magnetic field [see Figure 7.1(a)]
- the *cavity configuration* where the time-varying magnetic field encircles the beam [see Figure 7.1(b)].

The cavity configuration is the more used and of greater interest, but it is important to understand the betatron acceleration effect, which in fact also exists to a small degree in synchrotrons when the magnets are ramped and has therefore to be taken into account in accurate calculations.

7.1   BETATRON ACCELERATION[1-3]

As described in Chapter 1, the betatron resembles a cyclotron as regards its magnet, but the beam stays on a stationary orbit near to the pole edge rather than spiralling outwards from the centre as its energy increases. This stationary orbit is achieved by varying the guiding magnetic field to match the energy of the charged particle and at the same time to accelerate the particle via the change in the flux enclosed by the orbit. Clearly some definite relationship must be observed in order to match the induced voltage to the guide field.

The important equation is (7.6), which is simplified by the circular symmetry to,

$$E_s = -\frac{1}{2\pi\rho}\frac{\partial}{\partial t}\int_S \mathbf{B}\cdot d\mathbf{S}. \tag{7.7}$$

The force on the particle due to E_s will be azimuthal and equal to eE_s and the equation for the azimuthal motion becomes,

$$\frac{d(mv)}{dt} = eE_s. \tag{7.8}$$

For stable cyclotron motion (see Chapter 2), the governing relationship is,

$$mv = -eB\rho. \tag{7.9}$$

In order that the radius remains constant as the field varies, the equality,

$$\frac{d(mv)}{dt} = -e\rho\frac{dB}{dt} \tag{7.10}$$

must be respected. The rate of change of the guiding field in (7.10) can be related to the integrated flux enclosed by the orbit via (7.9) and (7.8) to give,

$$2\frac{dB}{dt} = \frac{1}{\pi\rho^2}\frac{\partial}{\partial t}\int_S \mathbf{B}\cdot d\mathbf{S}. \tag{7.11}$$

The surface integral of the field can be replaced by the average field, B_{av} multiplied by the area inside the orbit, so that,

$$2\frac{dB}{dt} = \frac{dB_{av}}{dt}. \tag{7.12}$$

Thus the average field inside the orbit must increase at twice the rate of the guide field. This is known as the *2-to-1 rule* or the *Wideröe condition*. In the schematic cross-section of a betatron shown in Figure 1.7 (Chapter 1), the slope towards the edge of the pole is not only made to provide focusing but also to reduce the field in order that the 2-to-1 rule can be respected. The betatron has the advantage of being simple and robust. It is limited in energy by saturation in the iron core and the size of the magnet required.

7.2   BASIC ACCELERATING CAVITY
FOR SYNCHROTRONS

In a synchrotron one or more accelerating gaps are fed with an rf voltage with its frequency and phase such that the particles get a positive energy increment each time they pass the gap. The magnetic guide field tracks the particles' energy to keep the equilibrium orbit unchanged. The frequencies are relatively high, so that the accelerating gaps can take the form of cavities.

The simplest cavity is the *pill-box cavity*, which can either be regarded as a shorted, radial transmission line (i.e. two parallel discs guiding waves radiating from the axis of the discs and reflecting them from a short circuit at a constant radius to form a standing wave mode), or as a section of cylindrical waveguide at cut-off (see Figure 7.2). In the terminology of the microwave engineer, the first case is called a radial line resonator and the second a $TM_{0,1,0}$ resonator. The letters TM indicate that the magnetic field is transverse to the axis of the cylinder. The suffixes indicate the number of nodes in the field pattern, first azimuthally, then radially and finally axially (note this does not correspond to the normal order of the coordinates).

If a small hole is made in each wall of the pill-box cavity on its axis, the wall currents are not significantly perturbed and the cavity will function normally. A charged particle can then be sent through the cavity at the correct moment to see the maximum electric field,

$$E_s \propto J_0(kr)\, e^{j\omega t} \qquad B_{cav} \propto j J_1(kr)\, e^{j\omega t}.$$

A practical design of a radial line resonator for a synchrotron would have drift tubes on the axis and in proton machines would often also be loaded with a ferrite core. The ferrite core reduces the quality factor, which increases the losses but at

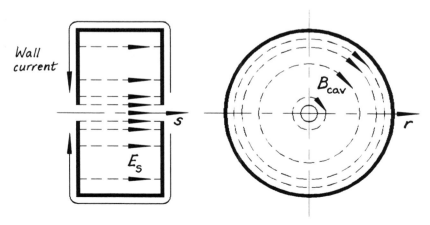

Figure 7.2. Pill-box cavity or radial line resonator (TM_{010} mode).

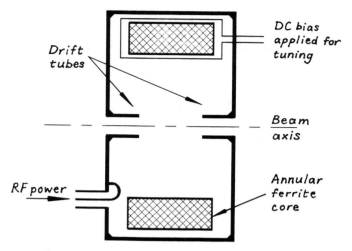

Figure 7.3. Foreshortened radial line resonator.

the same time makes the cavity less sensitive to beam loading. It also reduces the cavity size and it can be used for tuning (see Figure 7.3).

7.3   TRAVELLING WAVE REPRESENTATION OF THE ACCELERATING FIELD

Figure 7.4 shows schematically an accelerating gap with an applied accelerating voltage,

$$u(t) = \hat{u} \cos \int \omega(t) \, dt. \tag{7.13}$$

The frequency ω is assumed to be quasi-constant, but it is written in integral form in order to account for the slow variations needed during the acceleration process. For convenience, the longitudinal field is now expressed as a function of the azimuthal coordinate Θ ($= 2\pi s/C_0$), so that

$$E(t) = u(t)/L_g \qquad \text{for } |\Theta| < \pi L_g/C_0 \tag{7.14}$$

$$E(t) = 0 \qquad \text{for } (\pi L_g/C_0) < |\Theta| < \pi. \tag{7.15}$$

This field has a spatial periodicity of 2π in Θ and can be Fourier analysed with the result,

$$E(t) = \frac{\hat{u}}{C} \left[\cos \int \omega \, dt + \sum_{n=1}^{\infty} \left[\cos\left(\int \omega \, dt + n\Theta \right) + \cos\left(\int \omega \, dt - n\Theta \right) \right] \right] \tag{7.16}$$

where C is the machine circumference. This equation comprises two sets of rotating

143

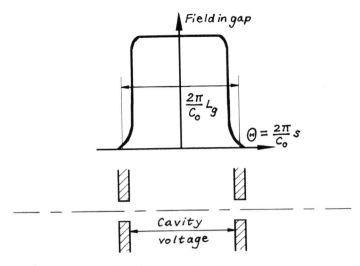

Figure 7.4. Electric field in an accelerating gap.

waves. All the wave components act as a.c. fields on the particles (with zero average effect) except the one that satisfies the condition,

$$\int \omega \, dt - h\Theta = \text{constant}, \quad \text{or } h\frac{d}{dt}\Theta = \omega \quad \text{written as } h\Omega_0 = \omega \quad (7.17)$$

where Ω_0 is the angular frequency of a particle with velocity v_0 running on a closed orbit of circumference C_0 according to the relationship,

$$\Omega_0 = 2\pi v_0 / C_0. \quad (7.18)$$

Such a particle is called an *equilibrium particle* or sometimes also a *phase stable* or *synchronous particle*. It is assumed that the guide field is increased to match the energy gain of the equilibrium particle from the rf gap. In this way, the equilibrium particle's closed orbit will remain constant. Thus the only component of interest for analysing the longitudinal motion is

$$E = \frac{\hat{u}}{C}\cos\left(h\Theta - \int \omega \, dt\right) \quad (7.19)$$

where h represents the number of rf cycles per particle revolution and is called the *harmonic number*.

It should be noted that although it was convenient to assume a short single gap for the acceleration, it is immaterial for the analysis how the field component (7.19) is set up. There could be one or many gaps or even travelling wave structures.

7.4 EQUATIONS FOR THE LONGITUDINAL MOTION

In analogy with the transverse motion, the problem is not only to provide the correct conditions for the equilibrium particle, but also to ensure stability for particles that deviate slightly from the motion of the equilibrium particle. It is useful to define a coordinate for the longitudinal deviation with respect to the travelling wave.

$$\theta = h\Theta - \int \omega \, dt. \tag{7.20}$$

The motion of a particle close to the equilibrium particle is then given by all the forces exerted on this particle equated to the rate of change of momentum, i.e.

$$\frac{d}{dt} p = \frac{e\hat{u}}{C} \cos \theta - \left[\frac{e}{C} \int_0^R 2\pi r \left[\frac{\partial B}{\partial t} \right] dr \right]. \tag{7.21}$$

The first term on the righthand side is the force exerted by the wave component (7.19). By considering this 'effective' force, the equation can be written as a differential equation rather than a difference equation. The differential form is easier to manipulate, but it does imply that many turns will be considered. This is a good approximation if there are many gap-crossings per radian of a synchrotron oscillation[4].

When deriving the wave component (7.19), it was assumed that the voltage across the gap (and hence the energy gained by the particle) was independent of radial position in the gap. It is normal for other reasons to design the lattice with dispersion-free regions around the rf cavities. This automatically ensures that particles pass on the same orbit and thus receive the same energy increment. However, a closer study of the acceleration mechanism shows that the energy increment is in any case quasi-constant for highly relativistic particles and it is not necessary to invoke the existence of dispersion-free regions.

The second term, which is highlighted by a box, is the betatron acceleration force due to the time variation (ramping) of the magnetic field. The minus sign arises from the sign convention adopted in Figure 2.1. This term is often neglected (correctly for constant-field machines such as Fixed-Field Alternating Gradient accelerators FFAGs), but should in general be present.

For the equilibrium particle, the conditions are by definition chosen such that it stays still with respect to the travelling wave at the point $\theta = \theta_0$ (see Figure 7.5) and therefore experiences the force,

$$\frac{d}{dt} p_0 = \frac{e\hat{u}}{C_0} \cos \theta_0 - \left[\frac{e}{C_0} \int_0^{R_0} 2\pi r \left[\frac{\partial B}{\partial t} \right] dr \right]. \tag{7.22}$$

145

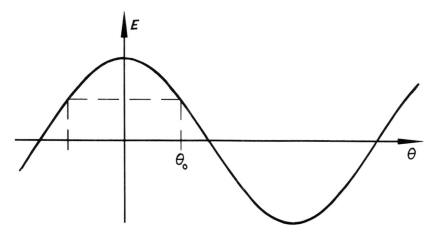

Figure 7.5. Travelling wave representation of the accelerating field.

Deviations from the motion of the equilibrium particle can then be written as the difference between (7.21) and (7.22)

$$\left[C \frac{d}{dt} p - C_0 \frac{d}{dt} p_0 \right] = e\hat{u}(\cos\theta - \cos\theta_0) - \int_{R_0}^{R} 2\pi r \left[\frac{\partial B}{\partial t} \right] dr.$$

This expression can be expanded to first order by putting $R = R_0 + \Delta R$, $C = C_0 + \Delta C$ and $p = p_0 + \Delta p$, so that

$$\boxed{\Delta C \frac{d}{dt} p_0} + C_0 \frac{d}{dt}(\Delta p) = e\hat{u}(\cos\theta - \cos\theta_0) - \boxed{eC_0 \left[\frac{\partial B}{\partial t} \right]_0 \Delta R}.$$

The second term on the righthand side contains the rate of change of flux between the equilibrium orbit and that of the test particle. This term gives a betatron acceleration effect. There is also some betatron acceleration associated with the equilibrium orbit, which can be seen more clearly in (7.22). The magnitude depends on the rate of change of the total flux inside the orbit. Since synchrotrons often have their return yokes evenly distributed between the inside and outside of the orbit (e.g. 'H' dipoles), this term is partially self-compensating. Both effects are always small compared to that of the rf voltage, even for machines like the Cosmotron where all magnet yokes were on the inside.

The basic cyclotron relationship (2.9) when applied to the equilibrium orbit and differentiated gives

$$\frac{d}{dt} p_0 = -e \left[\frac{\partial B}{\partial t} \right]_0 R_0,$$

so that the terms highlighted by boxes in the above relationship cancel, since

$\Delta C \equiv 2\pi \, \Delta R$, leaving

$$\left.\begin{aligned}
\frac{\mathrm{d}}{\mathrm{d}t}(\Delta p) &= \frac{e\hat{u}}{C_0}(\cos\theta - \cos\theta_0) \\[2mm]
&= \frac{e\hat{u}}{C_0}[\cos(\theta_0 + \Delta\theta) - \cos\theta_0].
\end{aligned}\right\} \tag{7.23}$$

In the literature, the reader may find a seemingly simpler derivation of (7.23) in which the betatron acceleration terms are firstly omitted when constructing (7.21), i.e.

$$\frac{\mathrm{d}}{\mathrm{d}t}p = \frac{e\hat{u}}{C}\cos\theta \tag{7.21}'$$

and then, when expressing the difference with the equilibrium orbit, a *second approximation is made that $C = C_0$*, so that

$$\frac{\mathrm{d}}{\mathrm{d}t}(\Delta p) = \frac{\mathrm{d}}{\mathrm{d}t}p - \frac{\mathrm{d}}{\mathrm{d}t}p_0 \simeq \frac{e\hat{u}}{C_0}[\cos(\theta_0 + \Delta\theta) - \cos\theta_0]. \tag{7.23}'$$

The end result (7.23)' is the same as (7.23). However, this result is only obtained if both approximations are made and one is left with the impression that the equation contains serious approximations when in fact it is correct. When betatron acceleration forces are not present, as in the FFAG machines, the two terms in the boxes are zero, so the final equation is again (7.23) without having had to make any approximations.

Now an important relationship is established between the particle's revolution frequency, its momentum and the lattice properties. The revolution frequency is given by $\Omega = 2\pi v/C$, which yields by logarithmic differentiation,

$$\Delta\Omega/\Omega_0 = \Delta v/v_0 - \Delta C/C_0. \tag{7.24}$$

Simple relativity theory gives,

$$\Delta v/v_0 = \gamma^{-2}\Delta p/p_0, \tag{7.25}$$

where γ now has the definition $\gamma = m/m_0$, and from the definition of the momentum compaction (2.41) in Section 2.6,

$$\Delta C/C_0 = \alpha \, \Delta p/p_0. \tag{7.26}$$

The combination of the above with (7.24) yields

$$\Delta\Omega/\Omega_0 = (\gamma^{-2} - \alpha)\,\Delta p/p_0. \tag{7.27}$$

This is frequently rewritten as,

$$\eta = \frac{\Delta\Omega}{\Omega}\bigg/\frac{\Delta p}{p_0} = (\gamma^{-2} - \alpha) \tag{7.28}$$

where η is the fractional change in revolution frequency per unit of fractional change in momentum spread.

In (7.27) $\Delta\Omega$ can be replaced by $\Delta\dot{\theta}/h$ from (7.20), if ω is constant, or slowly varying, compared to the particle oscillations about the equilibrium particle. In addition Ω_0 can be replaced by ω/h from (7.17).

$$\Delta p = p_0 \eta^{-1} \Delta\dot{\theta}/\omega. \tag{7.29}$$

The particle momentum is given by,

$$p_0 = mv_0 = mC_0\Omega_0/(2\pi) = m_0\gamma C_0\omega/(2\pi h). \tag{7.30}$$

The replacement of p_0 in (7.29) using (7.30) yields,

$$\Delta p = \frac{m_0\gamma C_0}{2\pi h\eta} \Delta\dot{\theta} \tag{7.31}$$

which when inserted into (7.23) gives,

$$\frac{\mathrm{d}}{\mathrm{d}t}\left[\frac{m_0\gamma C_0}{2\pi h\eta} \frac{\mathrm{d}}{\mathrm{d}t} \Delta\theta \right] = \frac{e\hat{u}}{C_0} [\cos(\theta_0 + \Delta\theta) - \cos\theta_0]. \tag{7.32}$$

This is the basic phase equation valid for all amplitudes and the equation which in particular should be used for the study of such problems as the *separatrices* (boundaries between stable and unstable regions).

Equation (7.32) is frequently approximated for small deviations in $\Delta\theta$, i.e. linearised, to the form,

$$\frac{\mathrm{d}}{\mathrm{d}t}\left[\frac{m_0\gamma C_0}{2\pi h\eta} \frac{\mathrm{d}}{\mathrm{d}t} \Delta\theta \right] + \frac{e\hat{u}}{C_0} \sin\theta_0 \, \Delta\theta = 0. \tag{7.33}$$

The derivation of this equation needs some care and Ref. 4 discusses the more subtle points in relation to the main papers in the literature. An alternative derivation using the Lagrangian formalism is given in Appendix D. As was stressed in the case of the transverse motion, the use of the Lagrangian/Hamiltonian formalism has the great advantage that the variables are automatically chosen such that the phase-space area of the beam is an invariant of the motion, i.e. the variables are canonically conjugate.

7.5   PHASE STABILITY

7.5.1   General considerations for phase stability

For the general consideration of longitudinal stability the linearised equation (7.33) is sufficient. Inspection of (7.33) shows that if the two coefficients have the same sign the solution will be oscillatory, whereas if the signs are opposite, the solution will be exponential in character and definitely unstable. This means that stability can only be obtained if:

$$0 < \theta_0 < \pi/2 \qquad \text{for } \gamma < \alpha^{-\frac{1}{2}} \tag{7.34}$$

or

$$0 > \theta_0 > -\pi/2 \qquad \text{for } \gamma > \alpha^{-\frac{1}{2}}. \tag{7.35}$$

Strictly, it would be more correct to put π rather than $\pi/2$ in the above limits, but since the user is usually interested in acceleration and not deceleration $\pi/2$ is more appropriate. At a given energy there will be a transition between the two conditions (7.34) and (7.35). Not surprisingly, this energy is called the *transition energy*,

$$\gamma_{\text{tr}} = \alpha^{-\frac{1}{2}} \qquad \text{or} \qquad E_{\text{tr}} = \alpha^{-\frac{1}{2}} m_0 c^2. \tag{7.36}$$

It is worth considering equation (7.36) for a moment. It can be seen from (7.28) that at low energies (i.e. below transition) an increase in energy leads to an increase in revolution frequency, whereas above this limit, it leads to a decrease. Consider a particle, which is below transition and is lagging in energy with respect to the equilibrium particle. This particle will be travelling more slowly than the accelerating wave (7.19). If the accelerating wave is phased such that by lagging the particle rides up the wave and thus receives more energy, then the particle will be able to catch up the equilibrium particle. This is the situation shown in Figure 7.6(*a*) and corresponds to θ_0 being between 0 and $\pi/2$. Clearly if the wave presented the opposite slope, the particle would progressively lag more and more until it was lost. Above transition the situation is reversed, as shown in Figure 7.6(*c*) and a particle which lags has too much energy. By lagging it slips back with respect to the wave as before,

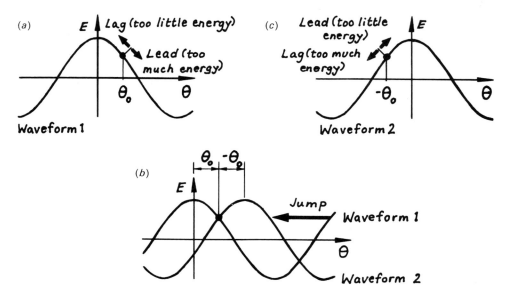

Figure 7.6. Phase stability and the phase jump at transition. (*a*) Below transition, (*b*) at transition, (*c*) above transition.

149

but now the phase has been changed and the particle receives less energy. This corresponds to θ_0 being between 0 and $-\pi/2$. In this way the particles are focused about the equilibrium particle. In order to maintain stability on either side of transition, the phase of the rf system has to be jumped by $2\theta_0$ just as the energy passes through γ_{tr}. By this means, the particle sitting at the stable phase θ_0 below transition will find itself at the new stable phase $-\theta_0$ above transition, as illustrated in Figure 7.6(b).

The momentum compaction α is a property of the lattice. In Section 2.6, it was shown that for weak focusing machines $\alpha = Q_x^{-2}$ and since for these machines $Q_x < 1$, γ_{tr} must in this case also be less than unity (which in fact means it is non-physical). Thus the stability condition (7.35) must always apply to weak focusing machines. It was also mentioned in Section 2.6 and illustrated in Section 4.7.1 that the expression $\alpha = Q_x^{-2}$ is a fair approximation for regular alternating-gradient structures. Since, in these structures, Q_x is normally well above unity, the transition energy will fall in the physical range. In electron machines, however, the injection energy is always above transition and again it is only (7.35) that applies, but for proton rings it often happens that the transition energy falls somewhere between the injection and top energies.

If the phase jump from θ_0 to $-\theta_0$ is done at the right moment, then (7.32) can be written in a form valid both below and above transition

$$\frac{d}{dt}\left[\frac{m_0\gamma C_0}{2\pi h\eta}\frac{d}{dt}\Delta\theta\right] = \frac{e\hat{u}}{C_0}[\cos(\pm\theta_0 + \Delta\theta) - \cos\theta_0] \tag{7.37}$$

where the upper sign applies below transition (i.e. $\eta > 0$) and the lower sign above transition (i.e. $\eta < 0$).

The linearised equation, valid for small values of $\Delta\theta$, can be simplified to,

$$\frac{d}{dt}\left[\frac{m_0\gamma C_0}{2\pi h|\eta|}\frac{d}{dt}\Delta\theta\right] + \frac{e\hat{u}}{C_0}\sin|\theta_0|\,\Delta\theta = 0. \tag{7.38}$$

This is equivalent to an oscillating mass point where the 'mass' goes to infinity when $E \to E_{tr}$. Figure 7.7 illustrates the kind of mechanical analogue described by (7.38) (this is also valid in the non-linear region).

In later sections the solution of the linearised equation (7.38) will be analysed in more detail both away from transition, where approximate solutions are simple, and close to transition where the solutions are more complicated.

7.5.2 Separatrices; the boundaries of stability

It is interesting to examine in more detail the boundaries between stable and unstable regions. These boundaries are normally called *separatrices*. For this purpose the conditions below transition will be analysed, i.e. (7.37) with the upper sign will be

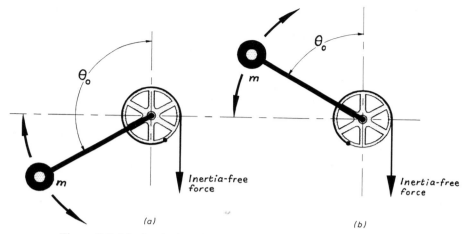

Figure 7.7. Mechanical analogue of the biased pendulum. (a) Stable phase point, (b) unstable phase point.

used. Since the phase plane trajectory for only one oscillation is needed, all the coefficients in (7.37) can be treated as constant, i.e.

$$\frac{\mathrm{d}}{\mathrm{d}t}\,\Delta\dot{\theta} = \frac{e\hat{u}}{C_0}\frac{2\pi h\eta}{m\gamma C_0}(\cos\theta - \cos\theta_0).\tag{7.39}$$

Multiplication of this equation by $\mathrm{d}\theta/\mathrm{d}t$ and integration gives,

$$(\Delta\dot{\theta})^2 = \frac{e\hat{u}}{C_0}\frac{4\pi h\eta}{m\gamma C_0}\int(\cos\theta - \cos\theta_0)\,\mathrm{d}\theta$$

$$= \frac{e\hat{u}}{C_0}\frac{4\pi h\eta}{m_0\gamma C_0}(\sin\theta - \theta\cos\theta_0 + \text{const.}).$$

The momentum deviation can now be introduced from (7.31),

$$(\Delta p)^2 = \frac{e\hat{u}m_0\gamma}{\pi h\eta}(\sin\theta - \theta\cos\theta_0 + \text{const.}).$$

At $\theta = \theta_0$ the restoring force in (7.37) is zero and therefore Δp has its maximum at this value of θ and the integration constant can be determined in terms of this maximum,

$$(\widehat{\Delta p})^2 = \frac{e\hat{u}m_0\gamma}{\pi h\eta}(\sin\theta_0 - \theta_0\cos\theta_0 + \text{const.})$$

so that,

$$\text{const.} = \frac{\pi h\eta}{e\hat{u}m_0\gamma}(\widehat{\Delta p})^2 - \sin\theta_0 + \theta_0\cos\theta_0$$

and

$$(\Delta p)^2 = (\widehat{\Delta p})^2 - \frac{e\hat{u}m_0\gamma}{\pi h\eta}[(\theta - \theta_0)\cos\theta_0 - \sin\theta + \sin\theta_0]\tag{7.40}$$

151

It should be noted that this expression is valid for any $\widehat{\Delta p}$ and θ and can therefore be used for plotting particle trajectories in a θ, Δp plane with $\widehat{\Delta p}$ as a parameter. For small values of $\widehat{\Delta p}$, which give small values of $\Delta\theta$, the trajectories are ellipses, given by,

$$(\Delta p)^2 + \frac{e\hat{u}m_0\gamma \sin\theta_0}{2\pi h\eta}(\Delta\theta)^2 = (\widehat{\Delta p})^2. \tag{7.41}$$

For large values of $\widehat{\Delta p}$ the trajectories deviate considerably from ellipses.

The trajectory representing the stability limit can be found in the following way. From Figures 7.5 or 7.6, it is seen that a particle below transition reaching $\theta = -\theta_0$ with a finite Δp will be unstable, since beyond that point it will have insufficient restoring force to be pulled back. A particle reaching this point with $\Delta p = 0$ will be metastable and a particle reaching $\theta = -\theta_0 + \varepsilon$ with $\Delta p = 0$ will be turned back for arbitrary small positive ε. This means that $\theta = -\theta_0$ (or $\Delta\theta_{\min} = -2\theta_0$) is the largest negative phase a particle can reach and still remain stable. When this value is substituted into (7.40), it gives the corresponding value of $\widehat{\Delta p}$, which is the largest attainable momentum deviation, i.e.

$$(\widehat{\Delta p_{\max}})^2 = \frac{2e\hat{u}m_0\gamma}{\pi h\eta}(\sin\theta_0 - \theta_0\cos\theta_0). \tag{7.42}$$

This is a very important expression, as it gives the absolute maximum momentum spread that can exist within the stable region. In more colloquial terms it gives the maximum momentum bite of the *bucket*. The expression for the separatrix can now be found by the substitution (7.42) into (7.40) to give,

$$(\Delta p)^2 = \frac{e\hat{u}m_0\gamma}{\pi h\eta}[\sin\theta + \sin\theta_0 - (\theta + \theta_0)\cos\theta_0]. \tag{7.43}$$

Modern accelerators are normally adjusted at injection to have no acceleration, i.e. $\theta_0 = \pi/2$. For this simple case, the momentum bite takes the simple form,

$$(\widehat{\Delta p_{\max}})^2 = \frac{2e\hat{u}m_0\gamma}{\pi h\eta} \tag{7.44}$$

and the separatrix of a stationary bucket (i.e. $\theta_0 = \pi/2$), takes the simplified form,

$$(\Delta p)^2 = \frac{e\hat{u}m_0\gamma}{\pi h\eta}(1 + \sin\theta)$$

or,

$$\Delta p = \pm\left[\frac{2e\hat{u}m_0\gamma}{\pi h\eta}\right]^{\frac{1}{2}}\cos(\Delta\theta/2). \tag{7.45}$$

From these expressions separatrices can be plotted for various conditions. Two typical examples have been drawn in Figure 7.8.

So far explicit expressions have been found for the three points of a separatrix

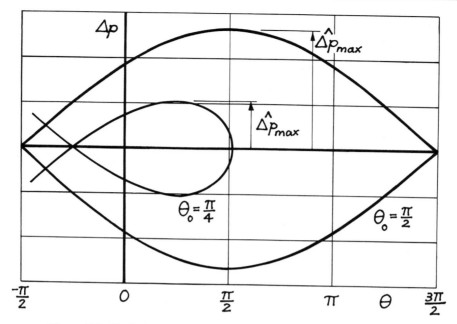

Figure 7.8. Typical separatrices.

at $(\theta = -\theta_0, \Delta p = 0)$ and $(\theta = \theta_0, \Delta p = \pm \widehat{\Delta p}_{max})$. The point limiting the separatrix to the right on Figure 7.8 is the fourth point that it is important to know for approximate calculations. This can be found by putting $\Delta p = 0$ in the equation for the separatrix, i.e.

$$(2\theta_0 + \Delta\theta) \cos\theta_0 - \sin(\theta_0 + \Delta\theta) - \sin\theta_0 = 0. \qquad (7.46)$$

It can be seen directly that $\Delta\theta = -2\theta_0$ is a solution to this equation but it is the solution that was found earlier by inspection. The desired solution of (7.46) giving the limit to the 'head', i.e. the righthand side of the separatrix in Figure 7.8, can only be found numerically and is drawn in Figure 7.9.

It can be noted from Figure 7.9 that $\Delta\theta_{max} \simeq \theta_0$ is a fairly good approximation to well above $\theta_0 = \pi/4$. This approximate solution, which can also be found analytically, is important for making simple estimates of the longitudinal stability of many accelerators. Above $\pi/4$, the solution rises sharply to $\Delta\theta_{max} = 2\theta_0$ at $\theta_0 = \pi/2$, which is the expected value, since this is the limiting case of no acceleration.

These calculations have been made for the situation below transition. Above transition, the equations are similar and yield mirror-image results with the axis $\theta = 0$ as the mirror-axis. This means that the phase limits for the stable region become (with the stable phase at $-\theta_0$):

$$\Delta\theta_{min} = -\theta_{max},$$

$$\Delta\theta_{max} = 2\theta_0$$

153

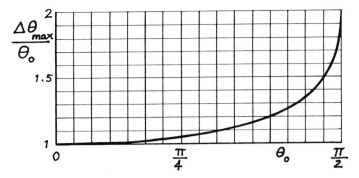

Figure 7.9. A universal curve for the limiting point in θ at the 'head' of the separatrix.

and

$$(\widehat{\Delta p}_{\max})^2 = \frac{e\hat{u}m_0\gamma}{\pi h(-\eta)}\left[\sin(-\theta_0) - (-\theta_0)\cos\theta_0\right]$$

or, written in a form valid both above and below transition,

$$(\widehat{\Delta p}_{\max})^2 = \frac{e\hat{u}m_0\gamma}{\pi h|\eta|}\left(\sin|\theta_0| - |\theta_0|\cos\theta_0\right). \tag{7.47}$$

A simple interpretation of the region of stability in the $(\Delta\theta, \Delta p)$ plane is that a particle that has its momentum and phase within the area enclosed by the separatrix will perform stable, but in general non-linear oscillations, whereas particles outside will be lost. Those below the stable area in Figure 7.8 will drift inwards, while those above will first swing round the stable region turning with the same sense as the stable particles, before also drifting inwards.

7.5.3 Longitudinal acceptance and emittance

It is natural to define a *longitudinal acceptance* as the area inside the separatrix. This area can in general only be found by numerical calculations and reference can be made to Symon and Sessler[5] for this. However, the very important special case of a stationary bucket can be found analytically from (7.45).

$$A_l(\text{stationary}) = 4\widehat{\Delta p}_{\max}\int_0^{\pi}\cos(\Delta\theta/2)\,\mathrm{d}(\Delta\theta)$$

$$= 8\widehat{\Delta p}_{\max} = 8\left[\frac{2e\hat{u}m_0\gamma}{\pi h|\eta|}\right]^{\frac{1}{2}}. \tag{7.48}$$

Although numerical calculations are needed for non-stationary buckets the following approximate formula is very handy for estimates

$$A_l \simeq \left[\pi + \frac{\Delta\theta_{\max}}{\theta_0} \right] \theta_0 \, \widehat{\Delta p}_{\max} \qquad (7.49)$$

where the values for $\Delta\theta_{\max}/\theta_0$ can be found from Figure 7.9. Again it should be remembered that for $\theta_0 \lesssim \pi/4$, $\Delta\theta_{\max}/\theta_0 \simeq 1$, giving the even simpler formula for small θ_0,

$$A_l \simeq (\pi + 1)\theta_0 \, \widehat{\Delta p}_{\max}. \qquad (7.50)$$

The formula (7.49) in fact is such a good approximation that it is hardly ever worth the effort of making further numerical computations to obtain the exact value. A demonstration test can be made for $\theta_0 = \pi/2$, giving,

$$A_l \simeq 0.5\pi(\pi + 2) \, \widehat{\Delta p}_{\max} = 8.08 \, \widehat{\Delta p}_{\max}$$

which is within 1% of the correct value [see (7.48)].

If an ensemble of particles, known as a *bunch*, occupies a certain area within A_l, this area is called the *longitudinal emittance*, $\mathcal{E}_l$. As in the transverse case, particles will be stable if $\mathcal{E}_l$ is entirely inside A_l and to avoid dilution of the emittance $\mathcal{E}_l$ should be centred around $\theta = \theta_0$, $\Delta p = 0$ and have the same shape as the trajectories calculated above. This is called matching the longitudinal emittance to the longitudinal acceptance, or to the *rf bucket*. The longitudinal emittance of the bunch is an invariant*.

For certain purposes it is important to have a feeling for how the bucket size changes with energy. For instance, if the bucket is filled at injection, it is important that the bucket size stays larger than the emittance at all later times to avoid the loss of particles. There are many parameters at disposal to achieve this, the main ones being $\hat{u}$ and θ_0. However, to see whether this situation is likely to occur, an analysis of the variation of the bucket size under the conditions of constant $\hat{u}$, θ_0 and h is of interest. For this special case (7.42) is rewritten in a simplified form,

$$(\widehat{\Delta p}_{\max}/\widehat{\Delta p}_0)^2 = \gamma/|\eta| = \gamma^3/|1 - (\gamma/\gamma_{\mathrm{tr}})^2| = \gamma_{\mathrm{tr}}{}^3(\gamma/\gamma_{\mathrm{tr}})^3/|1 - (\gamma/\gamma_{\mathrm{tr}})^2|.$$

Written in this form the expression is valid both below and above transition and $\widehat{\Delta p}_0$ has the simple physical interpretation of being the momentum bite at zero kinetic energy if $\gamma_{\mathrm{tr}} \gg 1$. The universal curve $(\widehat{\Delta p}_{\max}/\widehat{\Delta p}_0)\gamma_{\mathrm{tr}}{}^{-3/2}$ is plotted in Figure 7.10.

This expression increases with γ up to $\gamma = \gamma_{\mathrm{tr}}$. However, close to γ_{tr} the analysis loses its validity as the assumptions made break down. More will be said about this later. Above transition the expression diminishes with increasing γ to a minimum at $\gamma = \sqrt{3}\,\gamma_{\mathrm{tr}}$. This minimum is given by,

$$(\widehat{\Delta p}_{\max}/\widehat{\Delta p}_0)^2 = (\sqrt{3}\,\gamma_{\mathrm{tr}})^3/2$$

which means that an accelerator with injection well below transition (in principle

* Reference 5 introduces the abbreviations: $\Gamma = \sin\theta_0$ and $\alpha(\Gamma) = A_l/A_l(\text{stationary})$, where $A_l(\text{stationary})$ is given by (7.48), which are often found in the literature.

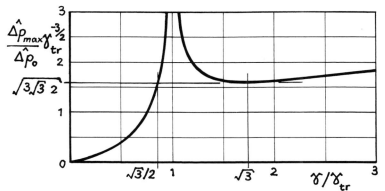

Figure 7.10. Bucket size (normalised by $\gamma_{\mathrm{tr}}^{3/2}$).

below $\gamma_{\mathrm{inj}} = \gamma_{\mathrm{tr}}\sqrt{3/2}$) and with little or no blow-up at transition will be in no danger of spilling out particles from the bucket at the '$\sqrt{3}$' point. This might look different for accelerators with injection above transition, but below the '$\sqrt{3}$' point, which is becoming common in modern machines (e.g. CERN-SPS with antiprotons and DESY-HERA). However, the injected beam in such cases is normally already tightly bunched and the reduction of the bucket height has little consequence.

7.6   SMALL-AMPLITUDE DEVIATIONS

For the purposes of finding the behaviour of particles that execute small oscillations around the phase stable point equation (7.38) will be used, which implies that the stable angle is jumped from θ_0 to $-\theta_0$ at the exact moment when γ reaches γ_{tr}. With this assumption, the adiabatic solution applies over the whole range of γ from injection to the top energy with the exception of a fairly narrow band on either side of transition. The criterion for the validity of this solution is that the coefficients change little over one radian of the oscillation. This breaks down in the neighbourhood of transition and this case will be treated separately in the next section.

The adiabatic solution of (7.38) can be written,

$$\Delta\theta = A\left[\frac{2\pi h|\eta|}{\gamma m_0 e\hat{u}\sin|\theta_0|}\right]^{\frac{1}{4}}\sin\int\omega_{\mathrm{s}}\,\mathrm{d}t \qquad (7.51)$$

where ω_{s} is called the *synchrotron oscillation frequency* and is given by,

$$\omega_{\mathrm{s}} = \left[\frac{2\pi h}{C_0{}^2}\frac{e\hat{u}\sin|\theta_0|}{\gamma m_0}|\eta|\right]^{\frac{1}{2}}. \qquad (7.52)$$

With the same approximation, the momentum deviation is given by,

$$\Delta p = A\frac{\sin\theta_0}{\sin|\theta_0|}\left[\frac{2\pi h|\eta|}{\gamma m_0 e\hat{u}\sin|\theta_0|}\right]^{-\frac{1}{4}}\cos\int\omega_{\mathrm{s}}\,\mathrm{d}t. \qquad (7.53)$$

156

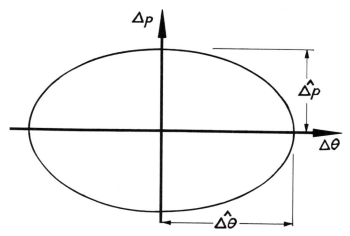

Figure 7.11. Small amplitude synchrotron oscillations in $(\Delta\theta, \Delta p)$ phase space.

It is now useful to 'forget' for a while that the solutions are not valid at, or close to transition. In the $(\Delta\theta, \Delta p)$ phase plane, the particles move on elliptical trajectories with the angular frequency ω_s (see Figure 7.11).

It can be seen from (7.51) and (7.53) that the area of the ellipse for a given particle stays constant as γ increases. This is a consequence of the invariance of the emittance. Below transition the particles move in the clockwise direction and above in the anticlockwise direction, with the frequency approaching zero at the transition point. It is important to notice that the frequency goes to zero because the 'equivalent mass' in (7.38) goes to infinity (rather than the 'equivalent spring force' going to zero). This makes the whole system stiff and slow. This greatly facilitates the 'gymnastics' needed for the phase jump at the transition point.

It is useful to introduce a variable $\omega_{s\infty}$, which would be the angular synchrotron oscillation frequency if $\gamma_{tr} = \infty$, i.e.

$$\omega_{s\infty} = \left[\frac{2\pi h\, e\hat{u}\, \sin|\theta_0|}{C_0{}^2} \frac{1}{\gamma m_0}\, \gamma^{-2}\right]^{\frac{1}{2}}. \tag{7.54}$$

The normalisation of ω_s by $\omega_{s\infty}$ leads to the formulae,

$$\frac{\omega_s}{\omega_{s\infty}} = |\eta|^{\frac{1}{2}}\gamma = \begin{cases} [1 - (\gamma/\gamma_{tr})^2]^{\frac{1}{2}} \text{ below transition,} \\ [(\gamma/\gamma_{tr})^2 - 1]^{\frac{1}{2}} \text{ above transition,} \end{cases} \tag{7.55}$$

which provide a convenient way of drawing a universal curve for the synchrotron oscillation frequency (see Figure 7.12). Furthermore this universal curve again illustrates the transition point very clearly.

It is of some interest to know the order of magnitude of this frequency and the CERN PS can be taken as a typical example. If the parameters for this machine are

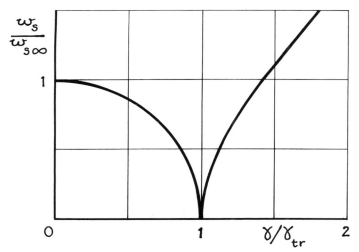

Figure 7.12. Universal curve for the phase oscillation frequency.

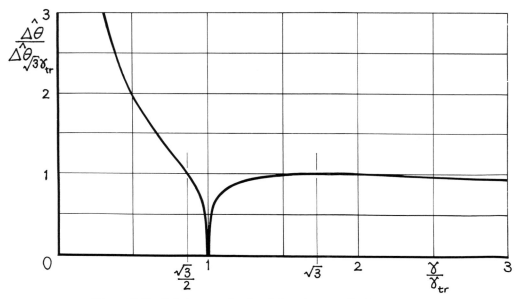

Figure 7.13. Relative amplitude of the phase oscillation (normalised to unity at $\gamma/\gamma_{tr} = \sqrt{3}$ and $\sqrt{3/2}$).

inserted into the above expressions the synchrotron oscillation frequency comes out in the range of 350–250 Hz for energies corresponding to $\sqrt{3}\gamma_{tr}$ to maximum[6].

The amplitude of the phase oscillation, given by (7.51), can also be represented by a universal curve, as drawn in Figure 7.13. Above transition this curve grows to a rather flat maximum at $\gamma/\gamma_{tr} = \sqrt{3}$, i.e. at the same point as the minimum for

$\widehat{\Delta p}_{\text{max}}$ in Figure 7.10. It is convenient to normalise the curve to unity at this maximum point as has been done in Figure 7.13. The normalised expression takes the form,

$$\widehat{\Delta\theta}/(\widehat{\Delta\theta})_{\sqrt{3}\gamma_{\text{tr}}} = \left[\frac{3\sqrt{3}}{2}[(\gamma/\gamma_{\text{tr}})^2 - 1](\gamma_{\text{tr}}/\gamma)^3\right]^{\frac{1}{4}} \tag{7.56}$$

and is of interest to notice that below transition the curve passes through unity at the point $\gamma/\gamma_{\text{tr}} = \sqrt{3}/2$.

It is necessary to check the phase oscillation amplitude under different conditions. For instance, in some accelerators the rf buckets are rather full at injection. If that is the case, it is important that the amplitude at the '$\sqrt{3}$' point is smaller than at injection in order not to spill particles out of the bucket. How far down on such a curve one chooses to be depends, amongst other things, on one's estimate of the beam's deterioration when going through the non-adiabatic transition zone (see next section).

The adiabatic solution for the momentum excursions was given by (7.53) and the associated average excursion is given by (7.26) and (7.36).

$$\frac{\Delta C}{C} = \alpha\frac{\Delta p}{p} = \gamma_{\text{tr}}^{-2}\frac{\Delta p}{p}. \tag{7.57}$$

With the assumption of adiabaticity all through the acceleration process, it is easy to evaluate how ΔC changes for a particular machine. The solution has a singularity at transition. However, it is clear that the approximation breaks down at this point and it is important to know how much the radial excursion increases in reality. How to calculate this is described in the next section.

It was stated earlier that it is desirable to match the shape of the longitudinal emittance to the shape of the trajectories inside the bucket. In this connection, the ratio of the axes of the trajectory ellipses (see Figure 7.11) is significant. This is obtained from (7.51) and (7.53) as,

$$\left[\frac{\widehat{\Delta p}}{\widehat{\Delta\theta}}\right] = \left[\frac{e\hat{u}m_0\gamma\sin|\theta_0|}{2\pi h|\eta|}\right]^{\frac{1}{2}} \tag{7.58}$$

which it is sometimes convenient to express in terms of the phase oscillation frequency (7.52),

$$\left[\frac{\widehat{\Delta p}}{\widehat{\Delta\theta}}\right] = \frac{m_0\gamma C_0}{2\pi h|\eta|}\omega_{\text{s}}.$$

This is important for the transfer of bunched beams from one machine to another. In order to avoid, or minimise, the dilution of the longitudinal emittance when transferring, it is necessary to:

- either pre-form the emittance to match the ratio (7.58) as defined by the second machine

159

● or to adapt the second machine to accept (7.58) as defined by the first machine.

Often a little of both is being done to achieve the match.

7.7   CLOSE TO TRANSITION[6]

Equation (7.33) will be taken as the starting point for the analysis of the behaviour of the particle very close to the transition point and as it is accelerated through this critical energy. For this purpose (7.33) is rewritten as,

$$\frac{m_0 C_0 \gamma}{2\pi h \eta} \frac{\mathrm{d}^2}{\mathrm{d}^2 t} \Delta\theta + \frac{m_0 C_0}{2\pi h} \frac{\mathrm{d}}{\mathrm{d}t} (\gamma/\eta) \frac{\mathrm{d}}{\mathrm{d}t} \Delta\theta + \frac{e\hat{u}}{C_0} \sin\theta_0 \, \Delta\theta = 0. \tag{7.59}$$

Let time be measured from the moment at which transition is passed and assume that γ varies linearly with time in the region of interest, i.e.

$$\gamma = \gamma_{\mathrm{tr}} + \dot{\gamma}t \tag{7.60}$$

where $\dot{\gamma}$ is the constant derivative of γ with time. For simplicity the abbreviation,

$$g = \gamma/\gamma_{\mathrm{tr}} = 1 + (\dot{\gamma}/\gamma_{\mathrm{tr}})t = 1 + \dot{g}t \tag{7.61}$$

is introduced. After a little manipulation this gives,

$$\gamma/\eta = \gamma_{\mathrm{tr}}{}^3 g^3/(1 - g^2) \tag{7.62}$$

so that

$$\frac{\mathrm{d}}{\mathrm{d}t} (\gamma/\eta) = \gamma_{\mathrm{tr}}{}^3 \frac{g}{(1 - g^2)} \left[3 + \frac{2g^2}{(1 - g^2)} \right] \dot{g}. \tag{7.63}$$

The above substitution can be made in the phase equation (7.59), but that gives unnecessarily complicated expressions for the present purpose. Since only a narrow band around the transition energy is of interest the approximation $g = 1$ can be introduced everywhere with the exception of,

$$1 - g = -\dot{g}t \tag{7.64}$$

when this appears in the denominator. With this approximation,

$$\gamma/\eta = \gamma_{\mathrm{tr}}{}^3/(-2\dot{g}t) \tag{7.65}$$

so that,

$$\frac{\mathrm{d}}{\mathrm{d}t} (\gamma/\eta) = \tfrac{1}{2}\gamma_{\mathrm{tr}}{}^3 \dot{g}/(\dot{g}t)^2. \tag{7.66}$$

When (7.66) is inserted into the phase equation (7.59) the result is,

$$\frac{d^2}{dt^2}\Delta\theta - t^{-1}\frac{d}{dt}\Delta\theta - \dot{g}t\frac{4\pi e\hat{u}h\sin\theta_0}{C_0{}^2\gamma_{tr}{}^3m_0}\Delta\theta = 0. \tag{7.67}$$

For convenience this is abbreviated to,

$$\frac{d^2}{dt^2}\Delta\theta - t^{-1}\frac{d}{dt}\Delta\theta + at\,\Delta\theta = 0 \tag{7.68}$$

which has Bessel-function solution*.

7.7.1   Above transition and after the phase has been shifted

In this case t and a are both positive and the Bessel-function solution can be written as,

$$\Delta\theta = t[A'J_{2/3}(\tfrac{2}{3}\sqrt{at^{3/2}}) + B'J_{-2/3}(\tfrac{2}{3}\sqrt{at^{3/2}}] \tag{7.69}$$

where A' and B' are constants.

7.7.2   Below transition and before the phase has been shifted

The solution is most easily found by the introduction of,

$$\tau = -t \tag{7.70}$$

which leads to,

$$\frac{d^2}{d\tau^2}\Delta\theta - \tau^{-1}\frac{d}{d\tau}\Delta\theta - a\tau\,\Delta\theta = 0. \tag{7.71}$$

Now that the equation is applied below transition a is negative and the solution becomes,

$$\Delta\theta = \tau[AJ_{2/3}(\tfrac{2}{3}\sqrt{(-a)\tau^{3/2}}) + BJ_{-2/3}(\tfrac{2}{3}\sqrt{(-a)\tau^{3/2}}]. \tag{7.72}$$

It is possible therefore to combine both solutions in the form,

$$\Delta\theta = |t|\left[\begin{matrix}A'\\A\end{matrix}\,J_{2/3}(\tfrac{2}{3}\sqrt{|at^3|}) + \begin{matrix}B'\\B\end{matrix}\,J_{-2/3}(\tfrac{2}{3}\sqrt{|at^3|})\right]. \tag{7.73}$$

7.7.3   Phase shift at the right moment

If the phase of the rf system is shifted at the right time, i.e. $t = 0$, it is readily seen that the two solutions match to give,

$$A = A' \qquad \text{and} \qquad B = B' \tag{7.74}$$

---

* See for example I. S. Gradshteyn, I. M. Ryzhik, *Table of integrals, series, and products* (Academic Press, Inc. 1980), 971, 8.491-12.

which means that the same solution applies on both sides of transition, i.e. the solution is symmetric to the approximation applied. This leads to an important conclusion:

When entering the transition region with an elliptical bunch in phase space, as given by the adiabatic solution, the bunch comes out on the other side with the same shape. After transition the particles have tracked the same phase-space trajectories backwards as they followed forwards before transition.

Under such ideal conditions the scales of the curves in Figure 7.13 for the adiabatic regions are the same on both sides of transition.

The expressions found in the transition region and the corresponding expressions for Δp, which have not been derived here, can be used to find the detailed behaviour of a particle, or more generally of a phase-space ellipse, as it goes through transition. This is tedious to do analytically, but numerical calculations are easy to perform for particular cases.

A last thing that can be done is to see what happens if the asymptotic approximation to the Bessel-function solutions is applied. This is simple and leads to the expected result (i.e. the adiabatic solution) providing the approximation,

$$|\gamma - \gamma_{tr}|/\gamma_{tr} \ll 1 \qquad (7.75)$$

is introduced. A further check ought to be carried out to see that the adiabatic solution is valid sufficiently close to transition that the approximation (7.75) is a valid one. This is for most practical cases true, but should perhaps not always be taken for granted, since the parameters depend strongly on the accelerator considered.

7.7.4 What can typically go wrong?

Two simple examples of situations where the above approach needs modifying will be described:

- mistiming of the transition jump,
- longitudinal space-charge forces.

In both cases the physics will be explained with little detailed mathematics.

(i) *Mistiming*

If the rf phase is shifted at the wrong moment, too early or too late, the beautiful symmetry is spoilt and the phase-space ellipse will emerge from the transition region with the wrong orientation. It will no longer match the adiabatic solution on the other side as indicated in Figure 7.14. This leads to an increase in the maxima of $\Delta\theta$ and ΔR of the synchrotron oscillation. Since only a small increase in the longitudinal emittance will be acceptable, there will be a corresponding tolerance on the allowed

162

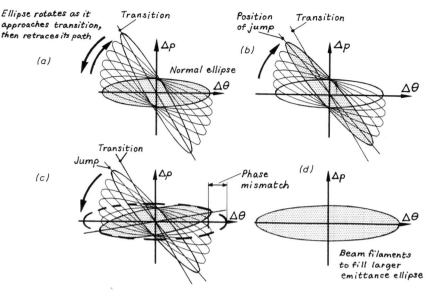

Figure 7.14. Mistiming transition. (*a*) Correct timing, (*b*) jump is made too early, (*c*) ellipse retraces its path but overshoots, (*d*) subsequent filamentation.

mistiming. The earlier phase equation is still valid for the 'error' region between transition and the phase shift.

$$\frac{d^2}{dt^2}\Delta\theta - t^{-1}\frac{d}{dt}\Delta\theta + at\,\Delta\theta = 0. \tag{7.76}$$

If the switch in phase is made too late then t is positive and a negative in the 'error' region, whereas if the switch is made too early the contrary is true. In either case, it is convenient to write the solution in terms of modified Bessel functions and to use a form valid on both sides of transition.

$$\Delta\theta = |t|[A''I_{2/3}(\tfrac{2}{3}\sqrt{|at^3|}) + B''I_{-2/3}(\tfrac{2}{3}\sqrt{|at^3|})]. \tag{7.77}$$

The procedure is to take a particle characterised by the two constants A'' and B'' and match the solution to the upper and lower edges of the time interval Δt to eliminate A'' and B'', and find A' and B'. With A' and B' the trajectory on the upper side of transition is known. The result will depend on Δt and hence a tolerance can be determined. Numerical calculations show that this tolerance is easy to satisfy. For instance, the timing tolerance for the CERN PS came out around 1 ms.

Non-linearities in the momentum compaction can cause a spread in the transition energy inside a bunch, which may also be of the order of the above tolerance[7].

163

(ii) *Longitudinal space-charge forces*[8]

So far only single particle dynamics have been considered, that is mutual forces between the beam particles have been neglected. When beam intensities become high this assumption is no longer valid. The self space-charge forces in a beam act in all directions, but only the longitudinal force inside a bunch will be considered here. This force is always repellent. It therefore counteracts the rf focusing force below transition and enhances the focusing above transition. Figure 7.15 shows the equilibrium bunch length with and without space charge forces as transition is crossed.

Consider a bunch which starts in an equilibrium (non-oscillating) condition far below transition. Due to space-charge forces its equilibrium bunch length will be larger than if there were no space charge. Outside the region around transition, the conditions change slowly in the machine compared with the period of the synchrotron oscillations and the bunch is able to adjust itself adiabatically to the changing conditions thus maintaining itself in equilibrium. At transition the synchrotron frequency goes to zero and all motion is 'frozen', so that just above transition the bunch will emerge with the larger bunch length left over from the influence of the space charge forces below transition. The true equilibrium bunch length is now even shorter than that without space charge and the bunch will start to oscillate (see Figure 7.16). The symmetry that was so important for a clean passage through transition has been broken.

Several remedies have been proposed and studied, the most successful of which has been the so-called γ_{tr}-*jump*[9]. A set of quadrupoles, with both focusing and defocusing lenses suitably located in the lattice, is pulsed at transition in such a way that γ_{tr} is *rapidly* reduced while holding the tunes constant, as shown in Figure 7.17. In the CERN PS this method is used daily to jump transition by several GeV within the order of one millisecond.

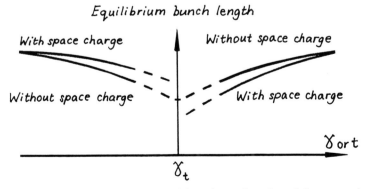

Figure 7.15. Equilibrium bunch length as a function of time assuming linear space charge forces (based on CERN/MPS/DL 73-9/Rev, March 1974).

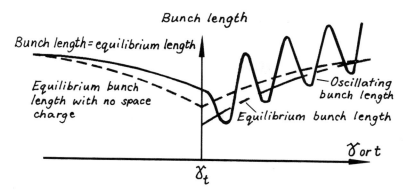

Figure 7.16. Bunch length oscillations as a function of time assuming linear space charge forces (based on CERN/MPS/DL 73-9/Rev, March 1974).

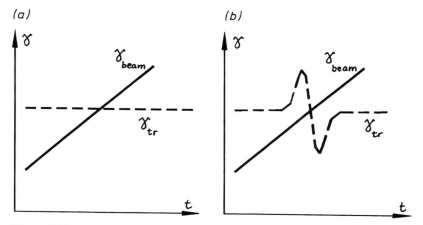

Figure 7.17. Speeding up transition crossing with a γ_{tr}-jump. (a) Without γ_{tr}-jump, (b) with γ_{tr}-jump (based on CERN/MPS/DL 73-9/Rev, March 1974).

7.7.5 Some practical examples

For proton synchrotrons it is often difficult and expensive to make the transition energy appear outside the energy range of the accelerator. It is normally not particularly difficult to pass transition, but the longitudinal dynamics is profoundly influenced by the features related with transition and the choice of parameters must be handled with care. A few examples can illustrate this.

In the original CERN PS design, the care taken for satisfying the phase jump timing when passing transition proved to be adequate to avoid difficulties at the relatively low beam intensities then achieved. The passing of transition became more of a problem as the intensity was increased and when, later still, the preparation of the PS as injector to the ISR imposed stricter demands on beam quality. Appropriate remedies were found as mentioned above. The AGS at Brookhaven was very similar, except that to date it has not faced the additional requirement of acting as an injector for a collider. The CERN SPS also crosses transition in its normal operation as a fixed-target accelerator and the difficulties encountered have been negligible. However, when the SPS operates in its collider mode, transition is lowered and the injection energy is raised to be always above transition. The same strategy is followed for the superconducting ring at Fermilab, which also operates as a proton–antiproton collider. The CERN ISR always operated above transition in its collider mode. Most other collider designs also aim at staying away from transition. For example, in the chain of proton accelerators in the DESY HERA project, the first injector ring (7.5 GeV PETRA III) has transition above its maximum energy, the second proton injector ring (40 GeV PETRA II) and the main proton ring (HERA) both have their injection energies above transition. Similarly the projects UNK (Russia), LHC (CERN), SSC (USA) and ELOISATRON (World Lab.) all plan their injection energies to be above transition. The RHIC project at Brookhaven is an exception. In this project, the main purpose is to collide relativistic heavy ions and it is difficult to avoid passing transition. The proposed superconducting Fermilab Main Injector, which will enable the luminosity in the Tevatron p–p̄ collider to be increased, will also cross transition.

In short, the transition energy does not cause grave difficulties for normal accelerators, whereas the more extreme requirements on beam quality for colliders makes it desirable to design the transition energy outside the operational range of these machines.

7.7.6   Special use of transition energy

It is interesting to look at what happens at transition when two different particle species with very similar mass, m, to charge, q, ratios are sitting in the same synchrotron (see Figure 7.18). This means that they have the same rf and therefore

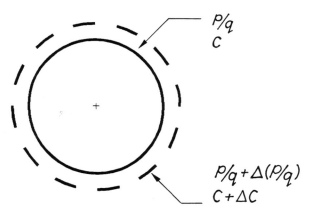

Figure 7.18. Orbits with particles of different species.

also the same revolution frequency and that their equilibrium orbits must differ by so little that they are both inside the vacuum chamber, i.e.

$$\Delta\Omega = 0 \quad \text{and} \quad \Delta C/C \ll 1$$

and the problem can be handled by differentiation.

$$\Omega/2\pi = \beta c/C$$

$$\frac{\Delta\Omega}{\Omega} = \frac{\Delta\beta}{\beta} - \frac{\Delta C}{C} = 0. \tag{7.78}$$

The two species of particles are distinguished by their slightly different β values.

$$\frac{p}{q} = \frac{m}{q} \frac{\beta c}{\sqrt{(1-\beta^2)}}$$

$$\frac{\mathrm{d}(p/q)}{p/q} = \frac{\mathrm{d}(m/q)}{m/q} + \mathrm{d}\left[\ln\frac{\beta}{\sqrt{(1-\beta^2)}}\right]. \tag{7.79}$$

The last term is the usual term present when accelerating a single species and was given in (7.25), so that

$$\frac{\mathrm{d}(p/q)}{p/q} = \frac{\mathrm{d}(m/q)}{m/q} + \gamma^2 \frac{\mathrm{d}\beta}{\beta}. \tag{7.80}$$

The substitution of (7.80) into (7.78) gives,

$$\frac{\Delta C}{C} = \frac{\Delta\beta}{\beta} = \gamma^{-2}\left[\frac{\Delta(p/q)}{p/q} - \frac{\Delta(m/q)}{m/q}\right].$$

From the definition of momentum compaction,

$$\frac{\Delta(p/q)}{p/q} = \alpha^{-1}\frac{\Delta C}{C} = \gamma_{\mathrm{tr}}^2 \frac{\Delta C}{C}$$

so consequently, the orbit separation for the two species becomes,

$$\frac{\Delta C}{C} = \frac{1}{\gamma_{\mathrm{tr}}^2 - \gamma^2} \frac{\Delta(m/q)}{m/q}. \tag{7.81}$$

There are two interesting examples of the possible uses of this equation. The first was proposed by Van der Meer[10] to investigate a possible mass difference between protons and antiprotons. In this case, the two charges are assumed equal but of opposite sign, so that (7.81) takes the form,

$$\frac{\Delta C}{C} = \frac{1}{\gamma_{\mathrm{tr}}^2 - \gamma^2} \frac{\Delta m}{m}. \tag{7.82}$$

If there is a mass difference, then there will be an enhanced orbit separation near transition. This was proposed as an experiment in the CERN SPS, but detailed analysis led to the conclusion that other methods were more accurate.

The second example[11,12] shows how this phenomenon can be used to produce a very pure beam of a single species. There is a minute difference in m/q for O^{8+} and S^{16+},

$$\frac{\Delta(m/q)}{m/q} = 0.54 \times 10^{-3}.$$

This was exploited in the CERN PS, where both species of ions have been accelerated, for driving out the O^{8+} ions that were contaminating the beams of S^{16+} and vice versa.

7.8   GENERAL REMARKS

This chapter dealt with the basic theory of phase stability and phase oscillations inside the stable region (rf bucket). The discovery of phase stability, due to McMillan and Veksler (see Section 1.2.5), led to the invention of the synchro-cyclotron and synchrotron. These were weak-focusing machines at that time, but the theory for the longitudinal motion remains essentially unchanged for alternating-gradient focusing and for linear accelerators. The main difference lies with the transition energy, which for weak-focusing rings is in the non-physical region below the rest energy and for alternating-gradient rings moves into the physical region. This complicates the theory and requires special care to be taken in the machine design for either crossing this singularity, or for choosing the parameters such that it is well outside the operational range. The special case of a linear accelerator can be treated by going to the limit of infinite radius, so that $\gamma_{\text{tr}} \to \infty$.

Alternating-gradient focusing is often called strong focusing, since the betatron oscillations are reduced in amplitude compared to those in weak-focusing machines (amplitude $\propto \sim 1/\sqrt{Q}$). The radial motion associated with the momentum excursions due to the phase oscillations, for instance, are even more compressed in alternating-gradient machines ($\Delta x \propto \sim 1/Q^2$). It was normal in weak-focusing machines for the transverse limits of the rf buckets to extend beyond the physical vacuum chamber, whereas in alternating-gradient rings the radial extent of a bucket is typically a fraction of the physical aperture. This particular feature was made use of in the ISR for stacking hundreds of injected pulses side by side in the momentum space.

For the betatron motion, the starting point is a linear equation and non-linearities are usually treated as perturbations leading to a general transverse stability limitation called the *dynamic aperture*. For phase oscillations, the starting equations are already non-linear, which naturally led in Section 7.5.2 to an analytical derivation of the extent of the stable region. This may be considered as an analogy to the transverse dynamic aperture. However, the longitudinal stability region is easier to determine because the non-linearity is simple and well defined, contrary to the transverse case where the non-linearities come largely from complicated field errors. This can be expressed semi-quantitatively that the longitudinal motion involves $1\frac{1}{2}$ degrees of freedom against $2\frac{1}{2}$ for the transverse motion (the time dependence of the Hamiltonian is said to add a $\frac{1}{2}$ degree of freedom).

Errors and disturbances also play a role in the longitudinal motion, often leading to growth of the effective bunch area (due to filamentation from non-linearities) and to particles being spilled over the edge of the bucket or hitting the aperture. A detailed treatment of this is considered to be outside the scope of the book, but the basic theory can be found in Refs 13 and 14. The early weak-focusing synchrotrons took their rf frequency directly from a frequency programme and the noise in the rf system caused a steady and substantial beam loss during the acceleration. Already in the early alternating-gradient machines, this was effectively suppressed by feed-back (phase lock and radial control from the beam, see Section 11.6.1), but in certain colliders, like the ones for protons and antiprotons, it was desirable to operate without such feedback. Fortunately the development of low-noise electronics over the last few decades, has made it possible to do this and to rely solely on a frequency programme without suffering from noise-induced degradation of the beam quality. For most machines, however, the use of feedback from beam signals continues to be standard practice.

CHAPTER 8

Image and space-charge forces
(transverse)

By the early forties, machine designers understood weak focusing and phase focusing well enough for their attention to be drawn away from single-particle phenomena towards the next limitation on performance, that of space charge. For example, Kerst[1] in 1941 was already discussing the self, or direct, space-charge force inside the beam, which was limiting the injection current and changing the optimum injection voltage into his betatron. Some years later the effects of image space-charge forces arising from the proximity of the conducting vacuum chamber walls were discussed by the MURA staff[2] in 1959. A comprehensive theory of these two mechanisms was published by Laslett[3] in 1963. Subsequent work has taken into account the penetration of a.c. fields through the vacuum chamber walls[4,5], the solution of image coefficients for more complicated boundary conditions[6-8], and some practical applications[9,10]. These ideas have led on naturally to the new fields of study of coherent instabilities and space-charge dominated beams.

8.1   THEORETICAL CONTEXT

The overall scheme for investigating space-charge effects is summarised in Table 8.1. This chapter will concentrate on how the local self and image fields manifest themselves as an incoherent tune shift for the motion of individual particles and as a coherent tune shift for the beam as a whole.

Since the effects of interest are integrated over complete turns in the machine, it is reasonable to build the theory on the smooth or weak approximation for the single-particle unperturbed betatron motion. This also assumes that the vacuum chamber does not have a particularly strange form, or that the optics are particularly exotic. The influence of the other particles in the beam on the test particle is then added as a transverse Lorentz force on the righthand side of (8.1). The form of (8.1) implies that the beam is uniform and parallel to the boundary wall.

Table 8.1. *Overall scheme for investigating transverse space-charge effects*

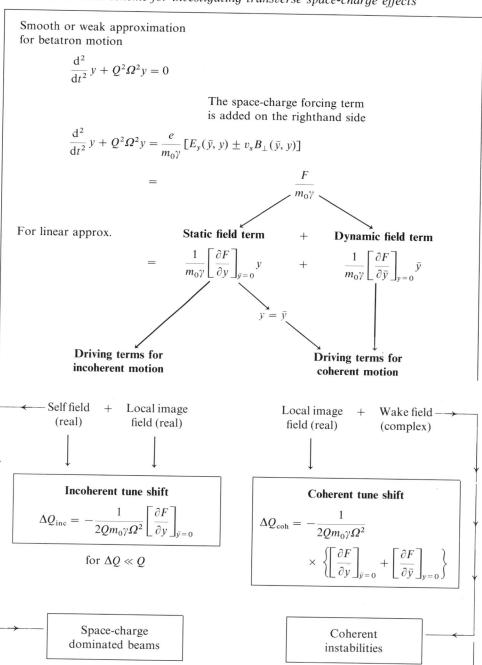

$$\underbrace{\frac{\mathrm{d}^2}{\mathrm{d}t^2}\, y + Q^2\Omega^2 y}_{} = \underbrace{e[E_y(\bar{y}, y) \pm v_s B_\perp(\bar{y}, y)]}_{} = F/m_0\gamma \qquad (8.1)$$

Unperturbed   Space-charge forcing term, which
test particle   depends on test particle's position
motion y and the average beam position $\bar{y}$

where the upper sign $(+)$ corresponds to $y = x$, $B_\perp = B_z$ and the lower sign $(-)$ to $y = z$, $B_\perp = B_x$. The suffix 'y' has been omitted from Q and F for brevity.

Depending on which contributions to F are considered, the analysis develops towards:

- **self fields** (incoherent tune shift, dominant at low energy)
- **local image fields** (incoherent and coherent tune shifts)
- **wake fields** (coherent transverse instabilities).

The function F can be expanded to first order in terms of the test particle's motion y and the average beam position $\bar{y}$ to give,

$$\frac{\mathrm{d}^2}{\mathrm{d}t^2}\, y + Q^2\Omega^2 y = \frac{1}{m_0\gamma} \left\{ \left[\frac{\partial F}{\partial y}\right]_{\bar{y}=0} y + \left[\frac{\partial F}{\partial \bar{y}}\right]_{y=0} \bar{y} \right\}. \qquad (8.2)$$

When $\bar{y}(t) = 0$ the beam and its associated fields are static and $(\partial F/\partial y)$ will be a real constant. This will be clear from the examples given later for the self and image fields. This term can therefore be assimilated into the lefthand side as an additional focusing term and it will cause a tune shift. This is the *incoherent tune shift*.

$$\frac{\mathrm{d}^2}{\mathrm{d}t^2}\, y + \left\{ Q^2\Omega^2 - \frac{1}{m_0\gamma}\left[\frac{\partial F}{\partial y}\right]_{\bar{y}=0}\right\} y = 0 \qquad \text{(for } \bar{y} = 0\text{)}. \qquad (8.3)$$

The new incoherent tune of the test particle is $Q_{\mathrm{inc}} = (Q + \Delta Q_{\mathrm{inc}})$, so that for small tune shifts,

$$Q_{\mathrm{inc}}^2 \simeq Q^2 + 2Q\,\Delta Q_{\mathrm{inc}} \qquad (8.4)$$

and

$$\Delta Q_{\mathrm{inc}} = -\frac{1}{2Qm_0\gamma\Omega^2}\left[\frac{\partial F}{\partial y}\right]_{\bar{y}=0}. \qquad (8.5)$$

In this linear approximation, $\bar{y}$ need not in fact be zero, but only constant in time, i.e. a static closed orbit distortion does not affect ΔQ_{inc}.

When $\mathrm{d}\bar{y}/\mathrm{d}t \neq 0$, the beam and its associated fields are time-varying and $(\partial F/\partial \bar{y})$ (for $y = 0$) will in general be complex, i.e. in addition to the real local image a phase factor will appear to account for local resistive wall effects and the wake fields left by other particles at other azimuthal positions in the accelerator. Only the real part will be considered here. The inclusion of the imaginary part leads to the complex coupling impedance, which is discussed in Chapter 9.

The coherent motion can be solved by choosing $y = \bar{y}$ in (8.2), i.e. the hypothetical particle at the beam centre with zero betatron amplitude, and making this substitution

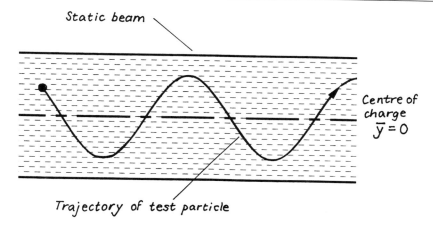

Figure 8.1. Incoherent motion of a test particle.

in (8.2) to give,

$$\frac{\mathrm{d}^2}{\mathrm{d}t^2}\,\bar{y} + \left(Q^2\Omega^2 - \frac{1}{m_0\gamma}\left\{\left[\frac{\partial F}{\partial y}\right]_{\bar{y}=0} + \left[\frac{\partial F}{\partial \bar{y}}\right]_{y=0}\right\}\right)\bar{y} = 0. \tag{8.6}$$

Providing the driving terms are small, the *coherent tune shift*, which is related to the average motion of the whole beam, can be written in the same way as the incoherent tune shift to give,

$$\Delta Q_{\text{coh}} = -\frac{1}{2Qm_0\gamma\Omega^2}\left\{\left[\frac{\partial F}{\partial y}\right]_{\bar{y}=0} + \left[\frac{\partial F}{\partial \bar{y}}\right]_{y=0}\right\}. \tag{8.7}$$

This coherent motion is called the *rigid dipole mode*. Higher order modes also exist such as the quadrupole and sextupole modes, which act on the beam's cross-sectional form. Figures 8.1 and 8.2 illustrate the incoherent and coherent oscillations.

8.2   COMPONENTS OF THE SPACE-CHARGE FORCE

8.2.1   Self fields

Consider first an elliptical beam in real space (see Figure 8.3). This is a standard model and the fields on the z-axis of the ellipse, inside the beam are:

$$E_z(x=0) = \frac{\lambda}{\varepsilon_0\pi b(a+b)}\,z \tag{8.8}$$

$$B_x(x=0) = \frac{\mu_0\lambda v_s}{\pi b(a+b)}\,z \tag{8.9}$$

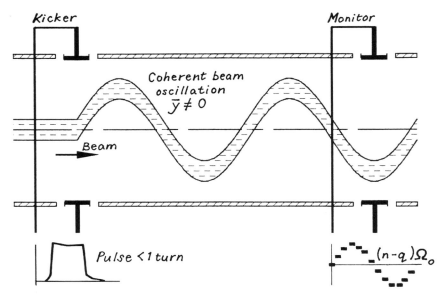

Figure 8.2. Excitation of a coherent oscillation by a kicker.

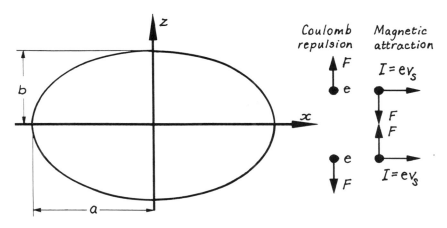

Figure 8.3. Elliptical beam.

where λ is the line charge density, a and b are the beam half width and height respectively and ε_0 and μ_0 are the permittivity and permeability of free space. The fields on the x-axis are similar with a sign change for B_z.

Equations (8.8) and (8.9) can be regarded as either the exact solutions for a uniform charge distribution[11], or the linear terms from the series solution for a bi-Gaussian charge distribution with half widths a and b equal to the standard deviation in the appropriate plane times the square root of two[12], i.e. $a, b = \sqrt{2}\,\sigma_{x,z}$.

174

Thus, for its vertical motion, a test particle will experience a Lorentz force due to all the other particles in the beam, of

$$F_z = e[E_z - v_s B_x] = e(1 - \beta^2) \frac{\lambda}{\varepsilon_0 \pi b(a + b)} z \qquad (8.10)$$

where

$\beta = v_s/c$ (note the relativistic use of β) and

$c = (\mu_0 \varepsilon_0)^{-\frac{1}{2}}$, the velocity of light.

An important feature of (8.10) is the internal compensation seen in the $(1 - \beta^2)$ term of the electric field by the magnetic field, which reduces the self force F to zero as $\beta \to 1$. For this reason self-field effects normally become unimportant at high energies and often it is only necessary to consider them at injection, but there is an exception to this when the beam is neutralised (see Section 8.3.2).

This model is simple and consequently is often used, although others exist for more realistic beams. Reference 13, for example, derives the incoherent tune shifts for the rather general case of a wide stack made from a series of elliptical beams with Gaussian distributions.

8.2.2   Image fields

The presence of a boundary distorts the self space-charge field. The strategy is to express this distortion as a difference term, which when added to the unperturbed self field gives the final perturbed field. Since an accelerator environment is characterised by conducting walls and ferromagnetic pole faces, often with simple parallel-plate, circular or elliptical geometries, it is natural to try and solve these cases using the technique of images. In turn, the image technique has the advantage of giving the final field as a superposition of the original unperturbed field and the fields from the images. Thus the self and image contributions are naturally separated as required by the above strategy. Concentrating now on the image contribution to the field (i.e. the field contribution from the boundaries), it is noticeable that the distances between a beam and its images are in general much greater than the transverse beam size. It is therefore usual and convenient to approximate the beam and its images by line charges or currents when calculating the image contributions (see Figure 8.4). An example calculation is given in Section 8.4.2.

8.2.3   Boundary conditions and coefficients

(i) *For incoherent tune shifts*

This case is straightforward since all the fields are static and the force F in (8.2) will reduce to the single term $[(\partial F/\partial y)y]$ (for $\bar{y} = 0$). The contributions to this term will

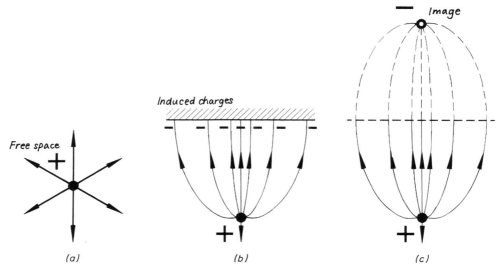

Figure 8.4. Effect of an infinite conducting plane on the electric field around a line charge and the image reconstruction. (a) Self field, (b) effect of an infinite conducting plane, (c) image charge reconstruction.

come from the self field, an electric image in the vacuum chamber and a magnetic image in the magnet pole faces. For the moment, the fields will be expressed in a general, linearised form by the introduction of self-field and image-field coefficients, $\varepsilon_{n,y}$, which contain the information about the geometry and the images. In the case of the vertical motion,

$$E_z(x = 0) = \frac{\lambda}{\varepsilon_0 \pi} \left[\underbrace{\frac{1}{b(a + b)} z}_{\{\text{self field}\}} + \underbrace{\frac{1}{h^2} \varepsilon_{1,z} z}_{\{\text{electric image in chamber}\}} \right] \qquad (8.11)$$

$$B_x(x = 0) = \frac{\lambda}{\varepsilon_0 \pi} \left[\underbrace{\frac{\beta^2}{b(a + b)} z}_{\{\text{self field}\}} + \underbrace{\frac{\beta^2}{g^2} \varepsilon_{2,z} z}_{\{\text{magnetic image in pole face}\}} \right] \qquad (8.12)$$

where, h is the half height of the vacuum chamber and g is the half gap of the magnet poles.

The components of the Lorentz force $(\partial F/\partial y)$ $(\bar{y} = 0)$ can be calculated from (8.11) and (8.12) and are quoted in Table 8.2 for both planes. Of course, the coefficients $\varepsilon_{n,y}$ may not be easy to calculate. One example is given in Section 8.4.2 and the known cases are quoted at the end of this chapter in Table 8.6. However, once these coefficients have been found, it is possible to calculate the space-charge tune shifts in a machine by combining the coefficients in a simple manner.

176

Table 8.2. *Boundary conditions and field coefficients for the incoherent tune shift*

Field source	Boundary condition	Components in $(\partial F/\partial y)_{\bar{y}=0}$
Self	Free space	$e\left\{\left[\dfrac{\partial E_y}{\partial y}\right]_{\bar{y}=0} \pm v_s\left[\dfrac{\partial B_\perp}{\partial y}\right]_{\bar{y}=0}\right\}_{\text{self}}$
		$= \varepsilon_{0,y}\left[\dfrac{\lambda e}{\pi\varepsilon_0}\right]\dfrac{1}{\gamma^2 b^2}$
d.c. electric image	$E_\parallel = 0$ on vacuum chamber	$e\left[\dfrac{\partial E_y}{\partial y}\right]_{\bar{y}=0} = \varepsilon_{1,y}\left[\dfrac{\lambda e}{\pi\varepsilon_0}\right]\dfrac{1}{h^2}$
d.c. magnetic image	$B_\parallel = 0$ on magnet poles	$ev_s\left[\dfrac{\partial B_\perp}{\partial y}\right]_{\bar{y}=0} = \varepsilon_{2,y}\left[\dfrac{\lambda e}{\pi\varepsilon_0}\right]\dfrac{\beta^2}{g^2}$

Note. y can be x or z and h and g change according to the plane. It is customary, however, to quote separate coefficients for the self field $\varepsilon_{0,x} = [(a/b)(1 + a/b)]^{-1}$ {horizontal} and $\varepsilon_{0,z} = (1 + a/b)^{-1}$ {vertical}.

(ii) *For coherent tune shifts*

This case is more complicated since $\bar{y}$ is varying in time, due to the coherent oscillation of the beam, and the boundary conditions depend upon whether the oscillating field of the beam is of a low enough frequency to penetrate the vacuum chamber and to reach the magnet poles, or not. In practice, a metallic vacuum chamber will always screen the electric field, but for the magnetic field all cases from full penetration to non-penetration are likely to occur. The two extremes will be dealt with here, and Zotter has treated the partial penetration of the magnetic field in Ref. 5. The expansion of the function F now contains both derivatives [see (8.6)], i.e.

$$F = \left\{\left[\frac{\partial F}{\partial y}\right]_{\bar{y}=0} + \left[\frac{\partial F}{\partial \bar{y}}\right]_{y=0}\right\}\bar{y}.$$

In an analogous way to that used for the incoherent tune shift, the fields can be expressed in terms of the image coefficients already introduced and four new ones $\xi_{n,y}$. The components of the force and the boundary conditions are summarised in Table 8.3.

In the second part of the table for non-penetrating fields, the magnetic image is decomposed into its d.c. part, which is bounded on the poles, and its a.c. part, which is bounded on the vacuum chamber. The d.c. part takes the same form as in Table 8.2, but the a.c. part is limited by induced currents in the vacuum chamber, so that the boundary condition is $B_\perp = 0$. A comparison of the scalar Poisson

Table 8.3. *Boundary conditions and field components for the coherent tune shift*

Field source	Boundary condition	Components in $(\partial F/\partial y)_{\bar{y}=0} + (\partial F/\partial \bar{y})_{y=0}$
	Penetrating magnetic fields	
Electric image	$E_{\parallel} = 0$ on vacuum chamber	$e\left\{\left[\dfrac{\partial E_y}{\partial y}\right]_{\bar{y}=0} + \left[\dfrac{\partial E_y}{\partial \bar{y}}\right]_{y=0}\right\}$ $= \xi_{1,y}\left[\dfrac{\lambda e}{\pi\varepsilon_0}\right]\dfrac{1}{h^2}$
Magnetic image	$B_{\parallel} = 0$ on magnet poles	$ev_s\left\{\left[\dfrac{\partial B_{\perp}}{\partial y}\right]_{\bar{y}=0} + \left[\dfrac{\partial B_{\perp}}{\partial \bar{y}}\right]_{y=0}\right\}$ $= \xi_{2,y}\left[\dfrac{\lambda e}{\pi\varepsilon_0}\right]\dfrac{\beta^2}{g^2}$
	Non-penetrating magnetic fields	
Electric image	$E_{\parallel} = 0$ on vacuum chamber	$e\left\{\left[\dfrac{\partial E_y}{\partial y}\right]_{\bar{y}=0} + \left[\dfrac{\partial E_y}{\partial \bar{y}}\right]_{y=0}\right\}$ $= \xi_{1,y}\left[\dfrac{\lambda e}{\pi\varepsilon_0}\right]\dfrac{1}{h^2}$
d.c. magnetic image	$B_{\parallel} = 0$ on magnet poles	$ev_s\left[\dfrac{\partial B_{\perp}}{\partial y}\right]_{\bar{y}=0} = \varepsilon_{2,y}\left[\dfrac{\lambda e}{\pi\varepsilon_0}\right]\dfrac{\beta^2}{g^2}$
a.c. magnetic image	$B_{\perp}$ on vacuum chamber	$ev_s\left\{\left[\dfrac{\partial B_{\perp}}{\partial \bar{y}}\right]_{y=0}\right\} = -(\xi_{1,y}-\varepsilon_{1,y})\left[\dfrac{\lambda e}{\pi\varepsilon_0}\right]\dfrac{\beta^2}{h^2}$

Note the electric field is always considered non-penetrating.

equation for electric charges and the vector Poisson equation for the currents, given in (8.13),

$$\left.\begin{aligned} \nabla^2\phi &= -\rho/\varepsilon_0 \quad \text{and} \quad \boldsymbol{E} = \boldsymbol{\nabla}\phi \\ \nabla^2\boldsymbol{A} &= -\mu_0\boldsymbol{j} \quad \text{and} \quad \boldsymbol{B} = \boldsymbol{\nabla} \times \boldsymbol{A} \end{aligned}\right\} \tag{8.13}$$

shows that the electric image coefficients will also apply to this magnetic case with the additional factor of β^2. Thus the component $(\partial F/\partial\bar{y})$ (for $y = 0$) can be expressed using $-(\xi_{1,y} - \varepsilon_{1,y})\beta^2$. The minus sign is needed because the magnetic component of the Lorentz force opposes the electric component.

8.2.4 Tune shifts in a costing beam

The combinations of Tables 8.2 and 8.3 together with equations (8.5) and (8.7) give the incoherent and coherent tune shifts to be expected in a coasting beam. These results are given in equations (8.14) to (8.17) and are collected in Table 8.4. For the combination of these expressions, use was made of:

$\lambda = Ne/(2\pi R)$ where N is the number of particles in the beam,

$\Omega = \beta c/R$ where β has its relativistic definition,

R/Q has been replaced by $\bar{\beta}_y$, the average value of β_y and

$$r_0 = \frac{1}{4\pi\varepsilon_0}\frac{e^2}{m_0 c^2}, \text{ the classical particle radius}$$

$= 1.5 \times 10^{-18}$ m for protons and 2.7×10^{-15} m for electrons.

The comparison of the final equations (8.14), (8.15) and (8.17) in Table 8.4 shows

Table 8.4. *Tune shifts in coasting beams*

Incoherent motion   Eqn. (8.14)

$$\Delta Q_{\text{inc}} = -\left[\frac{Nr_0}{\pi\beta^2\gamma}\bar{\beta}_y \right]\left[\frac{1}{\gamma^2}\frac{\varepsilon_{0,y}}{b^2} + \frac{\varepsilon_{1,y}}{h^2} + \beta^2\frac{\varepsilon_{2,y}}{g^2} \right]$$

  Beam   Self field   Electric image in   Magnetic image
  parameters   vacuum chamber   in magnet poles

Coherent motion. Penetrating fields. 'Integer formula'   (Eqn. (8.15)

$$\Delta Q_{\text{coh}} = -\left[\frac{Nr_0}{\pi\beta^2\gamma}\bar{\beta}_y \right]\left[\frac{\xi_{1,y}}{h^2} + \beta^2\frac{\xi_{2,y}}{g^2} \right]$$

  Beam   Electric image in   Magnetic image
  parameters   vacuum chamber   in magnet poles

Coherent motion. Non-penetrating fields. 'Half-integer formula'   Eqn. (8.16)

$$\Delta Q_{\text{coh}} = -\left[\frac{Nr_0}{\pi\beta^2\gamma}\bar{\beta}_y \right]\left[\frac{\xi_{1,y}}{h^2} + \beta^2\frac{\varepsilon_{2,y}}{g^2} - \beta^2\frac{(\xi_{1,y} - \varepsilon_{1,y})}{h^2} \right]$$

  Beam   electric image in   d.c. magnetic image   a.c. magnetic image
  parameters   vacuum chamber   in magnet poles   in vacuum chamber

Rearranging,   Eqn. (8.17)

$$\Delta Q_{\text{coh}} = -\left[\frac{Nr_0}{\pi\beta^2\gamma}\bar{\beta}_y \right]\left[\frac{1}{\gamma^2}\frac{\xi_{1,y}}{h^2} + \beta^2\frac{\varepsilon_{1,y}}{h^2} + \beta^2\frac{\varepsilon_{2,y}}{g^2} \right]$$

Note the electric field is always considered as non-penetrating.

that a similar form emerges as $\gamma \rightarrow \infty$ and that (8.14) and (8.15) converge to exactly the same equation, since as $\gamma \rightarrow \infty$ and $\beta \rightarrow 1$ the term $\gamma^{-2} = (1 - \beta^2) \rightarrow 0$.

Equation (8.14) applies to the incoherent motion of individual particles and is important for avoiding non-linear resonances in the tune diagram of high-energy machines (see Section 5.4). The most common concern is probably for the self-field effects which arise from the first term in (8.14) and dominate the low-energy performance of an injector chain. It is therefore interesting to examine the first term in isolation, i.e.

$$\Delta Q_{\text{self field}} = -\left[\frac{Nr_0}{\pi\beta^2\gamma} \bar{\beta}_y \right] \frac{1}{\gamma^2} \frac{\varepsilon_{0,y}}{b^2} = -\frac{Nr_0}{\beta^2\gamma^3} \frac{\bar{\beta}_y}{\pi b^2} \varepsilon_{0,y}.$$

In Section 8.2.1, it was stated that the half width and height a and b of a uniform elliptical beam were related to the standard deviations of an equivalent bi-Gaussian distribution by $a, b = \sqrt{2}\, \sigma_{x,z}$. The combination of these relations with the definition of emittance given in (4.31A) and the forms of $\varepsilon_{0,y}$ given in the notes below Table 8.2 lead to two expressions for the self-field tune shift in the two planes which are,

Horizontal plane: $$\Delta Q_{\text{self field}} = -\frac{1}{2} \frac{Nr_0}{\beta^2\gamma^3\varepsilon_x} (1 + b/a)^{-1}$$

Vertical plane: $$\Delta Q_{\text{self field}} = -\frac{1}{2} \frac{Nr_0}{\beta^2\gamma^3\varepsilon_z} (1 + a/b)^{-1}.$$

If the betatron modulation is ignored and a round beam is assumed, these expressions simplify to,

Round beam: $$\Delta Q_{\text{self field}} = -\frac{1}{4} \frac{Nr_0}{\beta^2\gamma^3\varepsilon_y}.$$

The last expression is of interest, since within the approximations made it is independent of the accelerator parameters and in particular of the focusing strength represented by $\bar{\beta}_y$.

Equation (8.15) applies to the coherent motion with fully penetrating fields and is important for changes in the closed orbit as the beam current is increased[14]. It also applies when a beam with a tune value close to an integer is kicked, since the frequency of its motion at a given azimuth will be low. This is the reason for its popular name *integer formula*.

Equation (8.17) applies to non-penetrating fields such as would occur when kicking a beam with a near half-integer tune value. Likewise this gives rise to its popular name *half-integer formula*.

In the conclusion of Ref. 5, Zotter gives the following criterion for deciding whether the fields will penetrate and hence which formula to apply.

If the skin depth, δ, of the vacuum chamber is less than the square root of the product

of the wall thickness, d, and the chamber half-height, h, the fields will be non-penetrating.

i.e. For non-penetration, $\qquad\qquad \delta_{\text{wall}} < (hd)^{\frac{1}{2}}$ $\qquad\qquad\qquad$ (8.18)

where,

$$\delta_{\text{wall}} = (2\rho_{\text{wall}}/\omega\mu_{\text{wall}})^{\frac{1}{2}}$$

$\omega = (n - Q)\Omega$, ρ_{wall} is the chamber's resistivity and μ_{wall} is its permeability.

8.3   BUNCHING, NEUTRALISATION AND PRACTICAL STRUCTURES

8.3.1   Bunching

Equations (8.14), (8.15) and (8.17) apply to storage rings with coasting beams, but space-charge problems first arose with bunched beams and accordingly account is taken of this in the early papers[15,3,4]. This was done by applying a bunching factor Λ, where Λ is the average current divided by the peak current. The idea was to calculate the worst-case conditions at the centre of the bunch.

In this model the self field and the electric field are simply enhanced by a factor $1/\Lambda$. The magnetic field is more complicated because the vacuum chamber acts as a filter, which stops the high frequency a.c. component from the bunching. The magnetic field has therefore now to be split into three parts: the d.c. term, one a.c. term from the transverse motion and a second a.c. term from the longitudinal bunching, which has an amplitude of $(\Lambda^{-1} - 1)$. As has already been explained the first a.c. term may or may not penetrate the vacuum chamber, but the second term from the axial bunching is always considered as non-penetrating as its frequency is high.

$$\text{Frequency of transverse motion} = (n - Q)\Omega$$

$$\text{Bunch frequency} = h\Omega.$$

There can be a factor of several thousand between these two frequencies, the bunch frequency always being the greater. Bunching is included in the equations (8.19) to (8.22).

8.3.2   Neutralisation

Neutralisation is not normally a problem in the bunched beam of an accelerator, since it has little time to build up, but it is of critical importance in storage rings with unbunched beams and appears very early in the corresponding literature[15]. In the case of a storage ring, a failure of the clearing electrode system can cause the loss of the beam. The typical loss mechanism would be an initial tune shift from neutralisation moving the beam edge into a resonance and so causing a beam loss, which then increases the outgassing from the vacuum chamber making the neutralisation and the consequent tune shift worse.

Neutralisation affects essentially only the electric field, since trapped particles may oscillate transversely, but they will only drift slowly longitudinally. The changes in the image terms ε_1 and ξ_1 are rarely important, but the balance of $(1 - \beta^2)$ between the electric and magnetic fields in the self-field term can easily be destroyed. In this way, the self-field term, which tends to disappear as the energy increases, can reappear and, since the beam size will be decreasing adiabatically as the energy increases (for protons), it can reappear strongly.

This effect is included in equations (8.19) and (8.22) by multiplying the electric field by a term $(1 - \eta_e)$, in which η_e is defined as the trapped charge in the beam normalised by the beam charge.

8.3.3   Practical structures

In order to take into account the different vacuum chambers around a machine and the distribution of magnets etc., the tune shifts can be expressed as a summation over the various components. By using the average betatron amplitudes in the localities of each type of component the beam characteristics can also be included. These modifications are implemented in the equations (8.19) to (8.22), which are grouped together in Table 8.5. The factors C_i represent the fractions of the circumference occupied by each type of component.

In Table 8.5 the electric field is always considered as non-penetrating, equally the a.c. magnetic field due to axial bunching is always non-penetrating and only the a.c. magnetic field from the transverse motion can penetrate under certain conditions (see Sections 8.2.4 and 8.3.1).

8.4   EVALUATION OF THE IMAGE COEFFICIENTS

8.4.1   General relationships

From Laplace's equation one finds that,

$$\varepsilon_{1,z} = -\varepsilon_{1,x} \quad \text{and} \quad \varepsilon_{2,z} = -\varepsilon_{2,x}.$$

The coherent image coefficients do not have any equally simple relationships. It is customary practice to divide the self-field term by the vertical beam height b, whereas for the other coefficients h and g represent the dimensions in the plane being considered.

182

Table 8.5. *Tune shifts with neutralisation, bunching and summation*

Incoherent tune shift   Eqn. (8.19)

$$\Delta Q_{\text{inc}} = -\left[\frac{Nr_0}{\pi\beta^2\gamma}\right]\left\{\underbrace{\frac{\bar{\beta}_y}{\Lambda}(\gamma^{-2}-\eta_e)\frac{\varepsilon_{0,y}}{b^2}}_{} + \sum_i C_i\bar{\beta}_{y,i}\left[\underbrace{\frac{(1-\eta_e)}{\Lambda}\frac{\varepsilon_{2,i}}{h_i^2}}_{} + \underbrace{\beta^2\frac{\varepsilon_{2,i}}{g_i^2}}_{} - \underbrace{\beta^2(\Lambda^{-1}-1)\frac{\varepsilon_{1,i}}{h_i^2}}_{}\right]\right\}$$

| Beam parameters | Self field | Electric image | d.c. magnetic image | a.c. magnetic image from axial bunching |

Rearranging,

$$\Delta Q_{\text{inc}} = -\left[\frac{Nr_0}{\pi\beta^2\gamma}\right]\left\{\frac{2\bar{\beta}_y}{\Lambda}(\gamma^{-2}-\eta_e)\frac{\varepsilon_{0,y}}{b^2} + \sum_i C_i\bar{\beta}_{y,i}\left[\left(\frac{(\gamma^{-2}-\eta_e)}{\Lambda}+\beta^2\right)\frac{\varepsilon_{1,i}}{h_i^2} + \beta^2\frac{\varepsilon_{2,i}}{g_i^2}\right]\right\}$$ Eqn. (8.20)

Coherent tune shift with penetrating magnetic field from transverse motion

$$\Delta Q_{\text{coh}} = -\left[\frac{Nr_0}{\pi\beta^2\gamma}\right]\sum_i C_i\bar{\beta}_{y,i}\left[\underbrace{\frac{(1-\eta_e)}{\Lambda}\frac{\xi_{1,i}}{h_i^2}}_{} + \underbrace{\beta^2\frac{\xi_{2,i}}{g_i^2}}_{} - \underbrace{\beta^2(\Lambda^{-1}-1)\frac{\xi_{1,i}}{h_i^2}}_{}\right]$$

| Electric image in vacuum chamber | d.c. magnetic image in magnet poles | a.c. magnetic image from axial bunching |

Rearranging

$$\Delta Q_{\text{coh}} = -\left[\frac{Nr_0}{\pi\beta^2\gamma}\right]\sum_i C_i\bar{\beta}_{y,i}\left[\left\{\frac{(\gamma^{-2}-\eta_e)}{\Lambda}+\beta^2\right\}\frac{\xi_{1,i}}{h_i^2} + \beta^2\frac{\xi_{2,i}}{g_i^2}\right]$$ Eqn. (8.21)

Coherent tune shift with non-penetrating a.c. magnetic fields

$$\Delta Q_{\text{coh}} = -\left[\frac{Nr_0}{\pi\beta^2\gamma}\right]\sum_i C_i\bar{\beta}_{y,i}\left[\underbrace{\frac{(1-\eta_e)}{\Lambda}\frac{\xi_{1,i}}{h_i^2}}_{} + \underbrace{\beta^2\frac{\xi_{2,i}}{g_i^2}}_{} - \underbrace{\beta^2\frac{(\xi_{1,i}-\varepsilon_{1,i})}{h_i^2}}_{} - \underbrace{\beta^2(\Lambda^{-1}-1)\frac{\xi_{1,i}}{h_i^2}}_{}\right]$$

| Electric image in vacuum chamber | d.c. magnetic image in magnet poles | a.c. magnetic images: from transverse motion | from axial bunching |

Rearranging

$$\Delta Q_{\text{coh}} = -\left[\frac{Nr_0}{\pi\beta^2\gamma}\right]\sum_i C_i\bar{\beta}_{y,i}\left[\frac{(\gamma^{-2}-\eta_e)}{\Lambda}\frac{\xi_{1,i}}{h_i^2} + \beta^2\frac{\varepsilon_{1,i}}{h_i^2} + \beta^2\frac{\varepsilon_{2,i}}{g_i^2}\right]$$ Eqn. (8.22)

8.4.2   Parallel-plate geometry

As an example of the calculation of the image coefficients, the parallel-plate geometry is taken here. In Figure 8.5 the beam is off-axis between two parallel conducting plates.

At a point y, the fields are summed from all the images of the line charge $+\lambda$ at the position $\bar{y}$. The field of the source line charge is not included, since this is already accounted for in the self-field term.

$$E_y^{\text{image}} = \frac{\lambda}{2\pi\varepsilon_0}\left[\frac{1}{2h-\bar{y}-y} - \frac{1}{2h+\bar{y}+y} + \frac{1}{4h-\bar{y}+y} - \frac{1}{4h+\bar{y}-y} + \frac{1}{6h-\bar{y}-y} - \cdots\right]$$

$$= \frac{\lambda}{\pi\varepsilon_0}\left[\frac{(\bar{y}+y)}{4h^2-(\bar{y}+y)^2} + \frac{(\bar{y}-y)}{16h^2-(\bar{y}-y)^2} + \frac{(\bar{y}+y)}{36h^2-(\bar{y}+y)^2} + \frac{(\bar{y}-y)}{64h^2-(\bar{y}-y)^2} + \cdots\right].$$

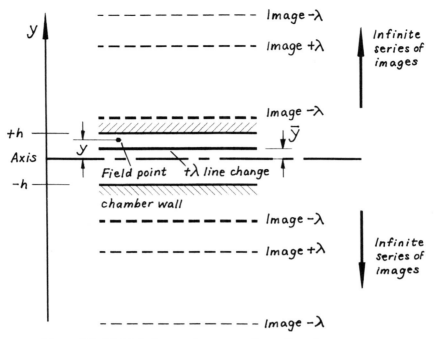

Figure 8.5. Multiple images due to an off-axis line charge.

The expansion of each term in the series and the retention of only the linear terms yields,

$$\simeq \frac{\lambda}{4\pi\varepsilon_0 h^2}\left[(\bar{y}+y)(1+3^{-2}+5^{-2}+\cdots)+\frac{(\bar{y}-y)}{4}(1+2^{-2}+3^{-2}+\cdots)\right]$$

$$=\frac{\lambda}{2\pi\varepsilon_0 h^2}\left[(\bar{y}+y)\frac{\pi^2}{8}+(\bar{y}-y)\frac{\pi^2}{24}\right]$$

$$=\left(\frac{\pi^2}{48}\right)\left(\frac{\lambda}{\pi\varepsilon_0}\right)\left(\frac{1}{h^2}\right)(2\bar{y}+y). \tag{8.23}$$

The comparison of (8.23) and Tables 8.2 and 8.3 yields,

$$\varepsilon_1=\frac{\pi^2}{48} \quad \text{and} \quad \xi_1=\frac{\pi^2}{48}+\frac{\pi^2}{24}=\frac{\pi^2}{16}.$$

These results and others are recorded in Table 8.6 of the next sub-section. The magnetic case with line currents sums in a similar way.

8.4.3   Coefficients for simple geometries

Many of the simple derivations for the image coefficients for centred beams are well described by Laslett[3]. These results are summarised in Table 8.6. For quick

Table 8.6. *Self-field coefficients for uniformly charged beams and image coefficients for centred beams*

Coefficients	Circular	Elliptical	Parallel plate	Comments
		Geometries		
$\varepsilon_{0,x}$ $\varepsilon_{0,z}$	$\frac{1}{2}$ $\frac{1}{2}$	$[(a/b)(1 + a/b)]^{-1}$ $(1 + a/b)^{-1}$	— —	$\Big\}$ Combined electric and magnetic self field
$\varepsilon_{1,x}$	0	$\dfrac{-h^2}{6d^2}[2(2 - k^2)K^2/\pi^2 - 1]$	$\dfrac{-\pi^2}{48}$	
$\varepsilon_{1,z}$	0	$\dfrac{h^2}{6d^2}[2(2 - k^2)K^2/\pi^2 - 1]$	$\dfrac{\pi^2}{48}$	Electric field and a.c. magnetic field with ξ_1
$\varepsilon_{2,x}$	†	‡	$\dfrac{-\pi^2}{24}$	
$\varepsilon_{2,z}$	†	‡	$\dfrac{\pi^2}{24}$	Magnetic field
$\xi_{1,x}$	$\frac{1}{2}$	$\dfrac{h^2}{4d^2}[1 - 4K^2(1 - k^2)/\pi^2]$	0	
$\xi_{1,z}$	$\frac{1}{2}$	$\dfrac{h^2}{4d^2}[4K^2/\pi - 1]$	$\dfrac{\pi^2}{16}$	Electric field and a.c. magnetic field with ε_1
$\xi_{2,x}$	†	‡	0	
$\xi_{2,z}$	†	‡	$\dfrac{\pi^2}{16}$	Magnetic field

† For circular magnet poles see text.
‡ Elliptical magnet poles are not treated in the literature.
In this table $K(k)$ is the complete elliptic integral of the 1st kind. K' is defined as $K(k')$ where $k' = \sqrt{(1 - k^2)}$. k and k' are defined by $\exp(-\pi K'/K) = [(v - h)/(v + h)]$, where v and h are the major and minor axes of the elliptical tube. Finally, $d = \sqrt{(v^2 - h^2)}$.

calculations, it is usually sufficient to represent a structure with just circular and parallel-plate geometries.

Table 8.6 contains some conspicuous omissions, such as the image coefficients $\xi_{2,H}$ and $\xi_{2,V}$ for the case of a circular magnetic yoke, which occurs in superconducting accelerator magnets. It is interesting to note that if $\mu = \infty$, any closed yoke around the beam will force $H_\parallel = 0$ and the beam will not be able to pass through

($\oint H_\parallel = I = 0$). The yoke becomes an infinite inductance. This can be remedied by putting gaps in the yoke. In superconducting magnets, however, the yoke is so saturated that μ approaches unity and the problem disappears. At low fields the permeability of the yoke will be strongly varying around its azimuth. This case has yet to be solved.

Derivations for the coefficients for off-centred beams and beams with finite dimensions in various geometries have been made by Zotter[5-8] and probably the most extensive collection of results has been made by Guignard[16].

8.5   MEASUREMENTS AND APPLICATIONS

The most direct way to measure tune is to kick the beam and to observe the resulting coherent oscillations (see Section 11.2.1). This method measures the coherent tune, but if the current in the beam is progressively reduced then the measured value will converge to the incoherent one. If the incoherent tune has to be measured in the presence of a large circulating beam, then some way has to be found of making only a very small fraction of the beam oscillate, or alternatively a method based on the Schottky scans can be used. These various methods are described in Section 11.2.

As mentioned above, the coherent tune can be measured by observing the coherent oscillation of a beam following a kick, but a large tune spread in the beam will quickly destroy the coherency, so this direct method is not always possible. When kicking a beam, the a.c. magnetic fields may, or may not, be penetrating the vacuum chamber, but one way of measuring with fully penetrating fields is to make measurements on the closed orbit. This is done by measuring the closed orbit distortion with a low intensity bunch and then again in the presence of a high intensity beam. The latter is achieved by passing an empty rf bucket through the stacked beam. The difference in the orbits will be due to the coherent tune shift ΔQ_{coh} for fully penetrating fields. By making a harmonic analysis of both closed orbits and examining the harmonic amplitudes close to the tune value, ΔQ_{coh} can be evaluated[14].

Measured before accumulation, $F_n = \dfrac{Q_{inc}^2}{Q_{inc}^2 - n^2} f_n$

Measured after accumulation, $F_n' = \dfrac{(Q_{inc} + \Delta Q_{coh})^2}{(Q_{inc} + \Delta Q_{coh})^2 - n^2} f_n$

where, F_n and F_n' are the Fourier amplitudes of the closed orbit, f_n are the Fourier amplitudes of the function containing the field errors, which are constant and n is an integer preferably close to Q for maximum sensitivity. The harmonic formulation of the closed orbit and the form of the function f_n have been described in Section 5.1.3.

The CERN ISR provided many examples of applications of the theory exposed in this Chapter. The accumulation of large stacks of tens of amperes for physics was only possible by making regular on-line corrections to the tune values based on detailed calculations of the incoherent space-charge tune shifts. In this way, it was possible to avoid non-linear resonances and to maintain a finite positive tune spread for stability[9,10]. Many of the physics experiments depended for their accuracy on the luminosity measurement, which in turn was based on making precision bumps in the closed orbits of the beams in the crossing regions[17]. The measurement of the coherent tune shift was important for the precision of these bumps and was used for determining one of the correction factors[14]. Reference 18 gives an interesting account of the investigation of a new working line in the tune diagram for the ISR. This line avoided certain resonances known as *rf overlap knockout resonances*[19] and had an intrinsically high stability against coherent transverse resistive-wall instabilities, but the coherent tune shift seen by the injected beam due to the influence of the nearby stack, so perturbed the orbit that injection became impossible with more than 22.8 A circulating in the ring. This promising line was never used operationally for this reason.

CHAPTER 9

Coherent instabilities

The first goal of any accelerator project is to reach the design energy. This battle is usually won rather quickly, only to be replaced by a longer struggle to reach the highest intensity possible. This latter phase can last for the whole of the accelerator's working life and, for the most part, will end up as a struggle against coherent instabilities. Both theoretically and experimentally the study of coherent instabilities is too advanced to be covered exhaustively in this book, but in line with the extreme importance of this subject an introduction is given in this chapter. More emphasis is put on physical description than the mathematical treatment and the latter is restricted to coasting beams. For more complete analyses, including bunched beams, the reader can refer to Hofmann[1], Zotter and Sacherer[2], Laclare[3] and Chao[4]. Many of the same phenomena were already known in electron tubes, where they were of interest as amplification mechanisms. A good account of the early work in this field can be found in Beck[5]. For those readers interested in the historical aspects, Maxwell[6] had already analysed the stability of a ring of weakly interacting particles in a prize-winning essay on Saturn's rings in 1856.

9.1   GENERAL DESCRIPTION

In all accelerators the beams are surrounded by cavities, pipes, plates, grids, etc., which for the most part are metal. Not surprisingly the surrounding equipment influences the fields around the beam, changing the spatial field pattern and phase with respect to what would exist if the beam were to be in free space. The beam will 'feel' the influence of these modified fields and under certain circumstances can become unstable under the action of their feedback.

The concept of a *beam–environment interaction* was first introduced in Chapter 8 with the analysis of space-charge image forces. The fields, however, were restricted to remaining in phase with either the single-particle motion or the average-beam motion. The results were simple tune shifts for the single-particle motion (incoherent) and the average-beam motion (coherent). Although bunched beams were introduced and some account was taken of the frequency of the oscillations to determine whether fields penetrated thin vacuum chamber walls, the problem was basically analysed with static charges, line currents and their images, so ensuring that the forces were always in phase with the motion.

In general, the fields created by the beam and then modified by the surrounding structure will be far more complex than those described in Chapter 8. In this wider description, they are referred to as *wake fields*. As an example, consider what happens inside a cavity when a beam performs a coherent motion. This can be a longitudinal or transverse oscillation as both illustrate the basic physics involved. The movement of charge will excite fields in the cavity and if the cavity has low losses it will 'ring'. The 'ringing' field, or wake field, will be seen by the upstream portions of the beam as they traverse the cavity shortly after. The consequences will depend on the situation. If the wake field in the cavity has a phase such that it reinforces the original oscillation, then the beam will be unstable. This means that the field around the beam must be 90° out of phase with the beam motion. Figure 9.1 illustrates this with a phase-space diagram and the simplified case of kicks applied to an oscillating system, first at the maximum excursion and then out of phase by 90°. In the first case, the kicks have little effect on the amplitude, but instead tend to affect the average angular velocity in the phase plane (i.e. they cause a tune shift) and in the second case, the kicks affect the amplitude directly causing either growth or damping.

It is understandable that a low-loss cavity can be excited by longitudinal bunching or transverse oscillations of the beam and that the cavity can store appreciable fields, which then apply feedback to the beam. It is less obvious that this effect can be significant in normal vacuum chambers and that it can in some cases become critical for tightly-bunched, high-intensity beams. Nevertheless, designers have learnt to spare no effort to keep changes in the chamber cross-section as small as possible and, when a change is unavoidable, to make it as smooth as possible. Tanks around monitors and even bellows are shielded and flanges are designed to be smooth. Without this

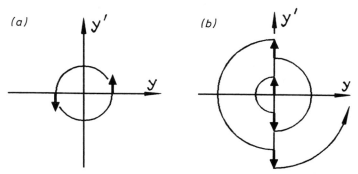

Figure 9.1. In-phase (*a*) and out-of-phase (*b*) excitation of an oscillator.

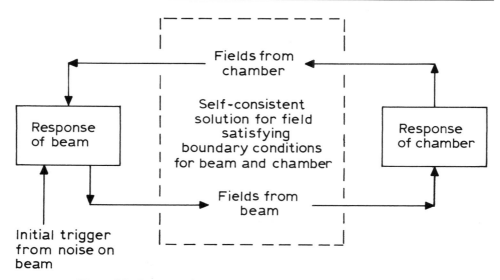

Figure 9.2. Beam–environment interaction in circular machines.

care, spontaneous oscillations or bunching of the beam at critical frequencies would grow out of the background noise and reinforce themselves via their wake fields.

The beam–environment interaction can be represented by the simple diagram of Figure 9.2, in which a disturbance on the beam radiates fields which excite the vacuum chamber. The responses of the chamber creates new fields which act back onto the beam causing a disturbance. If this disturbance enhances the original one, then the oscillation will grow and the solution is self-consistent. If such a solution exists, then the noise on the beam will provide the seed from which this instability can grow and the beam's longitudinal motion will provide the energy.

Figure 9.2 divides naturally into three regions or topics: the *beam*, the *chamber* and the *self-consistent solution for the field* linking the two. These will be discussed in the following sections.

- *Section 9.2* describes and classifies the coherent beam modes. This is the only section to include bunched beams.

- *Section 9.3* explains and compares the concepts of chamber impedance and wake fields using the example of a cavity.

- *Section 9.4* gives an analysis of the longitudinal stability of a coasting beam in a uniform circular vacuum pipe. The circular pipe is a natural choice for an accelerator, but it must be mentioned that it is one of the few cases for which the impedance can be solved analytically. The self-consistent solution is then found using the Vlasov equation.

- *Section 9.5* gives essentially the same analysis for the transverse stability, except that the opportunity is taken to derive the self-consistent solution from the single-particle equation of motion given in Chapter 8, rather than from the Vlasov equation. A version using the Vlasov equation can be found in Ref. 3.

The examples in Sections 9.4 and 9.5 employ a perturbation analysis, in which

- a perturbation on the beam is assumed
- Maxwell's equations are solved for the boundary conditions of the wall and the beam
- self-consistent solutions are sought, for which the field excited by the perturbation in the presence of the chamber matches the original perturbation thus enhancing it and leading to unstable growth.

The primary interest is in the stability threshold, so a perturbation analysis, which applies while the waves are still small, is ideal. As a consequence, the superposition principle holds and each frequency component can be studied separately.

9.2 CLASSIFICATION OF THE COHERENT BEAM MODES[7]

9.2.1 Classification

Before looking closely at the concepts of wake fields and chamber impedance and entering into the mathematical analysis, a phenomenological picture can be built up of the possible coherent modes of oscillation in the beam. The classification divides naturally into coasting beam and bunched beam cases. The former are simpler and are the only cases treated analytically in Sections 9.4 and 9.5.

Figure 9.3 illustrates the families of longitudinal and transverse modes (mode numbers n and k) in coasting beams and a similar summary is made in Figure 9.4 for the modes in bunched beams (mode numbers n, m and k). The bunched-beam situation is more complicated due to the extra boundary condition introduced by the external rf field. The bunched structure supports oscillation modes within the individual bunches, which are similar to the coasting beam case except that the bunch length sets the resonance condition for the standing wave rather than the machine circumference. In addition there are coupled-bunch modes, which are characterised by a definite phase relation between the oscillations from one bunch to the next.

As with all specialised subjects there is a certain amount of jargon. The longitudinal self-bunching is known generically as the *negative mass instability*[8,9]*. The longitudinal

---

* It is believed that the first accelerator paper to specifically use the term 'negative mass' was M. Q. Barton and C. E. Nielsen, *Int. Conf. on High Energy Accel.*, Brookhaven (1961), 163.

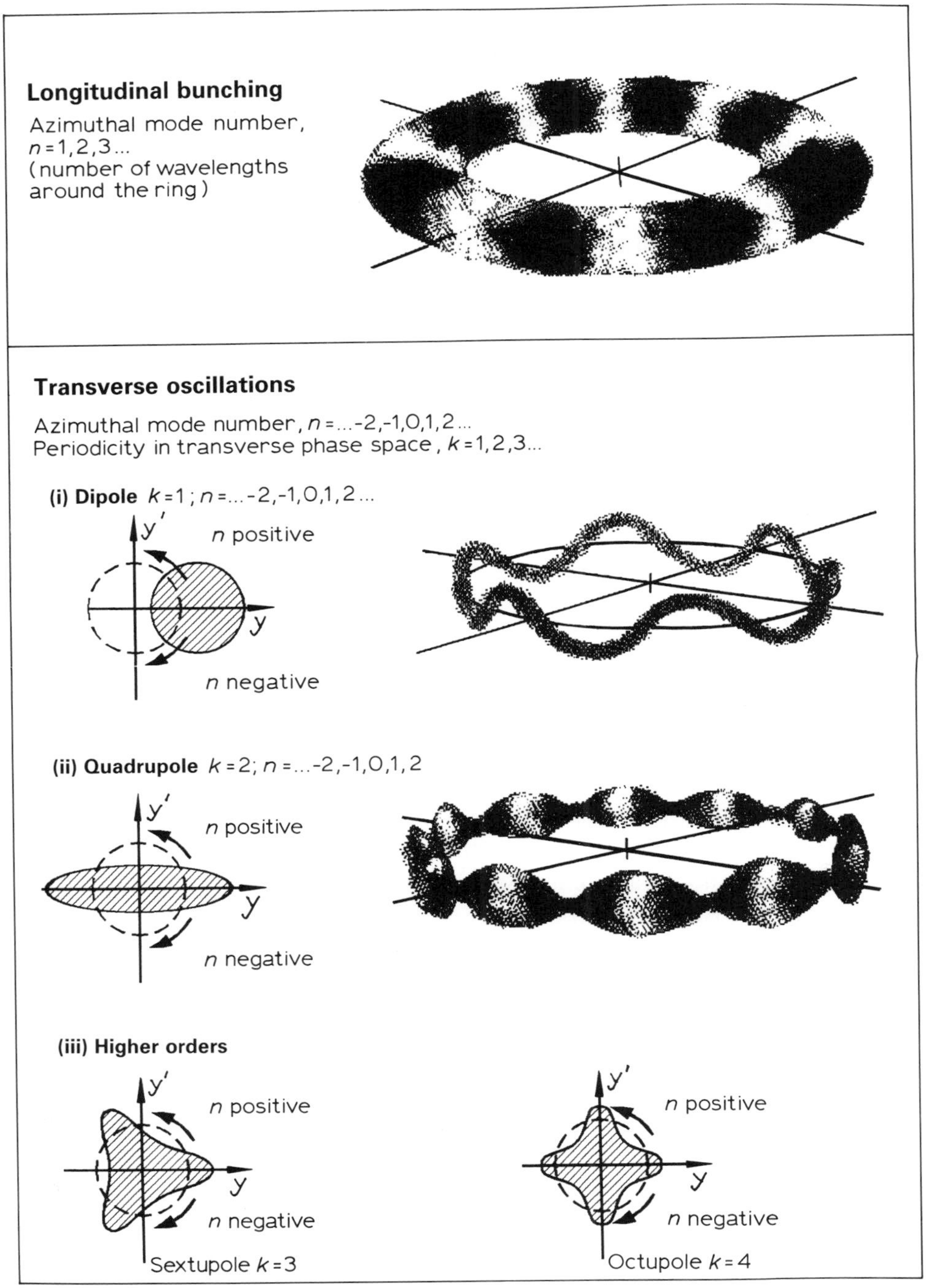

Longitudinal bunching

Azimuthal mode number,
$n = 1, 2, 3 ...$
(number of wavelengths
around the ring)

Transverse oscillations

Azimuthal mode number, $n = ... -2, -1, 0, 1, 2 ...$
Periodicity in transverse phase space, $k = 1, 2, 3 ...$

(i) Dipole $k = 1$; $n = ... -2, -1, 0, 1, 2 ...$

n positive

n negative

(ii) Quadrupole $k = 2$; $n = ... -2, -1, 0, 1, 2$

n positive

n negative

(iii) Higher orders

n positive

n negative

Sextupole $k = 3$

n positive

n negative

Octupole $k = 4$

Figure 9.3. Principal coherent beam modes in a coasting beam.

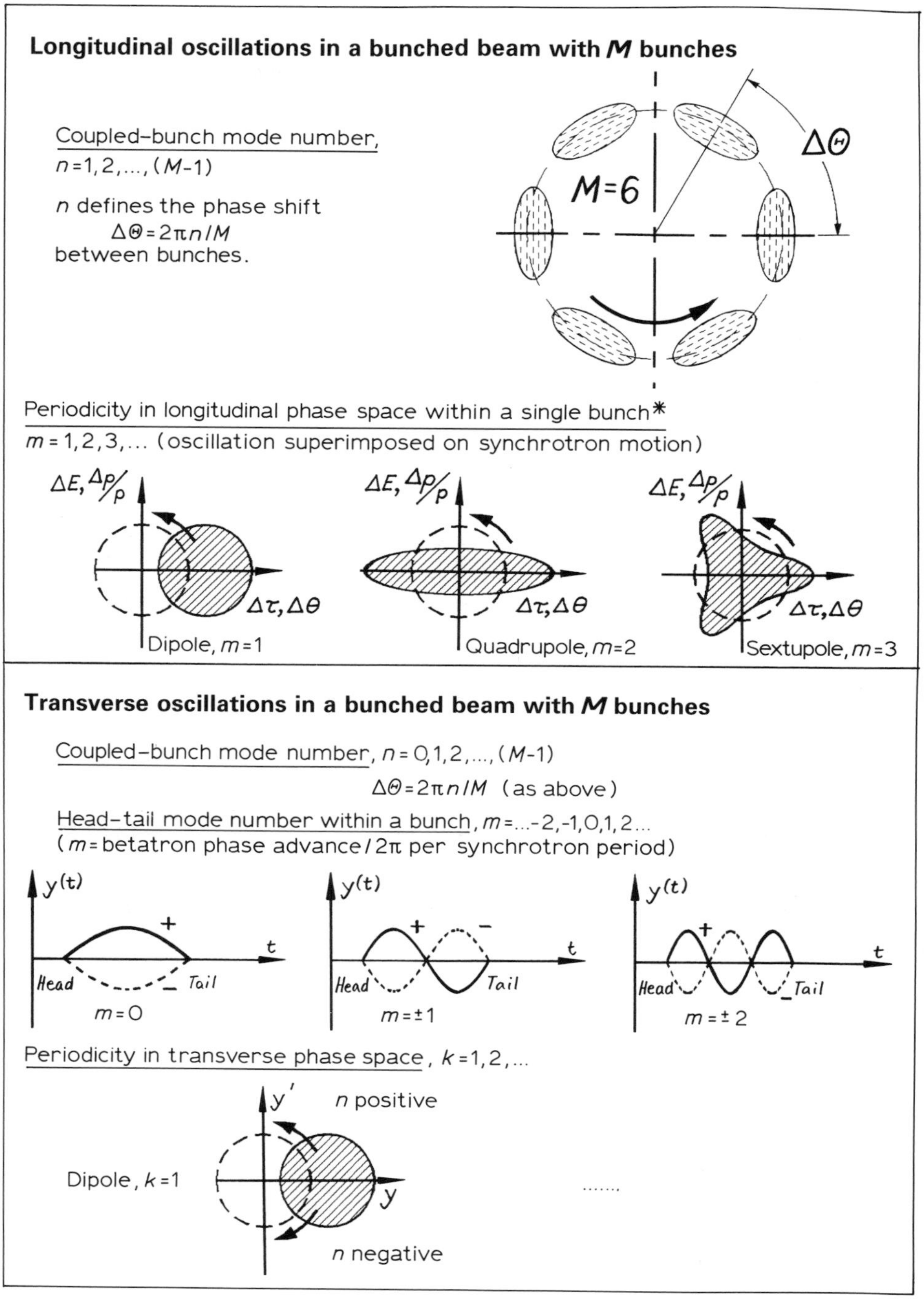

Longitudinal oscillations in a bunched beam with M bunches

Coupled–bunch mode number,
$n = 1, 2, ..., (M-1)$

n defines the phase shift
$\Delta\Theta = 2\pi n / M$
between bunches.

$M = 6$

$\Delta\Theta$

Periodicity in longitudinal phase space within a single bunch *
$m = 1, 2, 3, ...$ (oscillation superimposed on synchrotron motion)

$\Delta E, \frac{\Delta p}{p}$ $\Delta\tau, \Delta\theta$ Dipole, $m = 1$

$\Delta E, \frac{\Delta p}{p}$ $\Delta\tau, \Delta\theta$ Quadrupole, $m = 2$

$\Delta E, \frac{\Delta p}{p}$ $\Delta\tau, \Delta\theta$ Sextupole, $m = 3$

Transverse oscillations in a bunched beam with M bunches

Coupled–bunch mode number, $n = 0, 1, 2, ..., (M-1)$

$\Delta\Theta = 2\pi n / M$ (as above)

Head–tail mode number within a bunch, $m = ..., -2, -1, 0, 1, 2 ...$
(m = betatron phase advance / 2π per synchrotron period)

$y(t)$ Head   Tail $m = 0$

$y(t)$ Head   Tail $m = \pm 1$

$y(t)$ Head   Tail $m = \pm 2$

Periodicity in transverse phase space, $k = 1, 2, ...$

y' n positive

Dipole, $k = 1$ y

n negative

Figure 9.4. Principal coherent beam modes in a bunched beam. The modes marked * degenerate into radial modes[15,16].

193

and transverse instabilities in coasting beams are often known by reference to the wall characteristics, e.g. the *transverse, resistive-wall instability*. The *Robinson instability*[10] is the name given to the simplest bunched-beam instability, which is a dipole motion in phase space caused by the interaction of the bunch with a cavity, usually the rf cavity. The transverse oscillations due to the action of the leading part of a bunch on the trailing part of the same bunch is generically known as the *head–tail instability*[11]. If the wavelength of a perturbation is much shorter than the bunch length, then bunched-beam instabilities behave like those in coasting beams. Historically these very high frequency cases are called *microwave instabilities*.

When the whole beam, or all the bunches in the beam, move in synchronism the motion is known as a *rigid dipole mode*. Quadrupole modes are sometimes known as *breathing modes* and higher order modes as *throbbing modes*.

9.2.2   Fast and slow waves

Positive mode numbers indicate *fast waves* and negative mode numbers indicate *slow waves*. The fast wave propagates in the same direction as the beam in real space or in phase space, while the slow wave travels in the opposite sense.

This can be illustrated by dropping a stone into a trough of water. The disturbance will travel equally to the left and to the right. If a similar experiment is made in a trough of moving water both waves are carried along at the speed of the water flow. If the speed of flow is fast enough the observer will see both waves now travelling in the forward direction, but one will be slow and the other fast.

In order to extend this model to a circular accelerator, imagine an observer travelling around the machine in the same reference frame as the beam. A disturbance in the beam next to the observer will then be seen to propagate away from him in both directions. The resonant modes will be characterised by an exact number of wavelengths around the circumference, so that the upstream (slow) and downstream (fast) waves will encircle the machine in opposite directions, meet and reinforce to create a standing wave at the resonant frequency. If the observer now jumps into the laboratory frame of reference he will see this standing wave turning around the machine with the angular velocity of the beam. The circular boundary condition requires that a standing wave be formed, so that the observer will always see the sum of the fast and slow waves. In the longitudinal case, the fast and slow waves usually have approximately the same phase velocity, which is about equal to that of the beam. This makes them hard to distinguish experimentally and consequently it is usual to have only one mode number for longitudinal bunching as in Figure 9.3. A general and very readable account of slow and fast waves and other wave phenomena can be found in Ref. 12.

Identifying the various beam modes by the signals from a beam monitor (intensity, position, or shape) can be extremely confusing. References 13 and 14 should be of some help to experimenters.

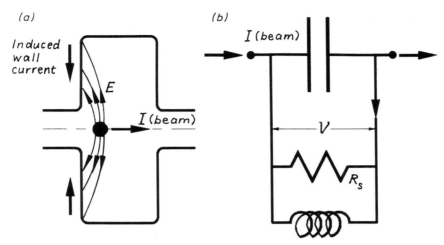

Figure 9.5. Analogue circuit for an rf cavity. (*a*) Cavity excited by beam, (*b*) beam current excites analogue circuit.

9.3   WAKE FIELDS AND COUPLING IMPEDANCE

A qualitative description of the beam–environment interaction was given in Section 9.1 and the example of a low-loss cavity, in which fields are long-lived and affect the beam over many traversals, was mentioned. This example was chosen as being the most easily understood from an intuitive standpoint. It can also be pursued further to illustrate two equivalent ways of analysing the *response of the chamber*. The treatment in this section in general follows that of Ref. 17, except that the Fourier transforms formalism is used.

The behaviour of an rf cavity, for example, is commonly studied using the equivalent circuit shown in Figure 9.5. The response of this circuit is well known and is governed by,

$$C \frac{d^2}{dt^2} V + \frac{1}{R_s} \frac{d}{dt} V + \frac{V}{L} = \frac{d}{dt} I, \tag{9.1}$$

where V corresponds to the gap voltage of the cavity and I corresponds to the current of the passing beam. In this example, the cavity is passive and is driven only by the beam current. The quality factor Q and the resonant angular frequency ω_r of the circuit are related to the lumped elements, L, C and R_s according to,

$$\omega_r = 1/\sqrt{(LC)} \tag{9.2}$$

and

$$Q = R_s \sqrt{(C/L)} = R_s C \omega_r. \tag{9.3}$$

The parameters Q and ω_r can be used to eliminate L and C, so converting (9.1) to,

$$\frac{d^2}{dt^2} V + \frac{\omega_r}{Q} \frac{d}{dt} V + \omega_r^2 V = \frac{\omega_r R_s}{Q} \frac{d}{dt} I. \tag{9.4}$$

195

Thus the gap voltage behaves as a driven, damped oscillator. It is convenient to solve this equation by the method of Fourier transforms, defined by*,

$$\tilde{f}(\omega) = \int_{-\infty}^{\infty} f(t) \, e^{-j\omega t} \, dt. \tag{9.5}$$

Contrary to normal mathematical practice, the Fourier transform is defined without the normalisation factor $1/\sqrt{(2\pi)}$. Consequently the factor $1/(2\pi)$ appears in the inverse transform [see later (9.12B)],

$$f(t) = \frac{1}{2\pi} \int_{-\infty}^{\infty} \tilde{f}(\omega) \, e^{j\omega t} \, d\omega. \tag{9.6}$$

The application of (9.6) to (9.4) gives,

$$-\omega^2 \tilde{V} + j\omega \frac{\omega_r}{Q} \tilde{V} + \omega_r^2 \tilde{V} = j\omega \frac{\omega_r R_s}{Q} \tilde{I}. \tag{9.7}$$

After some re-arrangements this can be written as,

$$\tilde{V} = \frac{R_s \tilde{I}}{1 - j[(\omega_r/\omega) - (\omega/\omega_r)]Q} = Z(\omega)\tilde{I}. \tag{9.8}$$

This means that if the frequency spectrum $\tilde{I}$ of the drive current is known then the spectrum of the induced voltage $\tilde{V}$ can be found by multiplying the current spectrum by the impedance $Z(\omega)$. In circuit theory, Z is known as the *generalised impedance*. Since in the present case the drive current is the beam, Z represents the coupling between the beam and its surroundings (represented by the simple equivalent circuit in Figure 9.5) and Z is therefore called the *coupling impedance*.

For further study of the solution to (9.8) there are two common approaches. One is to keep the solution in the form of (9.8) and to insert the Fourier transforms (i.e. spectra) of interesting driving currents. This is the *frequency domain* approach. The alternative is to take the inverse Fourier transform of (9.8) to find V as a function of time. This is the *time domain* approach.

The two approaches are, of course, fully equivalent, since they are both based on the same solution and it is a matter of convenience which to follow. Two examples will be given.

---

* Since $\sqrt{-1} = \pm j$, the transform could be defined with either sign. There is a tradition that physicists use $\exp(-i\omega t)$, while engineers use $\exp(j\omega t)$, so that $i = -j$ provides a confusion-free conversion, but this is not always respected. The 'engineers' convention, which is used in measuring instruments, is adopted here.

9.3.1 Delta pulse excitation and wake potential

Suppose a short beam pulse, represented by a delta function*, passes through the cavity close to the speed of light at $t = 0$, then

$$I(t) = q\,\delta(t) \tag{9.9}$$

and

$$\tilde{I}(\omega) = q. \tag{9.10}$$

Equation (9.8) then becomes

$$\tilde{V} = Zq. \tag{9.11}$$

It is now convenient to introduce a *wake potential*† w, whose Fourier transform is defined from (9.11) as,

$$\tilde{w} = \tilde{V}/q = Z. \tag{9.12A}$$

Thus the coupling impedance is equal to the Fourier transform of the wake potential, or as is more convenient in the present example, the wake potential is the inverse Fourier transform of the impedance,

$$w(t) = \frac{1}{2\pi} \int_{-\infty}^{\infty} Z(\omega)\,e^{j\omega t}\,d\omega. \tag{9.12B}$$

A more detailed discussion of this equivalence can be found in Ref. 18. If the coupling impedance is taken as given by (9.8), then after some manipulations the wake potential becomes,

$$
\left.
\begin{array}{l}
\text{For } t > 0,\ w(t) = \dfrac{V(t)}{q} = \dfrac{\omega_r R_s}{q}\,e^{-(\omega_r/2Q)t}\Big\{\cos[\omega_r\sqrt{(1 - 1/4Q^2)}\,t] \\[2ex]
\qquad\qquad - \dfrac{\sin[\omega_r\sqrt{(1 - 1/4Q^2)}\,t]}{2Q\sqrt{(1 - 1/4Q^2)}}\Big\}. \\[2ex]
\text{For } t < 0,\ w(t) = 0 \text{ (no field in front of pulse travelling at } c\text{)}.
\end{array}
\right\} \tag{9.13}
$$

If $Q \gg 1$ then (9.13) simplifies (see Figure 9.6) to,

$$w(t) = \frac{\omega_r R_s}{Q}\,e^{-\omega_r t/2Q}\cos(\omega_r t) \qquad \text{(for } t > 0\text{)}. \tag{9.14}$$

Thus a second charge q_2 in the beam, which is trailing the first, will gain or lose an

---

* The delta, or Dirac, function is frequently used to represent narrow pulses.

$$\int_{-\infty}^{\infty} \delta(t - t')\,dt = 1, \qquad \text{where } \delta(t - t') = 0 \text{ for all } t \text{ except } t = t'.$$

† Although w is known as the wake potential, it should be noted that its dimensions are voltage/charge.

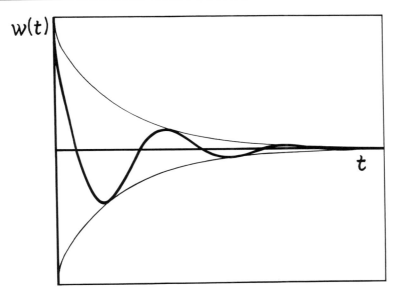

Figure 9.6. Wake potential following a delta current pulse crossing a high-Q cavity resonator.

energy from the wake field according to,

$$\Delta E = -w(t)q_1q_2. \tag{9.15}$$

The negative sign corresponds to the force being repulsive and causing deceleration and energy loss for a particle trailing directly behind the first particle (see also Figure 9.6).

A practical rf cavity, for example, would exhibit many sharp resonances other than that of the fundamental accelerating mode. These unwanted modes are called *parasitic modes*. They will normally be sufficiently separated to be treated individually. When the beam passes it excites the cavity and deposits energy into each mode as well as gaining energy from the fundamental mode. The energies deposited in the various modes decay and their wake fields combine to create the overall wake field seen by the tail of the bunch and sometimes by the following bunches if they are sufficiently closely spaced. The subsequent phase of this wake field with respect to the beam oscillation then determines whether the beam becomes unstable.

9.3.2   Sinusoidal excitation

Let it be assumed that the driving current is purely sinusoidal, given by,

$$I(t) = \hat{I}\,e^{j\omega_0 t}. \tag{9.16)*}$$

---

* In this chapter, $\hat{}$ will be specially re-defined as the complex amplitude (to include phase), rather than the peak value. This is known as a *phasor*. In the literature phasors may, or may not, include $e^{j\omega t}$.

The Fourier transform of this current is,

$$\tilde{I}(\omega) = 2\pi\hat{I}\,\delta(\omega - \omega_0) \tag{9.17}$$

which is seen most easily by performing the inverse transform on (9.17). When this is introduced into (9.8), the result is,

$$\tilde{V}(\omega) = Z(\omega)2\pi\hat{I}\,\delta(\omega - \omega_0)$$

and the inverse transform gives

$$V(t) = \frac{1}{2\pi}\int_{-\infty}^{\infty} Z(\omega)2\pi\hat{I}\,\delta(\omega - \omega_0)\,\mathrm{e}^{\mathrm{j}\omega t}\,\mathrm{d}\omega$$

$$= Z(\omega_0)\hat{I}\,\mathrm{e}^{\mathrm{j}\omega_0 t}\,.$$

Thus the response is, as expected, a sinusoidal oscillation and can be written as $\hat{V}\,\mathrm{e}^{\mathrm{j}\omega_0 t}$ and the ratio between the voltage amplitude (complex) and the current amplitude is the coupling impedance,

$$Z(\omega_0) = \hat{V}/\hat{I} \tag{9.18}$$

which for the particular circuit under consideration is given by (9.8).

The above definition of the impedance is based on the example of the lumped-element resonator in Figure 9.5. It can be generalised to a distributed structure, as the integral of *all the fields excited along the structure*, which may be one turn of an accelerator, normalised by the excitation current in the beam. It is easiest to assume a harmonic excitation current of amplitude $\hat{I}(\omega)$, which excites a harmonic field with complex amplitude $\hat{E}_s(\omega)$ and in analogy with (9.18) a *coupling impedance* is defined as,

$$Z_{\parallel}(\omega) = -\frac{\int \hat{E}_s(\omega)\,\mathrm{d}s}{\hat{I}(\omega)}\,[\Omega] \tag{9.19}$$

where the integral is taken over the structure or one full turn of an accelerator as applicable. $Z_{\parallel}$ indicates that this is the longitudinal impedance. The full significance of the wording 'all the fields along the structure' will be made clear in Section 9.4, where a space-charge term, which would be present with or without the chamber walls, will be included.

The principal results of the transverse case will be given without detailed justification. For a transverse oscillation, it is necessary to include the magnetic field, which is transverse to the beam, when expressing the deflecting force. In fact, the latter is usually the more important. Unlike the longitudinal case, the transverse wake field is zero directly after the charge and the sign is such that the sequel to (9.15) must be written with the opposite sign as,

$$\Delta E = w_{\perp}(t)\hat{y}q_1 q_2. \tag{9.20}$$

where yq_1 is the transverse dipole moment exciting the wake field. The transverse impedance is then defined as the *integral of the deflecting fields over one turn normalised by the dipole moment of the excitation beam current*,

$$Z_\perp(\omega) = j\frac{\oint [\hat{E}_r(\omega) + \beta c\hat{B}_\phi(\omega)]\,ds}{\beta I\hat{y}(\omega)} \, [\Omega\,m^{-1}]$$

(9.21)

where $\hat{y}(\omega)$ is the Fourier component of the vertical beam displacement and $Z_\perp$ indicates that it is the transverse impedance.

It would be correct to include the magnetic field in the definition of the longitudinal impedance (9.20), but it is common practice to neglect it, on the assumption that the product of the transverse field with the transverse motion is small compared with the electric field.

9.4   LONGITUDINAL INSTABILITY IN A COASTING BEAM

9.4.1   Description of the negative-mass instability[8,9]

Consider a beam in which a small spontaneous bunching, i.e. a fluctuation in the longitudinal density, has appeared. This will be accompanied by a longitudinal electric self field due to the changes in charge density. For the moment, there is no need to worry whether this field is enhanced or diminished by the presence of the chamber wall. The self field will exert a force on the beam particles directed from the high-density regions towards the low-density regions (see Figure 9.7). Beam particles such as P_1 in Figure 9.7 will lose energy due to the self field, while particles such as P_2 will gain energy. As was demonstrated in Chapter 7, the revolution frequency increases with energy gain below transition and decreases above it. Therefore, self fields tend to disperse density fluctuations below transition and enhance them above leading to a tighter and tighter bunching. This self-bunching phenomenon is known

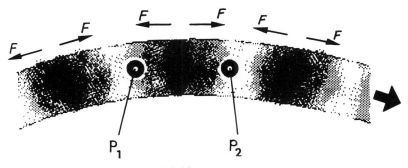

Figure 9.7. Longitudinal self field.

as the *negative-mass instability*. Thus beams are intrinsically unstable longitudinally above transition. In this case, some other mechanism is needed for damping.

The inclusion of the beam–environment interaction adds to the field seen by the beam. The response of the chamber may enhance the self-field effect, compensate it, or over-compensate it and make the beam unstable below transition rather than above, but in the event of self-bunching, this is still known as the negative-mass instability.

9.4.2   Specimen calculation of the longitudinal impedance[19]

In general it is difficult, if not impossible, to calculate analytically the impedance of the environment 'seen' by the beam. However, the case of a uniform circular vacuum pipe is both apt and analytically soluble. One turn in a smooth vacuum pipe can be represented by a lumped impedance, which is simple compared to the cavity example given earlier, but as indicated in the descriptive account above, there is also a self space-charge term to be taken into account. This term has to be found from first principles making the analysis rather more complicated.

Consider a mono-energetic beam of radius a centred in its vacuum chamber, a circular pipe of radius b (see Figure 9.8). Assume that the beam already has a longitudinal sinusoidal variation in the charge $\lambda(s)$ and that the wavelength of this fluctuation is much greater than b/γ. This is sometimes known as the *thin-beam approximation**.

The radial self fields of the beam due to the variation $\lambda(s)$ are given by[1],

$$
\left.
\begin{aligned}
E_r &= \frac{\lambda(s)}{2\pi\varepsilon_0}\frac{1}{r} \quad (b \geq r \geq a), \qquad
E_r = \frac{\lambda(s)}{2\pi\varepsilon_0}\frac{r}{a^2} \quad (r \leq a), \\[2mm]
B_\phi &= \frac{\lambda(s)\beta c\mu_0}{2\pi}\frac{1}{r} \quad (b \geq r \geq a), \qquad
B_\phi = \frac{\lambda(s)\beta c\mu_0}{2\pi}\frac{r}{a^2} \quad (r \leq a).
\end{aligned}
\right\}
\tag{9.22}
$$

It will be assumed that the characteristics of the pipe are such that the solution of the internal fields is completely independent of the external environment. In this case, the charge fluctuation in the beam will induce charges uniformly around the wall of the beam pipe that correspond to an equal and opposite alternating current to that on the beam. This wall current will see the wall's impedance and the electric field will be correspondingly modified. The longitudinal electric field seen by the beam can be found by equating the integral of the electric field around the loop shown in Figure 9.8 to the rate of change of the magnetic flux through that loop,

$$
[E_s(r=0) - E_w]\,\Delta s + \int_0^b [E_r(s+\Delta s) - E_r(s)]\,\mathrm{d}r = -\Delta s\,\frac{\partial}{\partial t}\int_0^b B_\phi\,\mathrm{d}r.
$$

* The 'thin-beam approximation' is explained in Ref. 20, pp. 322–8. In practice accelerator beams have little difficulty satisfying this criterion, especially with electron beams where γ can be very large.

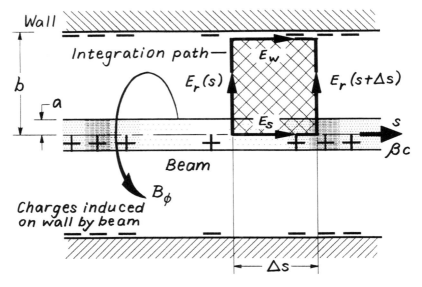

Figure 9.8. Coasting beam in a circular pipe.

After substitution from (9.22) for the field components, the integration gives,

$$[E_s(r = 0) - E_w] \Delta s + \frac{g_0}{4\pi\varepsilon_0} [\lambda(s + \Delta s) - \lambda(s)] = -\frac{\mu_0 g_0}{4\pi} \Delta s \beta_1 c \frac{\partial}{\partial t} \lambda \qquad (9.23)$$

where, $g_0 = [1 + 2 \ln(b/a)]$ and β_1 is the velocity of the perturbation. It is convenient to transform the time derivative in (9.23) to a spatial derivative by using $\partial/\partial t \equiv -\beta_1 c \, \partial/\partial s$, where $\beta_1 c$ equals the phase velocity of the wave (negative since positive time is equivalent to looking upstream), so that,

$$E_s(r = 0) = -\frac{g_0}{4\pi\varepsilon_0} (1 - \varepsilon_0\mu_0\beta_1^2 c^2) \frac{\partial\lambda}{\partial s} + E_w.$$

The bracket simplifies, since $1 = c^2\varepsilon_0\mu_0$, to give,

$$E_s(r = 0) = -\frac{g_0}{4\pi\varepsilon_0} (1 - \beta_1^2) \frac{\partial\lambda}{\partial s} + E_w. \qquad (9.24)$$

It is now necessary to make three assumptions.

- *Firstly*, E_s has been evaluated on the beam axis, but from here on, it will be assumed that this value is true for all positions across the beam, i.e. the longitudinal self fields are the same for all particles at the same azimuth in the machine.
- *Secondly*, the phase velocity of the wave is taken as being equal to the particle velocity in the unperturbed beam[20], i.e. $\beta_1 = \beta$. The fact that the

longitudinal fast and slow waves are both close to the beam velocity was mentioned earlier in Section 9.2.2 in connection with Figure 9.3.

- *Thirdly*, the perturbation in the line charge density and the resulting current are assumed to contain components with the forms,

$$\lambda(s) = \lambda_0 + \lambda_1\, e^{j(\omega t - n\Theta)} \qquad \text{and} \qquad I(s) = I_0 + I_1\, e^{j(\omega t - n\Theta)} \qquad (9.25)$$

where $\Theta = 2\pi s/C_0$ the mean azimuthal angle in the machine, n is the mode number already mentioned in Figure 9.3, and ω is the frequency of the perturbation. The line charge density and the current are related by, $I = \beta_1 c\lambda \simeq \beta c\lambda$.

With the above approximations the particles and the wave travel together, so that the voltage seen by a beam particle during a single turn will be $V_s = -C_0 E_s$, which becomes,

$$V_s \simeq \left[\frac{g_0 Z_0 C_0}{4\pi\beta\gamma^2}\right]\frac{\partial I}{\partial s} + V_w \qquad (9.26)$$

where $Z_0 = (\varepsilon_0 c)^{-1} = 377\ \Omega$, the impedance of free space. The voltage drop along the wall can be re-expressed as the wall current flowing into the wall impedance* Z_w. Since the wall current is equal in magnitude to the beam current, but flows in the opposite direction, this becomes

$$V_w = -Z_w I_w = Z_w I_1\, e^{j(\omega t - n\Theta)}. \qquad (9.27)$$

The substitution of (9.25) and (9.27) into (9.26) yields a simple expression for the beam-induced voltage experienced by a particle during one turn.

$$V_s = \left[Z_w - j\frac{n g_0 Z_0}{2\beta\gamma^2}\right]I_1\, e^{j(\omega t - n\Theta)} = Z_{\parallel} I_1\, e^{j(\omega t - n\Theta)} \qquad (9.28)$$

where $Z_{\parallel}$ is the total longitudinal coupling impedance. In accordance with normal a.c. circuit theory, real values of $Z_{\parallel}$ are resistive (negative resistance may occur in feedback systems) and imaginary values are reactive. The space charge term is negative and varies directly as ω (since $n = \omega/\omega_0$) and can therefore be considered as a negative inductance.

Three special cases will be quoted here: the smooth infinitely conducting wall[19]†, an inductive wall[21] and a resistive wall[22]†, which are close to practical situations. An analysis for a general wall impedance can be found in Ref. 23†.

---

* In circuit theory, the voltage drop round a loop is defined as $\sum ZI$, with I and V defined as positive in the direction of summation round the loop. Figure 9.8, however, imposes the convention that positive is in the coordinate directions of s and r. On axis the two conventions coincide, but they are opposed on the wall. Since impedance is defined by circuit conventions, $Z = +V/I$ on axis and $-V/I$ on the wall.
† These references refer to the average azimuthal field around the machine rather than the impedance, so that $2\pi R\langle E_\Theta\rangle \equiv -Z_{\parallel}\hat{I}$ and $+j$ should be replaced by $-i$ before comparing formulae.

(i) *Circular chamber with smooth infinitely conducting wall*[19]

In this case E_w must be zero. As long as the beam is centred, the field configuration is as if there were no walls and the beam were in free space.

$$Z_\parallel = -j \frac{n g_0 Z_0}{2 \beta \gamma^2}. \tag{9.29}$$

From the qualitative argument in Section 9.4.1, this reactive impedance makes the beam unstable above transition and stable below.

(ii) *Circular chamber with inductive wall*[21]

$$Z_\parallel = -j \frac{n g_0 Z_0}{2 \beta \gamma^2} + j\omega L' C_0 \tag{9.30}$$

where L' is the wall inductance per unit length of pipe. Briggs and Neil[21] concluded from (9.30) that a beam could be stabilised above transition by an inductive wall. In a footnote in Ref. 21, they refer to a private communication from Wideröe collaborating this from a chance observation. In 1949 during the development of a 30 MeV betatron, spiral windings were used to heat the glass vacuum envelope and later (1950) to provide an azimuthal magnetic field. When this winding was removed the accelerator would no longer work. Since a spiral winding presents an inductive impedance, it could have been the reason for the beam's earlier stability.

(iii) *Circular chamber with smooth resistive wall*[22]

When the wall thickness is greater than the skin depth, the wall impedance has equal resistive and inductive parts, so that

$$Z_\parallel = -j \frac{n g_0 Z_0}{2 \beta \gamma^2} + (1 + j) \frac{\mu_0 C_0}{2\pi b} \left[\frac{\omega}{2 \mu_r \mu_0 \sigma} \right]^{\frac{1}{2}} \tag{9.31}$$

where μ_r is the relative permeability of the wall and σ its conductivity. Equation (9.31) also contains the self-field term and the intrinsic stability depends on the relative magnitudes of the terms and whether the beam is above or below transition.

9.4.3   Longitudinal stability with a momentum spread

From the knowledge of the impedance seen by the beam, it is possible to predict whether the beam is stable or potentially unstable. It is only possible to say 'potentially unstable', since practical beams have a momentum spread and hence a revolution frequency spread, which may be sufficient to stabilise them. Such stabilisation is known as *Landau damping*[24]. Reference 24 is the original, but the interested reader may also like to read Hereward[25] and Kohaupt[26] who have written about Landau damping with accelerators in mind. It is not the intention to study Landau damping here, but in the following analysis it is automatically included when the Vlasov equation is applied. The Vlasov equation, which can be regarded as a formulation of Liouville's theorem (see Appendix E for a non-rigorous justification), affords the same advantage as the Hamiltonian mechanics formalism, it looks after the physics leaving only mathematical problems.

It is natural to describe particle distribution in a circulating beam in terms of the angular position and the angular momentum. This is done in some detail in Appendix D, where the average azimuthal angle Θ is used to define the conjugate general angular momentum. It is then shown that the longitudinal momentum p could be used as a conjugate variable in the linearised equations. Let therefore $f(\Theta, p, t)$ be a function describing the beam distribution. The substitution of the distribution function f into the Vlasov equation (see Appendix E) yields,

$$\frac{\partial}{\partial t} f(\Theta, p, t) + \dot{\Theta} \frac{\partial}{\partial \Theta} f(\Theta, p, t) + \dot{p} \frac{\partial}{\partial p} f(\Theta, p, t) = 0. \tag{9.32}$$

The rate of change of momentum, or force acting on a particle at position (Θ, t), is determined from (9.28) as,

$$\dot{p} = -\frac{eV_s}{C_s} = -\frac{eZ_\parallel}{C_s} I_1 \, e^{j(\omega t - n\Theta)}. \tag{9.33}$$

It is reasonable to make a trial solution with the same form as the current perturbation, i.e.

$$f(\Theta, p, t) = f_0(p) + f_1(p) \, e^{j(\omega t - n\Theta)} \tag{9.34}$$

in which $f_1 \ll f_0$. The total number of particles in the beam will be given by,

$$N = 2\pi \int_{\text{beam}} f_0(p) \, dp \tag{9.35}$$

and the d.c. current by,

$$I_0 = Ne\beta c/C_0 = Ne\Omega_0/(2\pi). \tag{9.36}$$

Similarly the amplitude of the a.c. component will be given by,

$$I_1 = e\Omega_0 \int_{\text{beam}} f_1(p) \, dp. \tag{9.37}$$

The substitution of the trial solution (9.34) and the expression (9.33) into (9.32) gives,

$$j(\omega - n\dot{\Theta}) f_1(p) \, e^{j(\omega t - n\Theta)} - \frac{eZ_\parallel}{C_s} I_1 \, e^{j(\omega t - n\Theta)} \frac{\partial}{\partial p} f(\Theta, p, t) = 0.$$

It will be assumed that the beam contains a relatively small momentum spread around the central orbit, so that C_s can be replaced by C_0. In the last term, f_1 will be omitted as small compared to f_0 and $\dot{\Theta}$ will be replaced by Ω to give,

$$j(\omega - n\Omega)f_1(p)\,e^{j(\omega t - n\Theta)} - \frac{eZ_\|}{C_0}I_1\,e^{j(\omega t - n\Theta)}\frac{d}{dp}f_0(p) \simeq 0. \tag{9.38}$$

An expression for the coherent behaviour of the beam is now obtained by integrating (9.38) over the momentum spread in the beam. The first term is evaluated with the use of (9.37).

$$I_1\,e^{j(\omega t - n\Theta)} = \frac{j}{C_0}e^2\Omega_0 Z_\| I_1\,e^{j(\omega t - n\Theta)}\int_{\text{beam}}\frac{df_0(p)/dp}{n\Omega - \omega}\,dp. \tag{9.39}$$

Equation (9.39) closes the 'loop' by giving the relationship between the current perturbation in the beam, I_1, and the voltage fed back to the beam, $-Z_\| I_1$, via the coupling impedance $Z_\|$.

The variable of integration is now changed to Ω using the definition of η from (7.28), which takes into account the optical properties of the lattice,

$$\Delta\Omega = \frac{\eta\Omega_0}{p_0}\Delta p. \tag{7.28}'$$

With (7.28)′, (9.39) simplifies in the linear approximation to,

$$\textit{Dispersion relation} \quad 1 = \frac{j}{C_0 p_0}e^2\Omega_0{}^2\eta Z_\|\int_{\text{beam}}\frac{df_0/d\Omega}{n\Omega - \omega}\,d\Omega. \tag{9.40}$$

The dispersion relation links the frequency of the disturbance ω with its wavelength represented by n.

A quasi-universal form of the dispersion relation, which depends only on the distribution, can be obtained by normalising the integral in (9.40). Two new variables are defined as,

$$\chi = \frac{1}{S}(\Omega - \Omega_0) \quad \text{(so that } d\chi = d\Omega/S) \tag{9.41}$$

$$\chi_1 = \frac{1}{S}\left(\frac{\omega}{n} - \Omega_0\right) \tag{9.42}$$

where S is the half-frequency spread at half maximum of the distribution $f_0(\Omega)$ (see Figure 9.9). If the full width of the momentum spread at half maximum is Δp_{fwhm}, then S becomes

$$S = \tfrac{1}{2}\eta\Omega_0\,\Delta p_{\text{fwhm}}/p_0. \tag{9.43}$$

Equation (9.35) for the total number of particles in the beam can be rewritten with the use of (9.41) as,

$$N = 2\pi\int_{\text{beam}}f_0(\Omega)\,d\Omega = 2\pi\int_{\text{beam}}f_0(\chi)S\,d\chi. \tag{9.44}$$

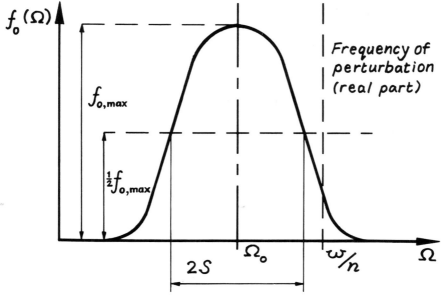

Figure 9.9. Beam distribution.

Equation (9.44) is then used to define a normalised distribution,

$$\psi(\chi) = \frac{2\pi S}{N} f_0(\chi) \qquad (\text{so that } \textstyle\int \psi(\chi)\, d\chi = 1). \tag{9.45}$$

The dispersion relation (9.40) can now be transformed using (9.36), (9.41), (9.42), (9.43), and (9.45) to give,

$$1 = \text{sign}(\eta)\text{j}\,\frac{Z_\parallel}{n}\,\frac{2eI_0}{\beta c p_0 \pi |\eta| (\Delta p_{\text{fwhm}}/p_0)^2} \int_{\text{beam}} \frac{d\psi/d\chi}{\chi - \chi_1}\, d\chi. \tag{9.46}$$

which for brevity is re-expressed as,

$$1 = \text{sign}(\eta)(U' - \text{j}V')\mathscr{I} \tag{9.47}$$

where $\mathscr{I}$ represents the integral*.

---

* The parameters U' and V' were introduced by Ruggiero and Vaccaro in Ref. 27. They are related to the parameters U and V introduced earlier by Neil and Sessler in Ref. 22. The inter-relationship is:

$$U - \text{j}V = [\tfrac{1}{2}n\pi\eta\beta c(\Delta p_{\text{fwhm}})^2/p_0](U' - \text{j}V'), \qquad \text{so that } Z_\parallel = \text{j}(U + \text{j}V)/eI_0.$$

The choice of U' and V' gives the most compact formulation of the dispersion relation with the integral in its normalised and quasi-universal form. The variables U and V, however, have the advantage of being closely related to the induced voltage. Both U, V and U', V' are widely used in the literature on instabilities.

When comparing references signs can be confusing. For example Refs 1, 2, 3 and 14 define $\eta = \gamma_{\text{tr}}^{-2} - \gamma^{-2}$ instead of $\eta = \gamma^{-2} - \gamma_{\text{tr}}^{-2}$, as used here. The terms in the denominators of the normalised and unnormalised dispersion integrals are also frequently inverted to adjust signs.

The solution of (9.46), or (9.40), is clearly going to run into difficulties when $\chi_1 = \chi$ (i.e. $\omega = n\Omega$). This particular problem was first treated correctly by Landau in his famous paper[24], in which he showed that the analytic solution to the integral in (9.46) when χ (i.e. ω) is real should be written as,

$$\int_{-\infty}^{\infty} \frac{d\psi/d\chi}{\chi - \chi_1} d\chi = \text{PV} \int_{-\infty}^{\infty} \frac{d\psi/d\chi}{\chi - \chi_1} d\chi \pm j\pi (d\psi/d\chi)_{\chi_1} \qquad (9.48)$$

where,

$$\text{PV} \int_{-\infty}^{\infty} \cdots \equiv \lim_{\varepsilon \to 0} \left\{ \int_{-\infty}^{\chi_1 - \varepsilon} \cdots + \int_{\chi_1 + \varepsilon}^{\infty} \right\} \cdots \textit{ The Cauchy principal value.}$$

The important term in (9.48) is the imaginary contribution from the pole, which gives rise to Landau damping. The sign of the imaginary term is often discussed[25] and should be negative for the above formulation of the integral. The positive sign is in fact admissible and corresponds to the man-made situation of a feedback system damping the oscillation.

The main goal here, however, is not to investigate analytical solutions of (9.48), but rather to predict when the beam will be stable. This can be elegantly done by a mapping technique developed by Ruggiero and Vaccaro[27], which has the advantage of giving a complete picture of the beam stability, while side-stepping the rather academic problem of the singularity in (9.48). This technique makes use of the fact that the angular frequency, ω, of the perturbation will in general be complex, so that,

$$\chi_1 = \text{Re } \chi_1 + j \text{ Im } \chi_1 \qquad (9.49)$$

and while the imaginary component exists, no matter how small it is, the singularity is removed from (9.46). The integral in (9.46) can then be evaluated numerically, and in some cases analytically, using

$$\text{Re } \mathscr{I} + j \text{ Im } \mathscr{I} = \int_{-\infty}^{\infty} \frac{(d\psi/d\chi)(\chi - \text{Re } \chi_1)}{(\chi - \text{Re } \chi_1)^2 + (\text{Im } \chi_1)^2} d\chi - j \int_{-\infty}^{\infty} \frac{(d\psi/d\chi) \text{ Im } \chi_1}{(\chi - \text{Re } \chi_1)^2 + (\text{Im } \chi_1)^2} d\chi \qquad (9.50)$$

so that (9.47) can be solved by,

$$U' = \text{sign}(\eta) \text{ Re } \mathscr{I}/|\mathscr{I}|^2 \qquad \text{and} \qquad V' = \text{sign}(\eta) \text{ Im } \mathscr{I}/|\mathscr{I}|^2. \qquad (9.51)$$

If this is done for a wide range of values, contour lines of constant $\text{Im } \chi_1$ can be plotted in the (V', U') plane, which can be related to $Z_\parallel$ by (9.46) and (9.47). The contours for negative $\text{Im } \omega$ (i.e. $\text{Im } \chi_1$ negative below transition and positive above) will correspond to growing solutions. Thus the area mapped in the (U', V') plane by $\text{Im } \omega < 0$ will be unstable and the area outside this mapping will be stable. The limit to the stable region can be approached by making $\text{Im } \chi_1$ very small. It can then be verified numerically that (9.46) converges on the analytic solution of (9.48) with the negative sign.

Figure 9.10 shows two typical stability plots. The first is for a beam with a

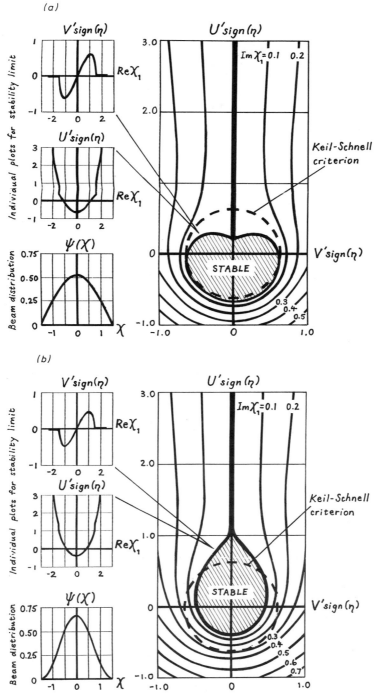

Figure 9.10. Longitudinal stability plots in (V', U') space. (a) Half-wave cosine distribution, (b) full-wave cosine distribution.

half-wave cosine distribution in momentum space and the second has a full-wave cosine distribution. The stable region is approximately circular, centred on the origin and has a long narrow 'neck' reaching up the U' axis. The exact shape of the contours of constant growth (Im $\omega < 0$) depends on the distribution ψ, but not critically. If there is a finite gradient $(d\psi/d\chi)$ at the edge of the distribution [e.g. Figure 9.10(a)] the stable region has a cusp. For smooth distributions with zero gradient at the edge [e.g. Figure 9.10(b)] the stable region extends further up the U' axis. Since it is the gradient of the beam distribution that appears in (9.46), it is the flanks of a distribution that contribute most to the dispersion integral and not, for example, the uniform central region of a stack. Broad smoothly sloping distributions have wider stability regions than narrow steep ones. In the extreme, a delta function is only stable on the positive U'-axis. Although it has large gradients, they are coincident and therefore self-compensating. The contours of constant Im $\omega < 0$ give only the initial growth rate of the instability. Perturbations on the beam inside the stable region remains as statistical noise on the beam. The dashed circles represent a stability criterion that will be explained in Section 9.4.4.

9.4.4 Longitudinal stability criteria

The first stability criterion appeared in Ref. 28 and was specifically derived for an rf cavity. It was expressed both as an upper limit on the quality factor for the resonant mode and as a maximum shunt impedance. The dispersion integral was solved analytically by assuming that the perturbation frequency had a small imaginary part and a Lorentz distribution was assumed for the beam. The result is quoted below for later reference,

$$Z_{\parallel \text{ cavity}} < \frac{nZ_0|\mathrm{d}f_{\text{rev}}/\mathrm{d}E|(\Delta E_{\text{hwhm}})^2}{m_0 c^2 r_0 N f_{\text{rev}}{}^2} \tag{9.52}$$

where Z_0 is the impedance of free space, r_0 the classical radius of the beam particle and ΔE is half of the energy spread at half maximum of the distribution.

Reference 27 gives a more extended study with analytic solutions for rectangular and Lorentz distributions, as well as several other distributions solved numerically. On the basis of these results a beautifully simple, but approximate, criterion was proposed[29] for beams with reasonable distributions (i.e. no discontinuities or a large central plateau). This criterion approximates the pear-shaped stability region with a circle centred on the origin, such that,

$$|U' - jV'| \lesssim 2/\pi. \tag{9.53}$$

Expressed with more practical parameters, found by equating (9.46) to (9.47), this inequality, known as the *Keil–Schnell criterion*, becomes

$$\left|\frac{Z_{\parallel}}{n}\right| \lesssim F_1 \frac{E_0}{e} \frac{|\eta|}{\gamma I_0} \left[\frac{\Delta p_{\text{fwhm}}}{\gamma m_0 c}\right]^2 \tag{9.54}$$

Table 9.1. *Form factor* F_1[30]

Distribution (shape factor*)	Form factor F_1					
	$-90°$	$-45°$	$0°$ V' axis	$45°$	$90°$ U' axis	$90°/-90°$
Triangular (2)	0	1.414	2	1.414	0	symmetric
Rounded triangle (2)	0.555	1.40	1.98	1.40	1.265	symmetric
Sawtooth (1.5)	0	1.768	2.25	1.768	0	0
Elliptic (1.115)	1.047	0.740	0	0	0	symmetric
Parabolic (1.414)	1.047	0.974	0.798	0.471	0	symmetric
Quartic (1.848)	1.073	1.084	1.194	1.407	2.183	symmetric
Circular limit† (1.64)	1.06	1.06	1.06	1.06	1.06	symmetric
Very smooth‡ (1.563)	1.089	0.908	0.693	0.644	1.694	symmetric

* 'Shape factor' = full width at half maximum/full width at base.
† 'Circular limit' is the distribution which gives exactly the circular stability region. It is very similar to the quartic distribution.
‡ 'Very smooth' is a distribution which has vanishing derivatives to all orders at its edges.

where E_0 is the rest energy of the beam particle and F_1 is a form factor. Equation (9.53) determines F_1 as unity, which is a fair approximation for well-behaved distributions. This criterion is included in Figures 9.10(a) and (b) as the dashed circles.

It is interesting to note that the criterion quoted earlier in (9.52) can be reduced to exactly the form of (9.54), with the form factor equal to π.

For those who want more than an approximate answer, but prefer to avoid mapping the (V', U') stability region in detail, a range of distributions is investigated in Ref. 30. The form factors obtained in this reference are reproduced in Table 9.1. In general, the presence of steep sides in a distribution reduces the stable area. The extreme case of a delta pulse is only stable on the positive U' axis. However, coasting beams usually have smooth profiles with tails and a form factor close to unity can be expected, but this is not always the case for bunched beams where the rf bucket can maintain sharp edges on the beam distribution. The results of the coasting beam theory can be sensibly applied to bunched beams providing the rise time of the instability is short compared to the synchrotron oscillation period. In this case, the current I_0 in (9.54) is taken as the peak current in one bunch.

The Keil–Schnell criterion has also been modified for stacks with a sharp, high-energy flank, a flat central region and a long, low-energy tail[31] (see Figure 9.11). This low-energy tail, which is created during the stacking process, gives the main contribution to the dispersion integral and the stability of the beam.

$$\left|\frac{Z_\|}{n}\right| \lesssim \frac{2E_0}{e}\frac{|\eta|}{\gamma I_0}\left[\frac{\Delta p_{\mathrm{fwhm}}}{\gamma m_0 c}\right]\left[\frac{\Delta p_{\mathrm{tail}}}{\gamma m_0 c}\right] \tag{9.55}$$

where Δp_{tail} is the full momentum width of the larger flank (on the low-energy side)

211

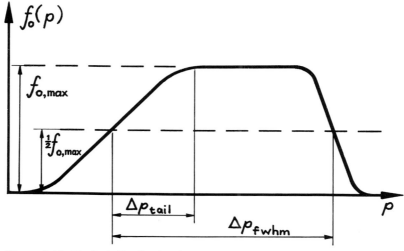

Figure 9.11. Typical distribution in a stacked beam.

at half maximum. Reference 31 also shows how dips or humps in the central plateau can drastically reduce the stability.

9.5 TRANSVERSE INSTABILITY IN A COASTING BEAM

The underlying physics for transverse instabilities is essentially the same as for the longitudinal case, although in detail the calculations may appear rather different. It is, for example, more complicated to calculate the impedance of the vacuum chamber, since magnetic forces must also be taken into account. There are of course many papers in the literature devoted to analytical solutions for various chamber geometries and algorithms for numerical computations. The uniform circular vacuum is again soluble analytically. However, at this introductory stage it is unnecessary to go through the detailed mathematics of this example, which can be found in Ref. 1 and many others, but it is useful to note that the longitudinal and transverse impedances of a uniform circular vacuum pipe of radius b with a resistive wall are related by,

$$Z_\perp = \frac{2c}{b^2 \Omega_0} \frac{Z_\parallel}{n}. \tag{9.56}$$

9.5.1 Transverse stability with a momentum spread

Once the impedance is known, the analysis can be continued as before with the Vlasov equation to find self-consistent solutions, as is done in Ref. 32. Readers will also find a slightly different approach in papers such as Ref. 2. Since this alternative method is a logical extension of the formalism developed in Chapter 8, the opportunity is taken to include it here for analysing the dipole mode of the transverse instability. In Chapter 8 the transverse motion of a test particle was expressed in linearised form using a space-charge forcing term F,

$$\frac{d^2}{dt^2} y + Q^2 \Omega^2 y = \frac{1}{m_0 \gamma} \left\{ \left[\frac{\partial F}{\partial y} \right]_{\bar{y}=0} y + \left[\frac{\partial F}{\partial \bar{y}} \right]_{y=0} \bar{y} \right\}. \tag{8.2}$$

The first driving term is evaluated with the static beam ($\bar{y} = 0$) and is therefore real. This term can be assimilated into the lefthand side as an incoherent tune shift (see Section 8.1) to give,

$$\frac{d^2}{dt^2} y + Q_{inc}^2 \Omega^2 y = \frac{1}{m_0 \gamma} \left[\frac{\partial F}{\partial \bar{y}} \right]_{y=0} \bar{y}. \tag{9.57}$$

This equation describes the motion of a single particle under the influence of a driving term derived from the average motion of the whole beam. This driving term can be shown to be directly proportional to the transverse impedance defined in (9.21). Thus (9.57) can be re-written as,

$$\frac{d^2}{dt^2} y + Q_{inc}^2 \Omega^2 y = \frac{e \beta I Z_\perp}{j m_0 \gamma C_0} \bar{y}. \tag{9.58}$$

A trial perturbation for the particle motion is assumed with the form

$$y = y_1 e^{j(\omega t - n\Theta)}$$

so that,

$$\frac{d}{dt} \equiv \frac{\partial}{\partial t} + \dot{\Theta} \frac{\partial}{\partial \Theta} = j(\omega - n\Omega). \tag{9.59}$$

By the use of (9.59), (9.58) can be re-written as,

$$[-(\omega - n\Omega)^2 + Q_{inc}^2 \Omega^2] y = \frac{e \beta I Z_\perp}{j m_0 \gamma C_0} \bar{y}$$

and then solved for y,

$$y = \frac{j e \beta I Z_\perp \bar{y}}{m_0 \gamma C_0} \frac{1}{[(\omega - n\Omega)^2 - Q_{inc}^2 \Omega^2]}. \tag{9.60}$$

All practical beams have a momentum distribution, which is related to the distribution of tune values via the chromaticity, Q',

$$\Delta Q_{inc} = \frac{Q'}{p_0} \Delta p \tag{9.61}$$

and to the revolution frequency by,

$$\Delta\Omega = \frac{\eta\Omega_0}{p_0}\Delta p. \tag{7.28}'$$

If $\psi(p)$ is the normalised momentum distribution in the beam, i.e. $\int \psi(p)\,\mathrm{d}p = 1$, then,

$$\bar{y} = \int y(p)\psi(p)\,\mathrm{d}p. \tag{9.62}$$

Thus by multiplying both sides of (9.60) by $\psi(p)$ and integrating, $\bar{y}$ can be eliminated to give,

$$1 = \frac{je\beta I Z_\perp}{m_0\gamma C_0} \int_{\text{beam}} \frac{\psi(p)}{[\omega - n\Omega(p)]^2 - Q_{\text{inc}}^2(p)\Omega^2(p)}\,\mathrm{d}p. \tag{9.63}$$

The integral in (9.63) can be split into partial fractions,

$$1 = \frac{je\beta I Z_\perp}{m_0\gamma C_0} \int_{\text{beam}} \frac{\psi(p)}{Q_{\text{inc}}\Omega}\left[\underbrace{\frac{1}{[(n - Q_{\text{inc}})\Omega - \omega]}}_{\text{slow wave}} - \underbrace{\frac{1}{[(n + Q_{\text{inc}})\Omega - \omega]}}_{\text{fast wave}}\right]\mathrm{d}p \tag{9.64}$$

which corresponds to the slow and fast waves. The frequency difference between these waves ($2Q_{\text{inc}}\Omega$) is much larger than the frequency spread that will exist in the beam, so only one type of wave will be able to contribute significantly to the integral for any one value of n. Since the fast wave is in any case difficult to excite, it is neglected in favour of the slow wave. The analysis for the fast wave, however, would follow the same steps with the difference of the minus sign. With the omission of the fast wave, (9.64) assumes a form analogous to longitudinal dispersion relation (9.40), i.e.

$$\textit{Dispersion relation} \qquad 1 \simeq \frac{je\beta I Z_\perp}{m_0\gamma C_0 Q_0 \Omega_0} \int_{\text{beam}} \frac{\psi(p)}{(n - Q_{\text{inc}})\Omega - \omega}\,\mathrm{d}p \tag{9.65}$$

where $Q_{\text{inc}} \to Q_0$ and $\Omega \to \Omega_0$ since the frequency spread in the beam will be small.

As before, the aim is to normalise the integral and to obtain a quasi-universal dispersion relation. First the denominator in (9.65) can be expressed to first order in momentum as,

$$[(n - Q_{\text{inc}})\Omega - \omega] = [n - (Q_0 + Q'\,\Delta p/p_0)][\Omega_0 + \eta\Omega_0\,\Delta p/p_0] - \omega$$

$$= \underbrace{[(n - Q_0)\eta - Q']\Omega_0\,\Delta p/p_0}_{} - \underbrace{[\omega - (n - Q_0)\Omega_0]}_{}.$$

Define,

$$[(n - Q_{\text{inc}})\Omega - \omega] = \qquad \Delta S\,\Delta p/\delta p \qquad - \qquad \Delta\omega \tag{9.66}$$

where ΔS is given by*,

---

* ΔS is analogous to S defined in (9.43) only inasmuch as it is a frequency spread needed for normalisation.

$$\Delta S = |(n - Q_0)\eta - Q'|\Omega_0\, \delta p/p_0 \tag{9.67}$$

and δp is the half-momentum spread at half-maximum of the distribution, i.e. $f(p_0 \pm \delta p) = \frac{1}{2}f(p_0)$. A new independent variable, which is given the same symbol as in the longitudinal case, is then defined as,

$$\chi = \frac{1}{\delta p}(p - p_0) \equiv \Delta p/\delta p. \tag{9.68}$$

With the use of (9.66), (9.67) and (9.68), the dispersion relation (9.65) can be re-written as,

$$1 = \frac{je\beta I Z_\perp}{m_0\gamma C_0 Q_0 \Omega_0\, \Delta S} \int_{\text{beam}} \frac{\psi(\chi)}{\chi - \chi_1}\, d\chi \tag{9.69}$$

where,

$$\chi_1 = \Delta\omega/\Delta S. \tag{9.70}$$

In analogy to the longitudinal case, (9.69) is further simplified using the same symbols* U', V' and $\mathscr{I}$ as before, to give,

$$1 = (U' + jV')\mathscr{I}. \tag{9.71}$$

The general remarks, made in Section 9.4.3 for the integral $\mathscr{I}$ of the longitudinal case, apply equally to the transverse case. The principal difference lies in the fact that while the derivative of a distribution function will assume both negative and positive values, the distribution function in (9.69) can only be positive or zero. This has the effect of limiting the stability plot in the (V', U') plane to positive values of V'. Two examples are given in Figure 9.12. The first is for a half-wave cosine distribution in momentum space and the second is for a full-wave cosine distribution. In this case the two curves are very similar. The dashed half circles represent a stability criterion that will be explained in Section 9.5.2.

9.5.2   Transverse stability criteria

As in the longitudinal case, the stability region for well-behaved beam distributions with neither discontinuities nor flat central regions can be approximated by a circle centred on the origin in the V', U' plot[33],

$$|U' + jV'| \lesssim 2/\pi. \tag{9.72}$$

---

* Although the same symbols are used, the definitions are different from those of U' and V' in Section 9.4.3. The analogy is carried further by also having the associated pair of variables U, V, but in this case, there are two definitions of U, V to be found in the literature, i.e.

$$(U + V) + jV = \Delta S(U' + jV') \quad \text{(introduced in Ref. 32)}$$

or

$$U + jV = \Delta S(U' + jV').$$

215

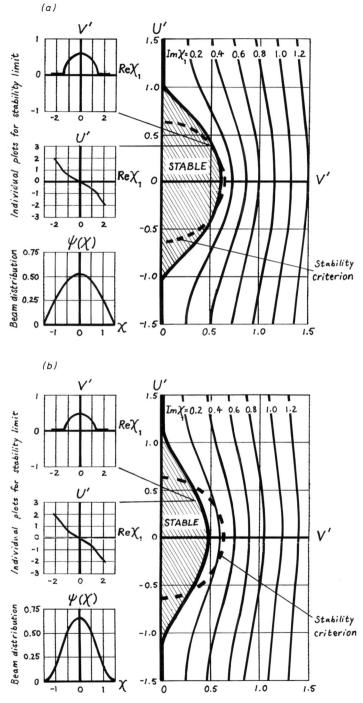

Figure 9.12. Transverse stability plots in (V', U') space. (a) Half-wave cosine distribution, (b) full-wave cosine distribution.

Table 9.2. *Form factor* F_2 [33]

Aspect ratio of stack*	0	1	10	100
Form factor F_2	1	0.89	0.65	0.45

* Aspect ratio = width of plateau/sum of edge widths at half maximum.

Expressed with more practical parameters, found by equating (9.69) and (9.71), this inequality becomes,

$$|Z_\perp| \lesssim F_2 \frac{E_0}{e} \frac{4Q_0\gamma\beta}{IR} \frac{\Delta p_{\text{fwhm}}}{p_0} |(n - Q_0)\eta - Q'| \qquad (9.73)$$

where R is the average machine radius and F_2 is a form factor which is again close to unity for well-behaved distributions. R/Q_0 is often replaced by the average betatron amplitude function in the appropriate plane, $\bar{\beta}_y$. This criterion is included in Figures 9.12(a) and (b) as dashed circles.

As before, the form factor can be adjusted to extend the criterion to wide stacks. Table 9.2 contains values for the form factor taken at the most unfavourable phases in the U', V' diagram. These 'worst case' values are quoted from Ref. 33 and were originally found by inspection of the plots in Ref. 34.

A more detailed study of form factors for the transverse case can be found in Ref. 35.

9.6 GENERAL REMARKS

The transverse and longitudinal planes have been treated in basically the same way and the results are very similar. Some differences should be noted. The longitudinal impedance depends predominantly on the induced electric fields, whereas the transverse impedance depends on both electric and magnetic fields and the latter are generally the more important. $Z_\parallel/n$ is analogous to $Z_\perp$ and they are usually quoted in this format. $Z_\parallel/n$ is typically of the order of ohms but with careful design it is possible to reach fractions of an ohm, while $Z_\perp$ is typically millions of Ω/m.

The impedance of a machine will vary with frequency. At low frequencies the impedance is dominated by the skin effect of the vacuum chamber, at medium and higher frequencies the impedance will behave like that of a broadband resonator ($Q \simeq 1$), which is due to the innumerable small resonators such as bellows, changes in cross-section, etc., and at certain frequencies there will be strong local resonances from individual pieces of equipment such as cavities, tanks, etc.

Defences available against coherent instabilities are limited. In the longitudinal case, it is important to design the chamber with a low coupling impedance in the first instance. For bunched beams, the Landau damping can be enhanced by introducing a spread in the synchrotron frequency with a so-called Landau cavity, but for coasting beams there is no such easy cure. Transversely there is some leeway for both coasting and bunched beams, inasmuch as the Landau damping can be enhanced by the use of sextupoles to increase Q', or octupoles to increase the tune spread with amplitude. Once Landau damping is insufficient, an active feedback system is the only solution in all cases.

One example of a coherent instability, which requires a small extension to the theory developed earlier, was a transverse, coherent instability that was observed in the ISR shortly after the machine was commissioned. At that time the phenomenon was not fully understood and it was dubbed the 'brickwall instability'. The characteristic behaviour of the 'brickwall' is shown in Figure 9.13(a). The accumulation of the beam by stacking in longitudinal phase space continued in a perfectly regular way until the intensity was about 3 A, at which point a *partial* loss would occur, usually from the top of the stack. After the loss the beam would stabilise, stacking would restart as before, until the intensity was again close to 3 A, whereupon a new partial loss would occur. This behaviour could be repeated several times giving rise to a characteristic sawtooth pattern.

The tune spreads across the stack had been calculated and the beam had been predicted to be stable. However, a further increase in these tune spreads stabilised the beam. In order to understand what was happening, it is necessary to study the shape of the working line in the tune diagram. As the line is loaded with the beam's space-charge, the incoherent tune increases in the horizontal plane and decreases in the vertical plane so progressively transforming the straight working line into the curved line as shown in Figure 9.13(b). Clearly the slope of the line is changed, the horizontal chromaticity is drastically reduced at the high-momentum end of the

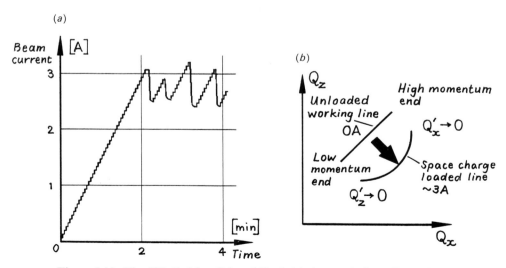

Figure 9.13. The ISR 'brickwall instability'. (a) Accumulation, (b) working line distortion.

line and the vertical chromaticity at the low-momentum end. In the theory presented in this chapter it is assumed that all regions of the stack 'see' essentially the same fields and that all parts of the beam contribute to the dispersion integral. However, in stacks, which are wide in momentum space, this is not necessarily true and the stability criterion should be satisfied locally in all regions of the stack. If this is not the case, a part of the beam may start to oscillate coherently and be lost.

CHAPTER 10

Radiating particles

A charge that is being accelerated radiates photons. In any kind of accelerator, the radiation due to the longitudinal acceleration is negligible. However, this is not always true for the radiation due to the transverse acceleration which is imposed by the guide field in circular machines. High-energy electron storage rings, for example, suffer severely from this fundamental energy loss by radiation and it is this fact that makes linear colliders the favoured electron machines of the future. While energy loss by radiation is a problem when considering the efficiency of an accelerator or the background caused in experimental physics detectors, it nevertheless does have certain redeeming features. Being a dissipative force, it is not Liouvillian and it is possible to exploit this to decrease the longitudinal and transverse phase space areas occupied by the beam. It can also provide some stabilisation against coherent oscillations. Best of all, this unwanted energy loss, or *synchrotron radiation* as it is called, can be turned into an intense source, which is typically in the visible to X-ray region with good directional properties and a well-defined polarisation. Accelerators dedicated to producing this radiation are called *synchrotron light sources* or simply *light sources*. Light sources have proved to be extremely valuable research tools and are a rapidly growing branch of the accelerator family.

The history of synchrotron radiation, with the theoretical and experimental development starting with Maxwell, can be found in a very readable account by Blewett[1]. The standard theory of radiating particles can be found in Refs 2 and 3 and reviews written primarily for accelerator physicists in Refs 4–6. Three of the main results of interest in this chapter, the radiated power and its angular and spectral distributions, will be quoted without derivation in the following sections. However, the derivations can be found in Refs 2–5 and the original references for the derivation of the Liénard–Wiechert potentials can be found in Ref. 1.

10.1   POWER RADIATED BY A RELATIVISTIC CHARGE

The total power radiated by an arbitrarily accelerated point charge was first derived by Liénard (1898) and is quoted below,

$$P_\gamma = \tfrac{2}{3}r_c m_0 c \gamma^6 [(\dot{\boldsymbol{\beta}})^2 - (\boldsymbol{\beta} \times \dot{\boldsymbol{\beta}})^2], \qquad (10.1)*$$

where $r_c = e^2/(4\pi\varepsilon_0 m_0 c^2)$, the classical radius of the particle. This can be found in a slightly different form in Ref. 2, p. 660 (cgs units), or in Ref. 3, p. 308. It is convenient to resolve (10.1) into components parallel to and perpendicular to the motion $\boldsymbol{\beta}c$, so that

$$P_\gamma = \tfrac{2}{3}r_c m_0 c \gamma^6 [\dot{\beta}_\parallel^2 + \dot{\beta}_\perp^2 - (\beta\dot{\beta}_\perp)^2] \qquad (\beta \equiv |\boldsymbol{\beta}|)$$
$$P_\gamma = \tfrac{2}{3}r_c m_0 c \gamma^4 (\gamma^2\dot{\beta}_\parallel^2 + \dot{\beta}_\perp^2). \qquad (10.2)$$

The relative importance of the two terms in (10.2) can be made apparent by translating acceleration into rate of change of momentum. The rate of change of the transverse momentum is simply

$$\dot{p}_\perp = m_0 \gamma c \dot{\beta}_\perp \qquad (10.3)$$

whilst the rate of change of the axial momentum is a little more complicated, since the energy of the particle changes and hence both its mass and velocity vary,

$$\dot{p}_\parallel = m_0 \gamma \dot{\beta}_\parallel c + m_0 \dot{\gamma} \beta_\parallel c$$
$$\dot{p}_\parallel = m_0 \gamma \dot{\beta}_\parallel c \gamma^2, \qquad (10.4)$$

which is obtained by the use of $\dot{\gamma} = \gamma^3\beta\dot{\beta}$, $\beta = \beta_\parallel$ and $\dot{\beta} = \dot{\beta}_\parallel$ when $\dot{\beta}_\perp = 0$. The substitution of (10.3) and then (10.4) into (10.2) gives the two limiting cases,

$$P_\gamma = \frac{2r_c}{3m_0 c} \dot{p}_\parallel^2 \quad \text{(axial)} \qquad P_\gamma = \frac{2r_c}{3m_0 c} \gamma^2 \dot{p}_\perp^2 \quad \text{(transverse)}. \qquad (10.5)$$

Equations (10.5) show directly that an accelerating force applied transversely to the particle motion will produce γ^2 times more radiation than the same force applied axially. In all synchrotrons and storage rings the radiation from transverse acceleration, known as *synchrotron radiation*, is by far the more important component and the axial component can be safely ignored.

The transverse acceleration is determined by the local curvature of the trajectory according to,

$$\dot{p}_\perp = m_0 \gamma \dot{\beta} c = m_0 \gamma (\beta c)^2/\rho, \qquad (10.6)$$

which when substituted into (10.5) and when $\dot{p}_\parallel$ is ignored gives,

$$P_\gamma = \tfrac{2}{3}r_c m_0 c^3 \beta^4 \gamma^4/\rho^2. \qquad (10.7)$$

---

* Here $\boldsymbol{\beta}$ is the relativistic β and the bold type indicates that it is a vector.

Equation (10.7) is usually applied to ultra-relativistic particles, in which case β is very close to unity and frequently it will be seen to be omitted. Equation (10.7) summarises the demise of the high-energy electron rings. The radiated power increases very rapidly indeed with the fourth power of γ, making these machines expensive to run. This power loss can be reduced by increasing the size of the machine, but this relief varies only as $1/\rho^2$ and although a large machine does mean a smaller power bill, it also means a higher construction cost.

Equation (10.7) also shows an important difference between protons and electrons. Consider beams of protons and electrons with the same total energy travelling on orbits with the same curvature, then

$$\frac{P_{\gamma \text{ electron}}}{P_{\gamma \text{ proton}}} = \frac{(\beta\gamma)^4_{\text{electron}}}{(\beta\gamma)^4_{\text{proton}}} \simeq \frac{m_0^4_{\text{proton}}}{m_0^4_{\text{electron}}} \simeq 10^{13}.$$

This illustrates why synchrotron radiation has been negligible so far in proton machines (although it cannot quite be disregarded for some future accelerators), whereas the role played by radiation in the beam dynamics and in the design of circular electron machines is of prime importance and is in fact the main limitation on the achievable energy.

Integration over the circumference of the machine gives the energy loss per turn, U_γ (in practice β can be taken as unity),

$$U_\gamma = \oint P_\gamma \, \mathrm{d}t = \oint \frac{P_\gamma}{\beta c} \, \mathrm{d}s \simeq \tfrac{2}{3} r_c m_0 c^2 \gamma^4 \oint \frac{\mathrm{d}s}{\rho^2}. \tag{10.8}$$

If the bending radius is the same in all dipoles, then $\rho = \rho_0$ and

$$U_\gamma \simeq \tfrac{4}{3} \pi r_c m_0 c^2 \gamma^4 / |\rho_0|, \tag{10.9}$$

since the straight sections do not contribute to the integral. In any case, this is usually a good approximation and is known as an *iso-magnetic lattice*. It is worth noting in the above equation, that the group of constants $(r_c m_0 c^2)$ is, in fact, *independent* of the type of particle.

From (10.9) the total power lost in synchrotron radiation by a beam is,

$$P_{\gamma \text{ tot}} \simeq \tfrac{4}{3} \pi r_c m_0 c^2 \gamma^4 f N / |\rho_0| = \tfrac{4}{3} \pi r_c m_0 c^2 \gamma^4 I / (e|\rho_0|) \tag{10.10}$$

where f is the revolution frequency, N is the number of particles in the beam and therefore $I = eNf$, the circulating current. The power loss has to be refurbished by the rf accelerating structure and in a large electron machine more rf voltage will be needed for this, than for acceleration. Some numerical examples for both electron and proton machines are given in Table 10.1.

As Table 10.1 shows, the large electron machines are prolific power consumers and at the higher power levels considerable care has to be taken to ensure that this energy is absorbed without damage to machine components.

Table 10.1. *Comparison of synchrotron radiation losses*

Machine	Beam energy (GeV)	Bending radius (m)	P_γ per particle (μW)	U_γ per particle (MeV)	Beam current (mA)	$P_{\gamma\,\text{tot}}$ per beam (MW)
ESRF e$^\pm$	6	23.37	1.60	4.91	100	0.491
LEP e$^\pm$	55	3096	0.646	261	3	0.784
LEP e$^\pm$	100	3096	7.06	2860	3	8.57
LHC p$^\pm$	8000	2669	3.4×10^{-5}	0.0119	865	0.0103*
SSC p$^\pm$	20 000	9098	1.2×10^{-4}	0.137	71.1	0.0097*

* Significant load for cryogenic system.

In addition to the radiated power loss, there is also an ohmic loss in the rf cavities, which is given by,

$$P_\Omega = \left[\frac{kU_\gamma}{e}\right]^2 \frac{1}{Z_s L_c} = \left(\frac{4}{3}\frac{k}{e}\,\pi r_e m_0 c^2\right)^2 \frac{\gamma^8}{Z_s L_c \rho^2}, \qquad (10.11)$$

where k is a factor representing the over-voltage required for phase stability and bunch shape, Z_s is the shunt impedance per unit length of the accelerating structure and L_c is the length of this structure. This equation assumes that the rf voltage is applied continuously, unlike the CERN LEP machine where the rf power is coupled into low-loss, spherical storage cavities between bunch passages. The ohmic power loss has the very strong energy dependence γ^8 and even if one lets L_c increase with the maximum energy this power loss will still dominate in large machines. Despite the energy-storage cavities used in LEP, the ohmic loss in the rf structure is about 12 MW at 55 GeV and starts to become prohibitively large soon after. This level of loss gives the main incentive for the development of superconducting rf structures. If successfully exploited this technology may raise the radiation limit to around 100 GeV per beam (LEP II upgrade). The justification for building circular electron accelerators for much higher energies seems doubtful, which explains the current interest in the development of linear colliders and the construction of the first prototype, the SLC at SLAC.

10.2   ANGULAR DISTRIBUTION

The angular distribution of the radiation is of importance for positioning absorbers, shielding experimental equipment, diagnostic applications and for designing secondary beam lines in light sources. The power radiated into a solid angle dΩ is quoted below for a relativistic charge ($\gamma \gg 1$) travelling instantaneously on a circular orbit (see Figure 10.1 for the definition of the geometry). This expression can be found,

223

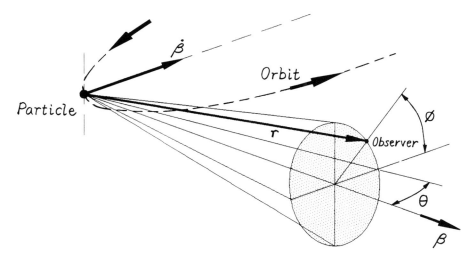

Figure 10.1. Distribution of synchrotron radiation.

for example, in Ref. 2, p. 665, in a slightly modified form where it is expressed in cgs units

$$\left[\frac{dP_\gamma}{d\Omega}\right]_{ret} \simeq \frac{2}{\pi} r_c m_0 c (\dot{\boldsymbol{\beta}})^2 \frac{\gamma^6}{(1+\gamma^2\theta^2)^3} \left[1 - \frac{4\gamma^2\theta^2 \cos^2\phi}{(1+\gamma^2\theta^2)^2}\right]. \tag{10.12}$$

The radiation pattern given by (10.12) is often described as a narrow spotlight beam with an opening angle of about $\pm 1/\gamma$ shining directly ahead of the particle as it sweeps round the synchrotron. The light is mainly polarised parallel to the median plane of the accelerator.

10.3   FREQUENCY SPECTRUM

The spectral distribution of the radiation is of especial interest to light source users and for diagnostic purposes. The main results are again quoted without derivation. This subject is treated in most of the references already mentioned, but Ref. 7 also includes many details concerning the properties of Airy functions and their use in the evaluation of the frequency spectrum.

The power radiated into the frequency range ω to $\omega + d\omega$ can be expressed in the form,

$$\frac{dP_\gamma(\omega)}{d\omega} d\omega = P_\gamma S\left(\frac{\omega}{\omega_c}\right) d\left(\frac{\omega}{\omega_c}\right) \tag{10.13}$$

where

$$\int_0^\infty \frac{dP_\gamma(\omega)}{d\omega} d\omega = P_\gamma \quad \text{and} \quad \int_0^\infty S\left(\frac{\omega}{\omega_c}\right) d\left(\frac{\omega}{\omega_c}\right) = 1. \tag{10.14}$$

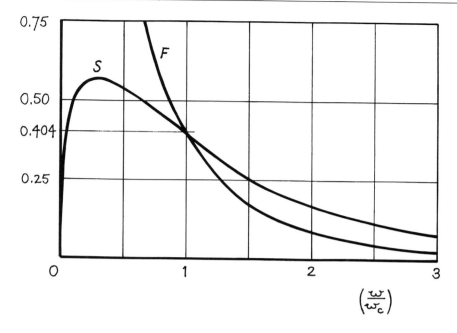

Figure 10.2. Universal functions $S(\omega/\omega_c)$ for the power spectrum and $F(\omega/\omega_c)$ for the photon number.

The function $S(\omega/\omega_c)$ is the *normalised spectral power density*. The analytic form of this universal function is given below and is sketched in Figure 10.2.

$$S\left(\frac{\omega}{\omega_c}\right) = \frac{9\sqrt{3}}{8\pi}\frac{\omega}{\omega_c}\int_{\omega/\omega_c}^{\infty} K_{5/3}\left(\frac{\omega}{\omega_c}\right) d\left(\frac{\omega}{\omega_c}\right) \tag{10.15}$$

where $K(\omega/\omega_c)$ represents the modified Bessel function of fractional order 5/3. The constant ω_c is known as the *critical frequency* and is chosen so as to divide the spectrum with equal powers being radiated above and below ω_c. The critical frequency and *critical wavelength* can be shown to be,

$$\omega_c = \tfrac{3}{2}c\gamma^3/|\rho| \qquad \text{and} \qquad \lambda_c = \tfrac{4}{3}\pi|\rho|/\gamma^3. \tag{10.16}$$

The behaviour of the Bessel function in (10.15) can be approximated for extreme arguments[7] to give,

$$\frac{\omega}{\omega_c} \ll 1; \qquad S \simeq 1.33\left(\frac{\omega}{\omega_c}\right)^{\frac{1}{3}}$$

$$\frac{\omega}{\omega_c} \gg 1; \qquad S \simeq \frac{9\sqrt{3}}{8\sqrt{(2\pi)}}\left(\frac{\omega}{\omega_c}\right)^{\frac{1}{2}} e^{-(\omega/\omega_c)}\left[1 + \frac{55}{72(\omega/\omega_c)} - \frac{10\,151}{2\times72^2(\omega/\omega_c)^2} + \cdots\right]$$

and the function S has a maximum at $\omega/\omega_c \simeq 1/3$ (see Figure 10.2).

225

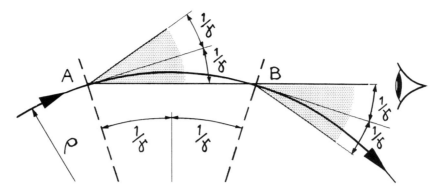

Figure 10.3. Observation of synchrotron light.

The rapid attenuation at low frequencies and the extended distribution at high frequencies in $S(\omega/\omega_c)$ can be understood qualitatively from the geometry of the curved beam trajectory. An observer, in the plane of the machine and outside the ring, will be periodically illuminated by narrow beams of light from the circulating particles (see Figure 10.3) and the duration of the flash from a single particle will be related to the frequency at which the spectral power starts to diminish.

The duration of the flash will be given by the difference between the time for the particle to pass from point A to B and the time for the first ray of light emitted at A to reach B. This will depend on the particle velocity, the machine radius and the opening angle of the cone of light, which will be taken as equal to the characteristic value $1/\gamma$.

$$\Delta t = t_{\text{part.}} - t_{\text{ray}} = \frac{2\rho}{c}\left[\frac{1}{\beta\gamma} - \frac{1}{\gamma} + \frac{1}{3!}\frac{1}{\gamma^3} - \cdots\right] \simeq \frac{4}{3}\frac{\rho}{c\gamma^3}.$$

If the time Δt is defined as corresponding to half an oscillation, then it defines the so-called *characteristic frequency*, which is a fixed point on the universal curve $S(\omega/\omega_c)$ in Figure 10.2. This point falls close to the critical frequency defined earlier in (10.16).

$$\omega_{\text{char}} = \pi c\gamma^3/|\rho|. \tag{10.17}$$

10.4   QUANTUM EMISSION

The synchrotron light is emitted in discrete quanta of energy u such that,

$$u = \hbar\omega \tag{10.18}$$

where $\hbar$ is Planck's constant divided by 2π. Equation (10.13) can be expressed in terms of the number of quanta emitted per unit time $\dot{n}(u)$ in the energy range u to

$u + du$,

$$\frac{dP_\gamma(\omega)}{d\omega} \, d\omega = u\dot{n}(u) \, du. \tag{10.19}$$

In analogy with (10.13), a universal function $F(u/u_c)$, the *normalised photon-number spectrum* can be defined which determines the number of quanta emitted in a certain energy range,

$$P_\gamma S(\omega/\omega_c) \, d(\omega/\omega_c) \equiv u_c^2 (u/u_c)\dot{n}(u) \, d(u/u_c)$$

so that,

$$\dot{n}(u) = \frac{P_\gamma}{u_c^2} \frac{u_c}{u} S\left(\frac{\omega}{\omega_c}\right) = \frac{P_\gamma}{u_c^2} F\left(\frac{u}{u_c}\right) \tag{10.20}$$

where u_c is the *critical energy* and is defined as,

$$u_c = \hbar\omega_c \tag{10.21}$$

and the two universal functions $F(\omega/\omega_c)$ and $S(\omega/\omega_c)$ are related by,

$$S\left(\frac{\omega}{\omega_c}\right) = \frac{\omega}{\omega_c} F\left(\frac{\omega}{\omega_c}\right). \tag{10.22}$$

The function $F(\omega/\omega_c)$ is also included in Figure 10.2. The rate at which quanta are emitted is,

$$\dot{N}_\gamma = \int_0^\infty \dot{n}(u) \, du = \frac{P_\gamma}{u_c} \int_0^\infty \frac{u_c}{u} S\left(\frac{\omega}{\omega_c}\right) d\left(\frac{u}{u_c}\right)$$

$$\dot{N}_\gamma = \frac{15\sqrt{3}}{8} \frac{P_\gamma}{u_c}. \tag{10.23}$$

The evaluation of this integral is described in Ref. 7. The actual rate of emission fluctuates and (10.23) gives the mean value. The average energy of the quanta is given directly by,

$$\langle u \rangle = P_\gamma/\dot{N}_\gamma = \frac{8}{15\sqrt{3}} u_c. \tag{10.24}$$

The variance of the photon energy distribution, which is needed for evaluating the quantum excitation later in this chapter, is given by[7],

$$\langle u^2 \rangle = \tfrac{11}{27} u_c^2. \tag{10.25}$$

The mean number of quanta emitted per turn (assuming that the bending radius ρ is approximately equal to the average radius R) is,

$$N_\gamma/\text{turn} = \dot{N}_\gamma 2\pi|\rho|/c \tag{10.26}$$

Table 10.2. *Comparison of synchrotron radiation characteristics*

Machine	Beam energy (GeV)	Bending radius (m)	λ_c Critical wavelength (Å)	u_c Critical energy (keV)	$\dot{N}_\gamma$ Rate of emission (s^{-1})
ESRF e$^\pm$	6	23.37	1.04	1.19	1.3×10^{15}
LEP e$^\pm$	55	3096	0.104	119.2	1.1×10^{14}
LEP e$^\pm$	100	3096	0.001 15	716.5	2.0×10^{14}
LHC p$^\pm$	8000	2669	180	0.069	1.0×10^{13}
SSC p$^\pm$	20 000	9098	0.393	0.315	7.4×10^{12}

$1 \text{ Å} \equiv 10^{-10}$ m.

which after substitution for $\dot{N}_\gamma$ leaves the simple approximation,

$$N_\gamma/\text{turn} \simeq \pi\gamma/50. \tag{10.27}$$

Use has been made of the fine structure constant $\alpha_f = e^2/(2h\varepsilon_0 c)$, which is very nearly equal to 1/137.

Some numerical examples concerning the characteristics of the synchrotron radiation in the machines featured in Table 10.1 are given in Table 10.2.

10.5   DAMPING OF SYNCHROTRON OSCILLATIONS

10.5.1   Damped phase equation

Intuitively it can be seen that an equilibrium is possible between the energy lost by radiation and the energy gained from an rf system. The mean energy U_γ that is lost per turn per particle is given by (10.8). This is usually far smaller than the particle's energy. Ignoring the statistical fluctuations in U_γ, the parameters can be set such that there will be a particle that receives exactly the right amount of energy to compensate its radiation loss. It then remains to be shown that other particles that differ slightly in energy from the synchronous particle are stable. In other words, it is necessary to show that phase stability is not destroyed by radiation. This is done below for the general case of an accelerated beam.

The energy loss by radiation will be taken as equivalent to a *continuous* 'resistive drag' on the particle, which implies that *several turns* are being considered. The problem can then be formulated with differential rather than difference equations and use can be made of the results already obtained in Chapter 7, Section 7.4. The 'resistive drag' force acting on a test particle will be,

$$\text{'Drag' force} = \frac{U_\gamma(p)}{C}. \tag{10.28}$$

This force, which is energy dependent, must be included in the balance of forces acting on both the test particle, see (7.21), and the synchronous particle, see (7.22). The difference between these two equations then yields a modified version of (7.23), which describes the deviations of the test particle with respect to the synchronous particle. The reader is left to refer back to Chapter 7 for explanations of the intervening steps.

$$\frac{\mathrm{d}}{\mathrm{d}t} \Delta p = \frac{e\hat{u}}{C_0}[\cos(\theta_0 + \Delta\theta) - \cos\theta_0] - \frac{\Delta U_\gamma(p)}{C_0}. \tag{10.29}$$

The approximation $C = C_0$ has been applied to the radiation term and other small amplitude approximations will be applied later, since in the present case the equations need only be accurate over one damping time.

A complementary first-order differential equation was derived in Chapter 7 linking Δp and $\mathrm{d}(\Delta\theta)/\mathrm{d}t$ [see (7.31)], which takes into account the properties of the lattice, i.e. how the changes in revolution frequency and equilibrium orbit, corresponding to a change in momentum Δp, affect $\Delta\theta$.

$$\Delta p = \frac{m_0 \gamma C_0}{2\pi h \eta} \frac{\mathrm{d}}{\mathrm{d}t} \Delta\theta \tag{7.31}$$

This pair of first-order equations is now used to obtain a second-order equation in Δp. A small amplitude approximation has been used to simplify the cosine terms.

$$\frac{\mathrm{d}^2}{\mathrm{d}t^2}\Delta p + \frac{2\pi h \eta e\hat{u} \sin\theta_0}{m_0 \gamma C_0{}^2}\Delta p + \frac{1}{C_0}\frac{\mathrm{d}}{\mathrm{d}t}\Delta U_\gamma(p) = 0. \tag{10.30}$$

The radiation loss term can be expanded, again to first order, as

$$\frac{1}{C_0}\frac{\mathrm{d}}{\mathrm{d}t}\Delta U_\gamma(p) \simeq \frac{1}{C_0}\left[\frac{\mathrm{d}U_\gamma}{\mathrm{d}p}\right]\frac{\mathrm{d}}{\mathrm{d}t}\Delta p. \tag{10.31}$$

The substitution of (10.31) into (10.30) and the use of (7.52) for replacing the group of constants by the synchrotron oscillation frequency ω_s yields,

$$\frac{\mathrm{d}^2}{\mathrm{d}t^2}\Delta p + \frac{1}{C_0}\left[\frac{\mathrm{d}U_\gamma}{\mathrm{d}p}\right]\frac{\mathrm{d}}{\mathrm{d}t}\Delta p + \omega_s{}^2\Delta p = 0. \tag{10.32}$$

In the literature on radiative effects, it is customary to use energy deviations, rather than momentum deviations, which can be done very simply by applying the conversion formulae in Appendix D. The use of (17D) gives directly,

$$\frac{\mathrm{d}^2}{\mathrm{d}t^2}\Delta E + \frac{1}{T_0}\left[\frac{\mathrm{d}U_\gamma}{\mathrm{d}E}\right]\frac{\mathrm{d}}{\mathrm{d}t}\Delta E + \omega_s{}^2\Delta E = 0, \tag{10.33}$$

where T_0 is the revolution period of the synchronous particle. This is the equation of a damped oscillator and its solution will be of the form

$$\Delta E = A\,\mathrm{e}^{-\alpha_1 t}\cos(\omega_s t + B) \tag{10.34}$$

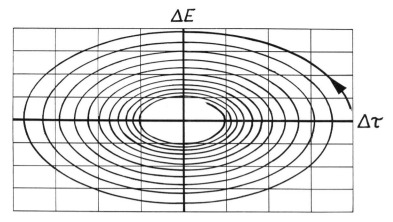

Figure 10.4. Motion in the (τ, E) phase space with damping.

where

$$\alpha_l = \frac{1}{2T_0}\left[\frac{\mathrm{d}U_\gamma}{\mathrm{d}E}\right].\tag{10.35}$$

The reader is left to verify that by returning to equations (10.29) and (7.31), it is possible to find a second-order differential equation for $\Delta\theta$ rather than Δp, which can be translated into a second-order equation for $\Delta\tau$ of exactly the same form as (10.33). The variable τ is defined in (16D) of Appendix D and (τ, E) would form a canonically conjugate pair in the absence of damping. The motion in the (τ, E) phase space is an inward elliptical spiral (Figure 10.4). Hence, all particles in the beam are continuously damped towards the synchronous particle. When considering the whole beam, it is more convenient to refer to the standard deviation (rms) of the energy distribution σ_l, which will also be damped with the damping constant α_l.

10.5.2   Evaluation of damping constant

In the previous section, the word 'damping' has been used on the tacit assumption that the constant α_l defined in (10.35) will be positive. The solution for the energy oscillations (10.34) was also given on the assumption that $\alpha_l < \omega_s$, which is equivalent to wanting the damping time to be greater than the synchrotron period. It is now necessary to evaluate the term $(\mathrm{d}U_\gamma/\mathrm{d}E)$ in order to clarify these points.

The mean energy loss U_γ was found from the integral of P_γ around the machine in (10.8). The integration in (10.8) must be made on the equilibrium orbit corresponding to the momentum of the particle and can be written as a first-order expansion about the orbit of the synchronous particle, as below

$$U_\gamma(p + \Delta p) = \oint \frac{P_\gamma(p_0 + \Delta p)}{\beta c}\,\mathrm{d}\sigma \simeq \oint \frac{P_\gamma(p_0 + \Delta p)}{c}\left(1 + \frac{D(s)}{\rho_0}\frac{\Delta p}{p_0}\right)\mathrm{d}s\tag{10.36}$$

where $d\sigma$ and ds correspond to the off-momentum and synchronous orbits respectively. The relationship between $d\sigma$ and ds was derived in Section 3.4.3 and is given in (3.52). Hence the difference in the energy loss per turn between a test particle and the synchronous particle will be to first order,

$$\Delta U_\gamma \simeq \oint \frac{1}{c} \left[\Delta P_\gamma + P_\gamma \frac{D(s)}{\rho_0} \frac{\Delta p}{p_0} + \cdots \right] ds.$$

In this context, it is usual to use energy rather than momentum as the independent variable when expressing the variation between two orbits. With the use of $cp \simeq E$ ($\beta \simeq 1$) and (17D) for Δp, the variation of U_γ with energy can be evaluated on the synchronous orbit, i.e. for $\Delta E \to 0$,

$$\left[\frac{dU_\gamma}{dE} \right]_0 = \oint \frac{1}{c} \left\{ \left[\frac{dP_\gamma}{dE} \right]_0 + \frac{P_\gamma}{E} \frac{D(s)}{\rho_0} \right\}_0 ds. \tag{10.37}$$

The radiated power P_γ can be written as,

$$P_\gamma = \frac{2r_c}{3m_0 c} \gamma^2 \dot{p}_\perp{}^2 \tag{10.38}$$

which comes from (10.5) with the $\dot{p}_\parallel$ term omitted for the reasons explained in Section 10.1. The transverse force is imposed by the guiding magnetic field according to $\dot{p}_\perp = -ecB_z$ from (2.5) with $\beta \simeq 1$ and γ can be re-expressed* as E/E_0, so that P_γ becomes,

$$P_\gamma = \frac{2r_c}{3m_0 c} \left[\frac{E}{E_0} \right]^2 (ecB_z)^2. \tag{10.39}$$

The derivative of P_γ with energy on the design orbit can therefore be written as,

$$\left[\frac{dP_\gamma}{dE} \right]_0 = 2\frac{P_\gamma}{E} + 2\frac{P_\gamma}{B_0} \left[\frac{dB_z}{dE} \right]_0.$$

The second term can be re-written with the help of the definition of the normalised gradient $k(s)$ given in (2.16).

$$\left[\frac{dP_\gamma}{dE} \right]_0 = 2\frac{P_\gamma}{E} + 2\frac{P_\gamma}{B_0} \left[\frac{dB_z}{dx} \frac{dx}{dE} \right]_0 = 2\frac{P_\gamma}{E} - 2\frac{P_\gamma \rho_0(s) k(s) D(s)}{E}. \tag{10.40}$$

The substitution of (10.40) into (10.37) leads to,

$$\left[\frac{dU_\gamma}{dE} \right]_0 = \oint \frac{1}{c} \left[2\frac{P_\gamma}{E} - 2\frac{P_\gamma \rho_0(s) k(s) D(s)}{E} + \frac{P_\gamma D(s)}{E\rho_0} \right] ds.$$

The third term in the square bracket arises from the change in orbit length with radial position. A closer inspection of Section 3.4.3 will show that this term is only

---

* Here E_0 refers to the rest mass, whereas the subscript $_0$ is used elsewhere in this section to indicate the central or synchronous orbit.

strictly true for sector dipoles ($\Lambda = 1$) and is zero for rectangular ones ($\Lambda = 0$).

$$\left[\frac{dU_\gamma}{dE}\right]_0 = \frac{U_\gamma}{E}\left\{2 + \frac{1}{U_\gamma}\oint \frac{P_\gamma D(s)}{c}\left[\frac{\Lambda}{\rho_0(s)} - 2\rho_0(s)k(s)\right]ds\right\}.$$

The energy is considered as constant during one turn for the evaluation of the integral, which then reduces to a form depending solely on the characteristics of the lattice,

$$\left[\frac{dU_\gamma}{dE}\right]_0 = \frac{U_\gamma}{E}\left\{2 + \oint \frac{D(s)}{\rho_0(s)}\left[\frac{\Lambda}{\rho_0^2(s)} - 2k(s)\right]ds\Bigg/\oint \frac{ds}{\rho_0^2}\right\}. \tag{10.41}$$

Thus the damping constant given in (10.35) becomes,

Radiation damping of synchrotron motion:
$$\alpha_l = \frac{1}{\tau_l} = \frac{1}{\sigma_l}\frac{d}{dt}\sigma_l = \frac{U_\gamma}{2T_0 E}(2 + \tilde{D}) \tag{10.42}$$

where

$$\tilde{D} = \oint \frac{D(s)}{\rho_0(s)}\left[\frac{\Lambda}{\rho_0^2(s)} - 2k(s)\right]ds\Bigg/\oint \frac{ds}{\rho_0^2(s)} \tag{10.43}$$

τ_l is the damping time of the synchrotron oscillations and $\Lambda = 1$ or 0 for sector and rectangular dipoles respectively. The quantity $\tilde{D}$ is known as the *lattice damping constant* and is one of the most important parameters of an electron machine lattice.

The second term in (10.43) is only non-zero when a gradient, a bending field and dispersion are all present. This can occur in combined-function magnets, off-axis in quadrupoles and in the end fields of dipoles with their end faces inclined with respect to the design orbit. The first term is likely to be small in a normal lattice. It is usual to design accelerators with all dipoles having the same bending radius (iso-magnetic lattice), which makes a useful simplification in (10.43),

$$\tilde{D} = \frac{\rho_0}{2\pi}\oint \frac{D(s)}{\rho_0(s)}\left[\frac{\Lambda}{\rho_0^2(s)} - 2k(s)\right]ds \qquad \text{(iso-magnetic lattice)} \tag{10.44}$$

where $\rho_0(s)$ inside the integral is a constant ρ_0 in the dipoles and ∞ elsewhere. In a combined-function lattice $\tilde{D}$ can be shown to have a value $\simeq 2$.

However, most modern accelerators and light sources are of the separated-function type, which makes it possible to further simplify (10.44) with the introduction of the momentum compaction defined in (3.53), i.e.

$$\tilde{D} = \frac{1}{2\pi\rho_0}\oint \frac{D(s)}{\rho_0}ds = \frac{C_0}{2\pi\rho_0}\alpha \qquad \text{(iso-magnetic, sep. function lattice)}. \tag{10.45}$$

The momentum compaction is usually well approximated by $1/Q_x^2$, which indicates that the natural value for $\tilde{D}$ in a separated-function machine is likely to be small so that the damping time for the energy oscillations from (10.42) will be approximately given by,

$$\tau_l = 1/\alpha_l \simeq T_0 E/U_\gamma. \tag{10.46}$$

This has the simple physical interpretation that the damping time is the time needed to radiate the equivalent of the particle's total energy.

It is now possible to comment on the assumptions mentioned at the start of this section, at least for separated-function machines. Firstly α_l is positive in (10.46), so damping does occur. Secondly the form of the solution for the energy oscillations presumes that α_l less than ω_s. This inequality can be checked by substitution from (10.46) and (7.52) to give,

$$\frac{U_\gamma^2}{T_0^2 E^2} < \frac{2\pi h |\eta| e\hat{u} \sin|\theta_0|}{C_0^2 \gamma m_0}.$$

At equilibrium, the energy $e\hat{u} \sin\theta_0$ delivered per turn by the rf system will just equal the energy lost by radiation U_γ, so that

$$U_\gamma/E < 2\pi h |\eta|.$$

On the lefthand side U_γ/E cannot exceed unity and is more usually a few percent. On the righthand side an approximation can be made for ultra-relativistic particles that $\eta \simeq \alpha$ and that $\alpha \simeq 1/Q_x^2$, so that

$$U_\gamma/E < 2\pi h/Q_x^2.$$

For LEP $Q_x \simeq 70$ and $h = 31\,320$ so the righthand side is $\simeq 40$ and the inequality is more than adequately satisfied.

10.6 QUANTUM EXCITATION OF SYNCHROTRON OSCILLATIONS

It was shown in the last section that as long as the effect of the radiation is regarded as a uniform 'drag' force on the particles the phase stability can be preserved and the synchrotron oscillations will be continuously damped towards the synchronous particle. It is unphysical to expect the beam to shrink to a singular point in phase space and indeed this can never occur. Firstly the photon emission from the beam is subject to random fluctuations, which introduce emittance growth and set a fundamental limit to the shrinking of the beam. This growth is known as *quantum excitation* and will be discussed in this section. Another limitation is the emittance growth due to intra-beam scattering. The latter is somewhat less fundamental, since if the beam is reduced to a single particle the effect disappears, whereas the quantum excitation remains. Several other mechanisms can also cause the beam to grow, for example, scattering on the residual gas, noise on the rf voltage, non-linear resonances and coupling, but none of these have the same fundamental character as the first two. The final beam size is the result of an equilibrium between the damping and the growth mechanisms.

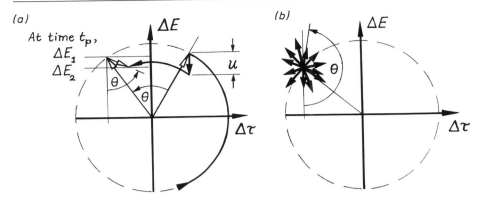

Figure 10.5. Effect of random photon emissions. (a) Single quantum, (b) many quanta projected to time t_p.

Consider the beam on a timescale which can resolve the emission of individual photons. The energy oscillation (10.34) with zero damping simplifies to,

$$\Delta E = A \cos(\omega_s t + B). \tag{10.47}$$

An instantaneous energy loss u by the emission of a photon at an arbitrary moment in time will change the amplitude and phase of the synchrotron oscillation as illustrated in Figure 10.5(a). The effects of many randomly distributed photon emissions at some later point in time t_p can be found by transporting all the kicks as vectors to the point t_p as shown in Figure 10.5(b). The kicks will appear randomly distributed in angle θ about this point. The expected average effect on the amplitude of the oscillation of a single photon emission can then be found by averaging over the distribution of random kicks. Consider first the projection on the energy axis of a single kick transferred to time t_p,

$$\Delta E_2 = \Delta E_1 - u \cos \theta \tag{10.48}$$

where the subscripts 1 and 2 represent the unperturbed and perturbed energies. The square of (10.48) gives,

$$(\Delta E_2)^2 = (\Delta E_1)^2 - 2\Delta E_1 u \cos \theta + u^2 \cos^2 \theta \tag{10.49}$$

and the average of many such kicks would be,

$$\langle (\Delta E_2)^2 \rangle = \langle (\Delta E_1)^2 \rangle - 2\langle \Delta E_1 u \cos \theta \rangle + \langle u^2 \cos^2 \theta \rangle \tag{10.50}$$

but since ΔE_1 and u are uncorrelated,

$$2\langle \Delta E_1 u \cos \theta \rangle = 2\langle \Delta E_1 \rangle \langle u \cos \theta \rangle = 0$$

which leaves,

$$\langle (\Delta E_2)^2 \rangle = \langle (\Delta E_1)^2 \rangle + \tfrac{1}{2}\langle u^2 \rangle. \tag{10.51}$$

The mean rate of quantum emission is $\dot{N}_\gamma$, given by (10.23), so there will be an increase in ΔE, given by (10.51), on average every $1/\dot{N}_\gamma$ seconds. Thus the growth rate of $(\Delta E)^2$ will be,

$$\frac{d}{dt}\langle(\Delta E)^2\rangle = \tfrac{1}{2}\dot{N}_\gamma\langle u^2\rangle. \tag{10.52}$$

The equations (10.51) and (10.52) are not only true for one particle in the beam, but for all particles in the beam. Once the whole beam is considered $\langle(\Delta E)^2\rangle$ is better known as σ_l^2 where σ_l is the standard deviation (rms) of the distribution. Thus (10.52) can be rewritten as,

Quantum excitation of synchrotron motion: $\qquad \dfrac{d}{dt}\sigma_l^2 = \tfrac{1}{2}\dot{N}_\gamma\langle u^2\rangle.$ (10.53)

Equation (10.53) shows an amplitude-independent growth of the longitudinal emittance, which depends on the fundamental characteristics of the radiation.

The quantum excitation described by (10.53) and the radiation damping described by (10.42) occur simultaneously. The two effects are opposed and lead to an equilibrium value for the rms of the energy distribution, which can be found by equating the two expressions to give,

Equilibrium between damping and quantum excitation: $\qquad \sigma_{l,\mathrm{eq}}^2 = \tfrac{1}{4}\dot{N}_\gamma\tau_l\langle u^2\rangle.$ (10.54)

The variance of the photon energy was quoted in (10.25).

10.7 DAMPING OF BETATRON OSCILLATIONS

The treatment of the synchrotron oscillations had two main features:

- *exponential damping* was obtained from the macro-effect of the radiation by considering it as a continuous process, and
- *amplitude-independent growth* was obtained by considering the micro-effects in the radiation known as quantum excitation.

A similar pattern re-appears in the analysis for betatron oscillations. The damping is a macro-effect depending on the energy loss over one turn and the energy gain in the rf cavities while the excitation is mainly due to the micro-effects of individual quanta. First the damping will be considered.

10.7.1 Zero dispersion at cavity

When replacing the energy loss due to radiation, the action of the rf cavity is to systematically add an increment to the particle's momentum *parallel to the axis*. RF cavities are usually installed in regions with zero dispersion, so this situation will be

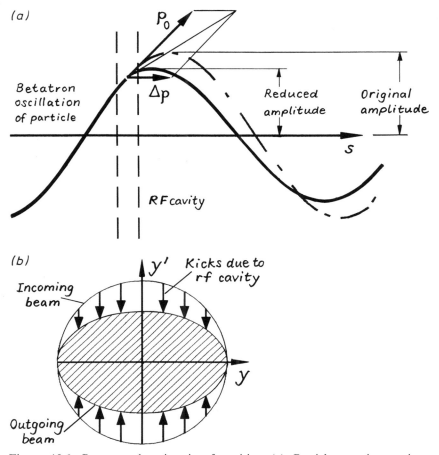

Figure 10.6. Betatron damping in rf cavities. (a) Particle crossing cavity, (b) reduction of phase-space area.

described first. In this case, it is easy to see that the change in momentum has no effect on the position of the equilibrium orbit, while the vector addition of the cavity's kick and the particle's motion leads to a reduction (damping) of the betatron oscillation in both planes [Figure 10.6(a)]. The slope of a betatron oscillation (horizontal or vertical) is simply $y' = p_\perp/p_\parallel$, so that the change in slope when crossing a cavity will be,

$$\Delta y' = \frac{p_\perp}{p_\parallel + \Delta p} - \frac{p_\perp}{p_\parallel} \simeq \frac{p_\perp}{p_\parallel}\left[\left(1 - \frac{\Delta p}{p_\parallel}\right) - 1\right] = -y'\frac{\Delta p}{p_0}. \tag{10.55}$$

Thus the incremental kick $\Delta y'$ is proportional to y' with the opposite sign. It therefore reduces the emittance of the beam as shown in Figure 10.6(b). The cavity is assumed to be short with respect to the betatron wavelength, so that the oscillation can be represented by a sinusoidal approximation and the displacement y can be taken as constant between entry and exit,

$$y = A \cos(s/\beta_y + \phi_0) \quad \text{and} \quad y' = -(A/\beta_y) \sin(s/\beta_y + \phi_0). \quad (10.56)$$

The amplitude A on entry to the cavity is directly obtained from (10.56) as

$$A^2 = y^2 + (\beta_y y')^2 \quad (10.57)$$

and the variation of A will be,

$$A \, \Delta A = y \, \Delta y + \beta_y^2 y' \, \Delta y'. \quad (10.58)$$

When passing through the cavity $\Delta y = 0$, but $\Delta y'$ is given by (10.55) so that (10.58) reduces to

$$A \, \Delta A = -(\beta_y y')^2 \, \Delta p / p_0,$$

which can be re-expressed using (10.56) for y' to give,

$$\frac{\Delta A}{A} = -\sin^2(s/\beta_y + \phi_0) \, \Delta p / p_0. \quad (10.59)$$

The phase ϕ_0 of the oscillation arriving at the cavity can be anywhere between 0 and 2π with equal probability. As usual, it is the average change in amplitude that is of interest, so (10.59) is averaged over all possible phases to give,

$$\frac{\Delta \langle A \rangle}{A} = -\langle \sin^2(s/\beta_y + \phi_0) \rangle \, \Delta p / p_0 = -\tfrac{1}{2} \Delta p / p_0. \quad (10.60)$$

Equation (10.60) is applied after each turn of the machine, so the average rate of change in A is determined by the revolution period T_0 and a damping constant α_β can be defined as,

$$\alpha_\beta = \frac{1}{A} \frac{\mathrm{d}}{\mathrm{d}t} \langle A \rangle = -\frac{1}{2T_0} \frac{\Delta p}{p_0}. \quad (10.61)$$

Equation (10.61) is not only true for one particle in the beam, but for all particles, so the damping constant describes the general behaviour. The fractional momentum error $\Delta p / p_0$, which appears after each turn, can be related to the energy loss per turn, U_γ, by

$$\frac{\Delta p}{p_0} = \frac{1}{\beta^2} \frac{\Delta E}{E} = \frac{1}{\beta^2} \frac{U_\gamma}{E} \simeq \frac{U_\gamma}{E} \quad (10.62)$$

so that the final expression for the betatron damping constant in either plane in the rf cavities is,

Damping by rf cavity of betatron motion: $\quad \alpha_\beta = \dfrac{1}{\tau_\beta} = \dfrac{U_\gamma}{2T_0 E}, \quad (10.63)$

where τ_β is the time constant of the damping. This is the same *adiabatic damping* that was first mentioned in Section 2.4 and always occurs when accelerating a beam.

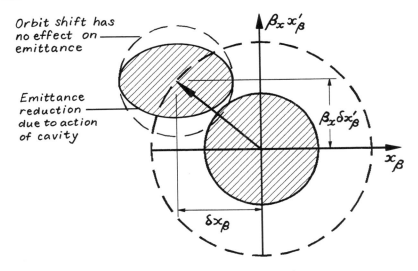

Figure 10.7. Effect of finite dispersion in a cavity.

10.7.2 Finite dispersion at cavity

In the case of finite dispersion at the cavity, the particles in a given slice of beam see a systematic step in their equilibrium orbit (see Figure 10.7). This does not affect the action of the cavity, which still reduces the emittance about the centre of the distribution and the shift equally has no direct effect on emittance. The closed orbit, about which the beam oscillates, will adjust so that the beam changes smoothly from oscillating about the equilibrium orbit on the upstream side of the cavity to the shifted equilibrium orbit on the downstream side. This becomes a little more complicated when the synchrotron motion is included, since the shift in the equilibrium orbit will be modulated at the synchrotron frequency. This modulation is slow and in most cases the closed orbit can adapt adiabatically, just as it would when ramping a dipole magnet. However, it is possible that the betatron and synchrotron frequencies will have an integer relationship and in this case *synchro-betatron resonances*[8] will occur. This is one reason why dispersion in cavities is avoided. Another reason is that an off-axis beam may excite an unwanted deflection mode in the cavity.

10.7.3 Anti-damping due to the macro-effect of the energy loss

For the macro-effect of the radiated energy loss on the betatron oscillations, it will be assumed that the random nature of photon emission can be disregarded, so that two equal particles on identical trajectories would experience the same macro-energy

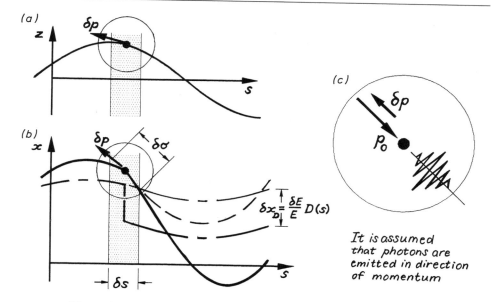

Figure 10.8. Macro-effects of energy loss on betatron oscillations. (a) No dispersion (vertical motion), (b) with dispersion (horizontal motion), (c) assumption for macro-effects.

loss* δE in an elementary segment δs of the lattice. The loss will be considered as a step loss inside δs.

Furthermore, it will be assumed that the photons are emitted exactly along the direction of motion of the particle. In this way, the recoil makes no difference to the real-space position and angle of the particle at the moment of emission. The only immediate effect is the reduction in the total momentum and any subsequent effect will depend upon the presence or absence of dispersion in the lattice. These two situations are illustrated in Figure 10.8.

In the absence of dispersion, the particle trajectory is unaffected to first order by its energy loss. This is usually the case for the vertical motion in a machine. However, in the presence of dispersion the equilibrium orbit about which the particle is oscillating is shifted in proportion to the energy loss. This shift in the equilibrium orbit changes the reference for the particle's motion, so modifying the amplitude and phase of the betatron oscillation. This is always the case for the horizontal motion in a machine and is analysed below.

The position of any particle can be expressed as the sum of its betatron and dispersion trajectories (see Section 3.4.2), i.e.

$$x = x_\beta + x_D. \tag{10.64}$$

---

* Δ has been used to indicate small changes such as the momentum loss Δp over one turn. The use of δ here indicates a smaller change, such as the radiation loss δE in an elementary segment of a dipole. It is assumed that average values still apply for this radiation loss.

If there is a small energy loss δE, then there will be a corresponding shift in the equilibrium orbit, which changes x_D, according to,

$$\delta x_D = D(s)\,\delta E/E, \tag{10.65}$$

where $\delta E/E$ is taken as equal to $\delta p/p_0$ for ultra-relativistic particles. Since the real-space position of the particle is unchanged, there must be a compensatory change in the betatron trajectory of,

$$\delta x_\beta = -D(s)\,\delta E/E. \tag{10.66}$$

When it is assumed that the recoil from the photon emission is exactly collinear with the particle momentum, there are analogous results for the angular changes, namely,

$$\delta x_D' = D'(s)\,\delta E/E \tag{10.67}$$

$$\delta x_\beta' = -D'(s)\,\delta E/E. \tag{10.68}$$

In the section of lattice $\delta\sigma$ where the energy loss δE occurs, the betatron oscillation can be taken as sinusoidal and the variation of its amplitude will then be given by (10.58), i.e.

$$A\,\delta A = x_\beta\,\delta x_\beta + \beta_x^2 x_\beta'\,\delta x_\beta'. \tag{10.58'}$$

The incremental changes δx_β and $\delta x_\beta'$ are related to $\delta E/E$ by (10.66) and (10.68), so that,

$$A\,\delta A = -[D(s)x_\beta + D'(s)\beta_x^2 x_\beta']\,\delta E/E. \tag{10.69}$$

The energy loss can be found by multiplying P_γ by the time taken to traverse the segment $\delta\sigma$, i.e.

$$\delta E = -P_\gamma(x_\beta)\frac{\delta\sigma}{\beta c}. \tag{10.70}$$

Equation (10.39) shows that the radiated power P_γ varies with the square of the guiding magnetic field, so that changes in radial position can be accounted for according to,

$$P_\gamma(x_\beta) = P_\gamma + 2\frac{P_\gamma}{B_0}\frac{dB_z}{dx}x_\beta. \tag{10.71}$$

The variation of arc length with radial position was discussed in Section 3.4.3 and is given by,

$$\delta\sigma = (1 + x_\beta/\rho_0)\,\delta s \tag{10.72}$$

where δs is measured along the design orbit. The energy loss δE in (10.70) can now be evaluated using (10.71) and (10.72), which to first order gives,

$$\delta E \simeq -\left[1 + 2\frac{x_\beta}{B_0}\frac{dB_z}{dx} + \frac{x_\beta}{\rho_0}\right]P_\gamma\,\delta s/\beta c. \tag{10.73}$$

Since the energy loss depends on x_β, it is different for the inner and outer halves of the oscillation. This introduces a bias when averaging over all phases of the betatron oscillation in the next step and the result is anti-damping. If the energy change were constant for all particles, as with an rf cavity, the overall effect would average to zero. The substitution of (10.73) into (10.69) yields,

$$A\,\delta A = [D(s)x_\beta + D'(s)\beta_x{}^2 x_\beta'] \left[1 + 2\frac{x_\beta}{B_0}\frac{\mathrm{d}B_z}{\mathrm{d}x} + \frac{x_\beta}{\rho_0}\right]\frac{P_\gamma}{E\beta c}\,\delta s.$$

The average over all the possible phases in the segment δs leaves only the terms in $x_\beta{}^2$,

$$A\langle\delta A\rangle = \frac{D(s)P_\gamma}{\beta c E}\left[\frac{2}{B_0}\frac{\mathrm{d}B_z}{\mathrm{d}x} + \frac{1}{\rho_0}\right]\langle x_\beta{}^2\rangle\,\delta s$$

and since $\langle x_\beta{}^2\rangle = \tfrac{1}{2}A^2$,

$$\frac{1}{A}\langle\delta A\rangle = \frac{D(s)P_\gamma}{2\beta c E}\left[\frac{2}{B_0}\frac{\mathrm{d}B_z}{\mathrm{d}x} + \frac{1}{\rho_0}\right]\delta s. \tag{10.74}$$

The integration of (10.74) around the whole machine gives,

$$\frac{1}{A}\Delta A = \frac{1}{2}\oint\frac{\rho_0 D(s)P_\gamma}{\beta c E}\left[\frac{1}{\rho_0{}^2} - 2k\right]\mathrm{d}s.$$

With the use of (10.43), (10.7) and (10.8), this is re-expressed as,

$$\frac{1}{A}\Delta A = \frac{U_\gamma}{2E}\tilde{D}. \tag{10.75}$$

Equation (10.75) is the expected effect on any single particle and is therefore also true for the beam as a whole. The sign is positive so this effect causes anti-damping. This effect can now be added to the damping obtained in the rf cavities. The combined action of these macro-effects is normally known as *radiation damping**. In the two final expressions, it is more general to replace the oscillation amplitude A by its rms value σ.

Radiation damping for horiz. β-oscillations (with dispersion):
$$\alpha_x = \frac{1}{\tau_x} = \frac{1}{\sigma_x}\frac{\mathrm{d}\sigma_x}{\mathrm{d}t} = \frac{U_\gamma}{2T_0 E}[1 - \tilde{D}] \tag{10.76}$$

Radiation damping for vert. β-oscillations (no dispersion):
$$\alpha_z = \alpha_\beta = \frac{1}{\tau_z} = \frac{1}{\sigma_z}\frac{\mathrm{d}\sigma_z}{\mathrm{d}t} = \frac{U_\gamma}{2T_0 E}. \tag{10.77}$$

---

* 'Radiation damping' may appear to be a misnomer, since the damping action of an rf cavity always occurs when accelerating a beam and has already been named 'adiabatic damping'. However, the radiation loss on each turn does allow this 'adiabatic damping' to be applied repeatedly at what is effectively constant energy and, in this way, the special name can be justified. Another interpretation (due to G. Rees), is that 'adiabatic damping' is in fact the misnomer, since the Liouvillian forces of an rf cavity conserve the more fundamental normalised phase-space area of the beam, whereas, while a particle radiates, it is the normalised phase-space area that is reduced. *Thus, the true damping is 'radiation damping'!*

10.8   QUANTUM EXCITATION OF THE
             BETATRON MOTION

In order to evaluate the damping in the earlier sections, it was assumed that the random nature of the photon emissions could be disregarded and that when emitting a photon the recoil of the particle was collinear with its momentum. It is now time to revise these approximations. Apart from being random, the emission of a photon now has two consequences for a particle:

- a small change in the direction of motion, and
- a small reduction in the total momentum.

In the *absence of dispersion*, only the *change in direction* will cause growth of the transverse emittance. This situation usually applies to the *vertical motion*. In the *presence of dispersion*, the *reduction in the total momentum* also causes emittance growth and its effect is by far the larger. This situation applies to the *horizontal motion*. The relative strengths of these two effects will be evaluated at the end of Section 10.8.2.

10.8.1   In the absence of dispersion (vertical motion)

The emission of a photon takes place directly ahead of the particle within a narrow cone with an opening angle of order $1/\gamma$. At the instant of emission the particle's position is unaffected, but in general its angle will be changed slightly due to the recoil (see Figure 10.9). The change in slope of the trajectory will be*,

$$\delta z' = \frac{\pm \delta p\, \theta_\gamma}{p_0 - \delta p} \simeq \pm \theta_\gamma \frac{\delta p}{p_0} \simeq \pm \theta_\gamma \frac{u}{E} \qquad (10.78)$$

where θ_γ is the projected angular deviation onto the z, s plane. The following analysis will make use of the sinusoidal approximation for the betatron motion given in (10.56), on the assumption that the effect is small. This will be justified in Section 10.8.2, where the comparison is made with the second effect related to the momentum change.

$$z = A \cos(s/\bar{\beta}_z + \phi_0) \qquad \text{and} \qquad z' = -(A/\bar{\beta}_z) \sin(s/\bar{\beta}_z + \phi_0). \qquad (10.56)'$$

The average value of the betatron amplitude is used in (10.56)' because it will be necessary to sum all the kicks from the photon emissions around the whole circumference. This is a less favourable situation for the sinusoidal approximation than in the earlier examples where the approximation was applied to short segments of structure, such as an rf cavity.

---

* The symbol δ is now used to indicate the effects of single photons and average values are no longer assumed.

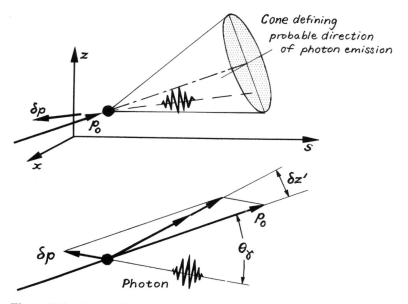

Figure 10.9. Micro-effect of emission of a photon.

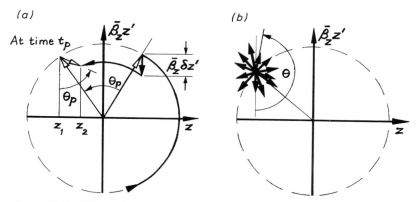

Figure 10.10. Effect of angular kicks from random photon emissions. (a) Single kick, (b) many kicks projected to time t_p.

Equation (10.56)′ describes a particle on a circular orbit in the z, $\bar{\beta}z'$ phase space. At random moments a particle will suffer kicks $\delta z'$ of the form (10.78). All the kick vectors can be transferred to a time t_p and summed to find the net or average value. This is basically the same model as used for the excitation of the synchrotron oscillations in Section 10.6, except that the kick $\delta z'$ now has an equal probability of being positive or negative. It is also more convenient on this occasion to project the kicks onto the horizontal axis before summing (see Figure 10.10). Thus the effect of a single emission transferred to time t_p is given by,

$$z_2 = z_1 + \bar{\beta}_z\,\delta z' \sin \theta_p \tag{10.79}$$

243

where the subscripts 1 and 2 represent the unperturbed and perturbed motions. The square of (10.79) averaged over many kicks becomes,

$$\langle z_2{}^2\rangle = \langle z_1{}^2\rangle + 2\langle z_1\bar{\beta}_z\,\delta z'\sin\theta_p\rangle + \langle(\bar{\beta}_z\,\delta z_1')^2\sin^2\theta_p\rangle$$

but since z_1 and $\delta z'$ are uncorrelated,

$$2\langle z_1\bar{\beta}_z\,\delta z'\sin\theta_p\rangle = 2\langle z_1\rangle\langle\bar{\beta}_z\,\delta z'\sin\theta_p\rangle = 0$$

which leaves,

$$\langle z_2{}^2\rangle = \langle z_1{}^2\rangle + \tfrac{1}{2}\bar{\beta}_z{}^2\langle\delta z'^2\rangle. \tag{10.80}$$

The mean rate of quantum emission per radiating particle is $\dot{N}_\gamma$, given by (10.23), so there will be a mean increase in $\langle z^2\rangle$ given by (10.80), on average every $1/\dot{N}_\gamma$ seconds. Thus the growth rate of $\langle z^2\rangle$ will be,

$$\frac{\mathrm{d}}{\mathrm{d}t}\langle z^2\rangle = \tfrac{1}{2}\dot{N}_\gamma\bar{\beta}_z{}^2\langle\delta z'^2\rangle. \tag{10.81}$$

Substitution for $\delta z'$ from (10.78) gives,

$$\frac{\mathrm{d}}{\mathrm{d}t}\langle z^2\rangle = \tfrac{1}{2}\dot{N}_\gamma\bar{\beta}_z{}^2\langle(\theta_\gamma u/E)^2\rangle. \tag{10.82}$$

The projected emission angle θ_γ and the quantum u are uncorrelated so,

$$\langle(\theta_\gamma u/E)^2\rangle = \langle\theta_\gamma{}^2\rangle\langle u^2\rangle/E^2.$$

The characteristic opening angle of the emission cone will be taken as $1/\gamma$, so that the mean square of the projected angle will be,

$$\langle\theta_\gamma{}^2\rangle = \tfrac{1}{2}\gamma^{-2}.$$

Furthermore, since (10.82) is true for each individual particle in the beam, it is also true for the whole beam and it is then more convenient to refer to the standard deviation (rms) of the vertical distribution σ_z and hence to the emittance $\mathcal{E}_z$,

$$\langle z^2\rangle \equiv \sigma_z{}^2 \simeq \bar{\beta}_z\mathcal{E}_z/\pi.$$

With these changes, (10.82) transforms to

Quantum excitation of vertical beam emittance:
$$\frac{\mathrm{d}\mathcal{E}_z}{\mathrm{d}t} = \frac{\pi\dot{N}_\gamma\langle u^2\rangle}{4E^2\gamma^2}\bar{\beta}_z. \tag{10.83}$$

10.8.2 In the presence of dispersion (horizontal motion)

The effect of an incremental energy loss on a betatron oscillation in the presence of dispersion has already been described in Section 10.7.3. In that particular case, all particles in the beam were subjected to the same average energy loss. In the present section, account will be taken of the statistical nature of the photon emission which makes the situation different for each particle. The incremental changes in the amplitude and slope of the oscillation have been taken from Section 10.7.3 and rewritten below for a single photon.

$$\delta x_\beta = -D(s)u/E \qquad (10.66)'$$

$$\delta x_\beta' = -D'(s)u/E. \qquad (10.68)'$$

The kicks (10.66)' and (10.68)' will be averaged along the lattice and over the beam, except that on this occasion full account will be taken of the lattice functions. This extra care will be justified at the end of this Section by showing that the effect of the momentum change in the presence of dispersion dominates the angle change. The result is also the key to high-brilliance lattices discussed in Section 10.11.

The unperturbed motion in the normalised coordinates $(Y, dY/d\mu)$, is a circle (see Section 4.1). The sudden loss of a photon kicks the particle by δY and $\delta(dY/d\mu)$. These increments depend on the local lattice functions and can be evaluated in terms of the real space increments δx_β and $\delta x_\beta'$ by the use of (4.3) and (4.22) from Chapter 4.

$$Y = A\cos[\mu(s) + \phi_0] = x(s)\beta_x^{-\frac{1}{2}}(s) \qquad \left.\right\} \qquad (10.84)$$
$$dY/d\mu = -A\sin[\mu(s) + \phi_0] = x(s)\alpha_x(s)\beta_x^{-\frac{1}{2}}(s) + x'(s)\beta_x^{\frac{1}{2}}(s)$$

The increments δY and $\delta(dY/d\mu)$ can therefore be expressed as,

$$\delta Y = \beta_x^{-\frac{1}{2}}\delta x \qquad \left.\right\}$$
$$\delta(dY/d\mu) = \alpha_x\beta_x^{-\frac{1}{2}}\delta x + \beta_x^{\frac{1}{2}}\delta x' \qquad (10.85)$$

and the modulus χ of the kick can be found from,

$$\chi^2 = (\delta Y)^2 + [\delta(dY/d\mu)]^2. \qquad (10.86)$$

Consider first one segment of lattice. Transfer a single kick from this segment forward to some time t_p [see Figure 10.11(a)],

$$Y_2 = Y_1 + \chi\cos\theta \qquad (10.87)$$

where the subscripts 1 and 2 represent the unperturbed and perturbed projections of the particle amplitude on the Y-axis. The square of (10.87) gives,

$$Y_2^2 = Y_1^2 + 2Y_1\chi\cos\theta + \chi^2\cos^2\theta. \qquad (10.88)$$

The average of many such kicks given over several turns to the same particle would introduce all phases so that,

$$\langle Y_2^2\rangle = \langle Y_1^2\rangle + 2\langle Y_1\chi\cos\theta\rangle + \langle\chi^2\cos^2\theta\rangle \qquad (10.89)$$

but since Y_1 and χ are uncorrelated,

$$2\langle Y_1\chi\cos\theta\rangle = 2\langle Y_1\rangle\langle\chi\cos\theta\rangle = 0$$

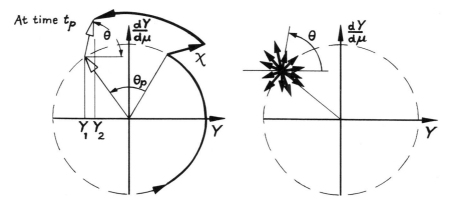

Figure 10.11. Effect of random photon emissions with dispersion. (a) Single quantum, (b) many quanta projected to time t_p.

which leaves

$$\langle Y_2^2 \rangle = \langle Y_1^2 \rangle + \tfrac{1}{2}\langle \chi^2 \rangle. \tag{10.90}$$

This produces the familiar 'star' vector diagram [Figure 10.11(b)]. The kick amplitude is now evaluated by the substitution of (10.66)′ and (10.68)′ into (10.85) and (10.86) to give,

$$\langle \chi^2 \rangle = \frac{\langle u^2 \rangle}{E^2} [\gamma_x D_x^2 + 2\alpha_x D_x D_x' + D_x'^2 \beta_x]. \tag{10.91}$$

The square bracket in (10.91) is the *dispersion* or *lattice invariant* $H(s)$ from (4.50), which was derived in Section 4.6.3. This function depends solely on the lattice and is constant in non-bending regions. Thus the probable increase in the square of the normalised betatron amplitude due to the emission of a single photon in a given segment of lattice can be found from (10.90) and (10.91) and written as,

$$\delta\langle Y^2 \rangle = \frac{\langle u^2 \rangle}{E^2} H(s). \tag{10.92}$$

The probable number of photons that will be emitted by the single radiating particle in a given segment will be $\dot{N}_\gamma(s)\,\delta s/c$, where $\dot{N}_\gamma$ is given by (10.23) with (10.7), so that,

$$\delta\langle Y^2 \rangle = \frac{\langle u^2 \rangle}{cE^2} \dot{N}_\gamma(s)H(s)\,\delta s. \tag{10.93}$$

Equation (10.93) can now be integrated around the ring. Only dipoles need to be included, since photon emission is zero elsewhere. The integral can be further simplified for an iso-magnetic lattice.

$$\delta\langle Y^2 \rangle = \frac{\langle u^2 \rangle}{cE^2} \oint_{\text{dipoles}} \dot{N}_\gamma(s)H(s)\,\mathrm{d}s = \frac{\langle u^2 \rangle}{cE^2} \dot{N}_\gamma \oint_{\text{dipoles}} H(s)\,\mathrm{d}s. \quad \text{(iso-magnetic)} \tag{10.94}$$

Since (10.94) is quite general, it applies equally to all particles in the beam and it is thus again more convenient to refer to the standard deviation of the distribution (rms) σ_x and hence to the emittance ε_x, as was already done for the vertical motion,

$$\langle Y^2 \rangle \equiv \langle x^2/\beta_x \rangle = \sigma_x^2/\beta_x = \varepsilon_x/\pi$$

so that finally,

*Quantum excitation
of horizontal beam
emittance:*

$$\frac{d\varepsilon_x}{dt} = \frac{\pi \dot{N}_\gamma \langle u^2 \rangle}{C_0 E^2} \oint_{\text{dipoles}} H(s)\, ds. \qquad \text{(iso-magnetic)} \qquad (10.95)$$

Equation (10.95) shows an amplitude-independent growth of the transverse emittance, which depends on the fundamental characteristics of the radiation and the lattice functions. This excitation and the damping described earlier by (10.76) will be opposed and lead to an equilibrium horizontal beam emittance, which can be found by equating the two expressions to give,

*Equilibrium between
damping and quantum
excitation:*

$$\varepsilon_{x,\text{eq}}^2 = \frac{\pi \dot{N}_\gamma \tau_x \langle u^2 \rangle}{4 C_0 E^2} \oint_{\text{dipoles}} H(s)\, ds. \qquad \text{(iso-magnetic)} \qquad (10.96)$$

It was mentioned in Section 10.8.1 that the horizontal excitation was far more important than the vertical. This can now be verified by comparing (10.95) with (10.83),

$$\frac{d\varepsilon_x/dt}{d\varepsilon_z/dt} = \frac{4\gamma^2}{C_0} \bar{\beta}_z^{-1} \oint_{\text{dipoles}} H(s)\, ds \simeq 4\gamma^2 \bar{H}(s)/\bar{\beta}_z.$$

The dominant term in $H(s)$ is D_x^2/β_x, and by the use of the approximations:

$$\bar{D}_x \simeq \rho_0/Q_x^2, \qquad \bar{\beta}_x \simeq R/Q_x, \qquad R \simeq \rho_0 \qquad \text{and} \qquad \bar{\beta}_z \simeq \bar{\beta}_x$$

this becomes, $\bar{H}(s)/\bar{\beta}_z \simeq 1/Q_x^2$, so that,

$$\frac{d\varepsilon_x/dt}{d\varepsilon_z/dt} \simeq 4\gamma^2/Q^2. \qquad (10.97)$$

Thus the natural vertical beam size is very much smaller than the horizontal one. For this reason, it is common in electron–positron colliders to increase the vertical emittance by coupling from the horizontal plane, in order to ensure that the beam–beam tune-shift limit is reached in both planes simultaneously.

10.9   DAMPING PARTITION NUMBERS

The three results (10.42), (10.76) and (10.77) for the damping in the three planes can be conveniently collected together as,

$$\tau_i = \frac{2ET_0}{U_\gamma J_i} \qquad (i \equiv l, x, z)$$

and

$$J_l = 2 + \tilde{D}, \qquad J_x = 1 - \tilde{D}, \qquad J_z = 1 \quad \text{and} \quad \sum J_i = 4.$$

(10.98)

The J_is are known as the *damping partition numbers*. The condition that the partition numbers add to four is a simple and general result[9].

There are two ways of influencing the damping rates.

10.9.1 Increase of U_γ

An increase in U_γ will increase the magnitude of all three rates. This is achieved by adding an element called a *dipole wiggler* to the lattice in a region of zero dispersion. A dipole wiggler comprises a series of dipoles of alternating sign, which add to give no net bending. In this case $\tilde{D}$ is essentially unchanged, but the radiated power can be increased considerably.

10.9.2 Design a suitable value of $\tilde{D}$

In combined-function lattices, $\tilde{D}$ can be large ($\simeq 2$ in an iso-magnetic ring) and consequently the horizontal oscillations are anti-damped. Since many of the early machines were of the combined-function type, this problem was met early on and several solutions were proposed[9,10]. The simplest was to displace the synchronous orbit inwards, which leads to unequal curvatures in the F and D units, which modifies $\tilde{D}$. It was found, for example, that in the CERN PS electrons could be made stable in all three degrees of freedom by an inward shift of 12 mm[10]. An alternative method which avoids any loss in aperture is the use of a *dipole–quadrupole* or *Robinson wiggler*[9]. This device comprises a series of magnets with alternate gradients, which give no net deviation but strongly affect the function $\tilde{D}$. This solution was first used for the By-pass at CEA[11] and later in the PS[12] when this machine became part of the injection chain for LEP. In modern, dedicated electron rings, the lattice is usually of the separated-function type and damping already occurs in all planes, but there is still a need to adjust the balance between the partition numbers and dipole–quadrupole wigglers are used.

10.10 RADIATION INTEGRALS

The six integrals defined below are specific to planar machines and are widely used in the literature on light sources. The reader will find that they are needed to form a bridge between accelerator-styled texts and light-source texts. These integrals were first introduced in Ref. 13, which also gives advice on their evaluation.

$$
\left.
\begin{aligned}
&I_1 = \oint \frac{D(s)}{\rho(s)}\, ds \qquad I_2 = \oint \frac{1}{\rho^2(s)}\, ds \qquad I_3 = \oint \frac{1}{|\rho^3(s)|}\, ds \\
&I_4 = \oint \frac{D(s)}{\rho^3(s)}\, [\Lambda - 2\rho_0{}^2(s)k(s)]\, ds \qquad I_5 = \oint \frac{1}{|\rho^3(s)|}\, H(s)\, ds
\end{aligned}
\right\}
\tag{10.99}
$$

where $\Lambda = 1$ for sector dipoles and $\Lambda = 0$ for straight dipoles. The principal results obtained in this chapter in terms of these integrals are,

$$
\left.
\begin{aligned}
&\alpha = I_1/C_0 \qquad\quad U_\gamma = \tfrac{2}{3} r_e m_0 c^2 \gamma^4 I_2 \qquad \tilde{D} = I_4/I_2 \\
&J_x = 1 - I_4/I_2, \qquad J_z = 1, \qquad\qquad J_l = 1 + I_4/I_2 \\
&\sigma_{l.\text{eq}}{}^2 = \frac{55}{32\sqrt{3}} \frac{\hbar}{m_0 c} \left[\frac{E^2}{m_0 c^2}\right]^2 \left[\frac{I_5}{2I_2 + I_4}\right] \\
&\varepsilon_{x,\text{eq}} = \frac{55\pi}{32\sqrt{3}} \frac{\hbar}{m_0 c} \left[\frac{E}{m_0 c^2}\right]^2 \left[\frac{I_5}{I_2 - I_4}\right].
\end{aligned}
\right\}
\tag{10.100}
$$

10.11 HIGH-BRIGHTNESS LATTICES

The most important parameter for light-source users is the *spectral brilliance* defined as,

$$
B_\lambda = \frac{d^4 N}{dt\; dS\; d\Omega\, (d\lambda/\lambda)}
\tag{10.101}
$$

where dS is the area of the source, $d\Omega$ is the solid angle and λ is the wavelength. The brilliance is usually expressed in the units of,

[photons, s^{-1}, mm^{-2}, mrad^{-2}, 0.1% relative bandwidth].

Machine designers, however, prefer to use *brightness* defined as,

$$
B = I/\varepsilon_x \varepsilon_z,
\tag{10.102}
$$

which has long been used in electron optics. Since the product of the source size dS and the solid angle $d\Omega$ is proportional to the product of the emittances $\varepsilon_x \varepsilon_z$ and the number of photons is proportional to the current I, the integral with respect to the wavelength of the spectral brilliance* defined in (10.101) is closely related to the brightness defined in (10.102).

Ultimately the brightness is limited by emittance growth mechanisms and, in particular, by the quantum excitation of the horizontal emittance as described in Section 10.8.2 and by equation (10.95). Since the excitation depends upon $H(s)$, the lattice designer can influence the final beam emittance in a fundamental way. This

---

* The integral over wavelength of the spectral brilliance as defined for a light source in (10.101) is identical to the *brightness* in classical optical photometry.

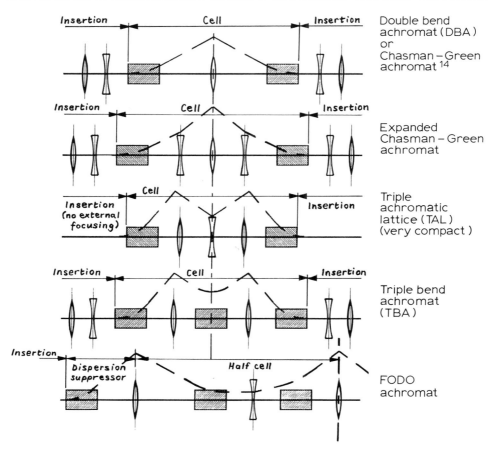

Figure 10.12. High-brightness achromats. The 'Double Bend Achromat' (DBA) is also referred to in the literature as a 'Double Focusing Achromat' (DFA), which appears to be a misnomer. The dispersion, D_x, is shown by dashed lines.

implies minimising D_x and optimising β_x to reduce $H(s)$ in the lattice dipoles. The standard solution is to create matched achromatic cells with short strong dipoles at the entry and exit and strong focusing in between. The layout is symmetric with the general feature that the dispersion builds up from zero at the entry to the first dipole, is focused back towards the axis in the central section and is reduced to zero by the final dipole. There are two basic categories of cells, those with and those without additional bending in the central dispersive region of the cell. Outside the cells there may be additional focusing or insertion devices such as wigglers and *undulators**.

---

* The angular divergence of the radiation from a wiggler is greater than the 'natural' value $1/\gamma$, whereas in an undulator it is of the same order. As a consequence, the radiation from a wiggler is incoherent while that from an undulator is coherent. Wigglers have few poles and are used for beam control, while undulators have many poles and are used as high-brilliance sources.

Several designs of high-brightness cells are summarised in Figure 10.12. These designs maintain the dispersion small, but only at the expense of very strong focusing, which in turn makes it necessary to add sextupoles. Insertion devices must also be carefully matched into the structure so that the overall optimising of the lattice is far from trivial.

CHAPTER 11

Diagnosis and compensation

There is much truth in the statement that a machine is only as good as its diagnostic equipment, but paradoxically it often happens that this is the first area to feel the cold draught of budget cuts. The careful preparation of algorithms for controlling machine parameters and plentiful instrumentation saves time and avoids the random chance of trial-and-error adjustments. Time spent in observation and understanding is always well spent and leads to future improvements in performance.

When commissioning an accelerator ring the first problem is to get a circulating beam. Apart from the jubilant press release announcing the first injection (of importance for funding agencies) little can be done until the beam makes at least a full circuit of the machine. This is primarily a problem of the closed orbit. The first step is taken during the machine design when a prognosis is made of the expected closed-orbit distortion, which is then used to set tolerances. Later algorithms are developed for measuring and correcting the orbit under various conditions as well as some tools for diagnosing the errors to guide the survey team. Orbit measurements can also be used for checking many optics parameters. Next on the list is the tune of the machine. The tune can be deduced from orbit measurements by noting the beam positions on four consecutive passages, but unless the beam is being lost after a few turns, it is more convenient and accurate to apply other techniques. Once the tune measurement is fully operational, attention will turn to measuring the chromaticities. It is difficult to ascribe an exact order to events when commissioning a machine but the measurement of profiles and emittances will also be of early interest. Several aspects of machine control will be covered in this chapter while luminosity is left for Chapter 12.

For those readers intending to specialise in diagnostics, the book in Ref. 1 is entirely devoted to this topic. References 2 to 4 are pedagogic papers on beam instrumentation. The first contains many instruments specifically for transfer lines, while the next two are more general and are particularly well furnished with references.

11.1   CLOSED ORBIT

11.1.1   Prognosis for random errors

The machine will inevitably contain many errors affecting the closed orbit. The steel quality in the magnets and the packing factor of the laminations will not be entirely constant and the machining and alignment will not be perfect. All these errors will contribute to the closed-orbit distortion according to (5.4). Most will be random in nature, while others will be systematic and some may be time or field dependent. There are many computer codes that can calculate exact solutions for any desired configuration of errors, but it is useful to have some means of estimating the expected orbit distortion from random errors for the purpose of determining tolerances.

Once estimates of the various random errors have been made, the expected root mean square (rms) amplitude of the orbit distortion averaged around the machine and over many sample machines can be found. In one hypothetical machine, the square of the distortion at any point [by the use of (5.4)] would be

$$[y(s)]^2 = \frac{\beta(s)}{4\sin^2 \pi Q}\left[\sum_n \beta_n^{\frac{1}{2}}\delta_n \cos[|\mu_n - \mu(s)| - \pi Q]\right]^2,\tag{11.1}$$

where δ_n is the angular kick due to the errors in the nth magnet and is given by,

$$\delta_n = \frac{l\,\Delta B}{B_0\rho_0}\text{ (dipoles)}\qquad \delta_n = -lk\,\Delta y\text{ (quadrupoles).}$$

It is reasonable to assume that the errors between magnets are uncorrelated so that only the δ_n^2 terms need to be retained in the sum in (11.1), which can then be averaged around the ring ($\langle\cos^2\rangle = \frac{1}{2}$) and over many sets of δ_n error values ($\sqrt{\langle\delta^2\rangle} = \delta_{\text{rms}}$) to give,

$$\langle y^2\rangle^{\frac{1}{2}} = \begin{cases} \dfrac{\bar\beta^{\frac{1}{2}}}{2\sqrt{2}|\sin\pi Q|}\left[\sum_n \beta_n\langle\delta_n^2\rangle\right]^{\frac{1}{2}} & \text{replace }\bar\beta\text{ by }\beta(s)\text{ for specific position,} \\[3mm] \dfrac{\bar\beta n^{\frac{1}{2}}\delta_{\text{rms}}}{2\sqrt{2}|\sin\pi Q|} & \text{simple case for a single class of errors.} \end{cases}\tag{11.2}$$

In the simple form of (11.2), the average value of β around the ring has been assumed equal to the average β at the errors. In general, the more important contributions to the closed orbit distortion are due to the quadrupole alignment errors, which are typically 0.1 mm rms.

11.1.2   Obtaining the first turn

(i) *Using beam position only*

Establishing the first turn relies on the same steering techniques that will already have been used in the injection transfer line, except that a ring is usually better

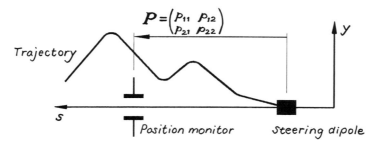

Figure 11.1. Point-to-point steering.

equipped with beam position monitors. The simple, but effective, method is *point-to-point steering*. The beam is centred on each monitor in the line in succession. This only makes use of the p_{12} term in the transfer matrix P between the corrector and the pickup.

$$\begin{pmatrix} \Delta y \\ \Delta y' \end{pmatrix} = \begin{pmatrix} p_{11} & \vdots\, p_{12}\, \vdots \\ p_{21} & p_{22} \end{pmatrix} \begin{pmatrix} 0 \\ \delta \end{pmatrix} \qquad (11.3)$$

where δ is the angular kick given by the corrector dipole. Clearly the sensitivity is improved by maximising p_{12}. This means that the monitor should be about $\pi/2$ in betatron phase downstream of the corrector (see Figure 11.1). The next corrector would then be placed near this monitor and so on. In practice, this simple layout with a corrector and monitor every quarter betatron wavelength falls part way between two situations. In a transfer line, it is permissible to omit many of these monitor/corrector pairs, especially in long-straight sections, whereas in a ring, only four monitors and correctors per betatron wavelength is considered as a minimum.

There are many types of beam position monitor. The semi-destructive types such as luminescent screens and secondary emission monitors can be useful in transfer lines, but in a ring the monitors must be non-destructive. There are three principal types: electrostatic, magnetic and electromagnetic, which all work with bunched beams. The electrostatic monitors, or *pickups*, sense the charges induced by the bunch on two plates on either side of the chamber. The difference of the induced signals is proportional to the transverse position of the beam, whereas the sum is proportional to the total beam current and is independent of position. The sum signal can be used as an additional diagnostic tool for localising beam losses. Magnetic detectors have the useful feature of not being affected by secondary particles, since they do not rely on measuring charge. In single-ring colliders, the electromagnetic directional-coupler monitors make it possible to detect one beam in the presence of the counter-rotating beam. The details of these monitors and others can be found in Refs 1–4.

(ii) *Extension of point-to-point steering*

Once a corrector is set, its effect at all other position monitors downstream is known. The overall response of the correction array can be summarised in matrix notation as,

$$
\begin{pmatrix} \Delta y_1 \\ \Delta y_2 \\ \vdots \\ \Delta y_n \end{pmatrix} = \begin{pmatrix} t_{11} & 0 & 0 & \cdots & 0 \\ t_{21} & t_{22} & 0 & \cdots & \vdots \\ t_{31} & t_{32} & t_{33} & & 0 \\ \vdots & \vdots & \vdots & \cdots & t_{nn} \end{pmatrix} \begin{pmatrix} \delta_1 \\ \delta_2 \\ \vdots \\ \delta_n \end{pmatrix}
\tag{11.4}
$$

where t_{ij} is the 'p_{12}' matrix element from the transfer between the ith monitor and the jth corrector. Very often the beam will survive a large fraction of the first turn, or perhaps a few turns, but with large excursions. It is necessary to correct these excursions in order to increase the likelihood of the beam surviving once the trajectory is closed after the first turn by using the position and angle steering described in the next section. Rather than laboriously steering step-by-step through all the monitors, it may be quicker and more efficient to correlate the observed orbit excursions with the calculated orbit deviations from each of the correctors in turn with the help of (11.4), in search of one which matches the observed trajectory the closest and can therefore be used to compensate it. This simple extension of the basic point-to-point steering can save considerable time and is a standard technique in transfer lines.

(iii) *Using position and angle*

The use of two adjacent monitors makes it possible to measure both beam position and angle and similarly, two adjacent magnets can be used to adjust position and angle (see Figure 11.2). For example, the ideal case of two monitors with a position resolution of 0.1 mm and separated by a drift space of 10 m can resolve angles of the order of 0.01 mrad. This slightly more sophisticated method might be used first at injection and later to close the first turn.

(iv) *Extension of position/angle steering*

The simple strategy described in (iii) can be extended to a global correction technique, which has become known as the *beam threader*. The principle is to search the measured orbit for the characteristic form of an unperturbed betatron oscillation, such as would follow a large dipole error. It is unlikely that any single corrector will be positioned so as to give an exact correction, but with two correctors the compensating oscillation can be launched with exactly the right amplitude and phase. The precision and flexibility of such a correction is best when the corrector magnets are orthogonal (i.e. separated by $\pi/2$ in betatron phase). If several such regions

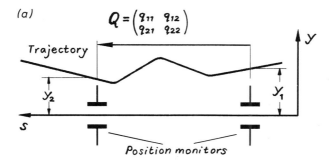

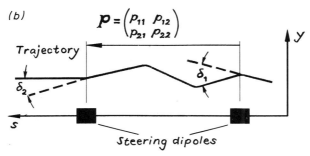

Figure 11.2. Position and angle measurements and steering. (a) $y_2 = y_2$ (measured), $y_2' = (q_{21} - q_{11}q_{22}/q_{12})y_1 + (q_{22}/q_{12})y_2$; (b) $\Delta y_2 = p_{12}\delta_1$, $\Delta y_2' = p_{22}\delta_1 + \delta_2$.

can be identified in the measured orbit, then compensating oscillations can be launched and stopped sequentially around the machine.

This technique was found to be well suited to a rather specific problem encountered in the CERN ISR[5]. When the superconducting low-β insertion was first introduced, it was feared that the combination of the strong gradients with small unavoidable misalignments would cause unacceptable orbit distortion and the above method was prepared as a supplement to the normal orbit correction programs. First the orbit was corrected in the normal way. The insertion was then powered and the orbit remeasured, *but only for the first turn*. In this way, the beam could be dumped before it returned to the insertion and possibly quench the magnets. The difference orbit showed an unperturbed betatron oscillation propagating around the machine with its source in the insertion. This could then be corrected as described above, so that the beam would re-enter the insertion on axis. Of course, there was still some distortion inside the insertion, but by establishing the first passage through the insertion as part of a corrected orbit for the whole machine, the danger was removed that on subsequent turns the insertion would be hit and quenched.

(v) *Injection optimisation*

The beam threader technique is ideally suited to injection optimisation. First the injection error is found by subtracting the average closed orbit from the first-turn orbit and then the inverted oscillation is launched using the last correction elements in the transfer line[6]*.

11.1.3   Algorithms for correcting circulating beams

Over the years considerable ingenuity has been put into closed-orbit correction algorithms and only the main points can be discussed here.

(i) *Overlapping orbit bumps*

The first reflex is to imitate point-to-point steering by applying bumps individually to each peak in the orbit distortion. For example, in a perfectly regular lattice with a phase advance of μ_0 per period, let the beam position monitors and the correction magnets be co-incident in each period. Any consecutive group of three magnets can be used to excite a '3-magnet bump' (see Section 5.1.2). Re-arrangement of the equations in Section 5.1.2 gives three expressions linking the strengths of the kicks to the amplitude of the bump at the centre monitor.

$$\delta_1 = \frac{1}{\beta \sin \mu_0} y_2; \qquad \delta_2 = -\frac{2 \cos \mu_0}{\beta \sin \mu_0} y_2; \qquad \delta_3 = \frac{1}{\beta \sin \mu_0} y_2. \qquad (11.5)$$

In this idealised lattice each bump affects only its central monitor, but each corrector is part of three separate bumps. Clearly the distortion measured at each monitor can be made zero by applying the appropriate bump (if errors in both monitors and correctors are neglected). In this situation, the net kick on the nth corrector will be,

$$\delta_n = \frac{1}{\beta \sin \mu_0} y_{n-1} - \frac{2 \cos \mu_0}{\beta \sin \mu_0} y_n + \frac{1}{\beta \sin \mu_0} y_{n+1}. \qquad (11.6)$$

In a region where y_n is an unperturbed betatron oscillation (between errors), it will have the form $y_n = \sqrt{\beta} \cos(n\mu_0 + \phi)$ where ϕ is a simple constant. This expression can be substituted into (11.6) and a little algebraic manipulation will show that the net kick given by (11.6) in such a region will be zero. Thus the straightforward correction of the orbit by setting all values of y_n to zero will not lead to unnecessary excitation of magnets outside the region of the field errors†. This idealised model is

---

* The term 'beam threader' is relatively recent and is not used in Refs 5 and 6. Reference 6 also describes a method which detects only the amplitude of the injection oscillation and by adding known vectors to the injection orbit searches for a minimum in the injection error.
† This analysis is attributed to T. Collins in *Accélérateurs circulaires de particles* by H. Bruck (Presses Universitaires de France), 116–17.

generalised in Ref. 7 into a practical method which takes into account the different positions of monitors and correctors.

A variant of the above, which is described in Ref. 8, relates the beam displacements Δy_n to the single free parameter of each bump by a matrix $A(n, m)$. The free parameter could be the kick amplitude in the first corrector δ_m; the others then being determined. Since any one bump will only cover two, or three monitors the matrix A will contain mainly zeros, thus easing the computation. The problem can then be treated directly as a least squares problem solved by,

$$[\delta_m] = -(A^{\mathrm{T}}A)^{-1}A^{\mathrm{T}}[\Delta y_n] \tag{11.7}$$

where T indicates the transpose of a matrix.

(ii) *Use of harmonics*

It has already been shown in Section 5.1.3 that field errors which are distributed around the ring as a pure sinusoidal harmonic of the function $\beta^{3/2} \Delta B/(B_0\rho_0)$ will excite the same harmonic in the closed orbit except that its amplitude is modified by the amplification factor $Q^2/(Q^2 - n^2)$ where n is the order of the harmonic. The amplification factor enhances those harmonics which are close to Q. This fact can be turned into a correction strategy for the closed orbit[9]. With m correctors (preferably equally distributed in betatron phase), it will be possible to excite $m/2$ sine harmonics and $m/2$ cosine harmonics with a reasonable purity provided that m is large enough to control the harmonics close to Q. Once these distributions are calculated, it only remains to Fourier analyse the observed closed orbit distortion and to sum over the contributions of each harmonic to find the corrector strengths. The method works well, but it is best adapted to regular structures as found in accelerators, as opposed to storage rings. The Fourier analysis is also more accurate when the monitors are equi-spaced in betatron phase. The need for purity in the harmonics and the order of the Fourier analysis implies that several units are required per betatron wavelength. If strong localised corrections should be avoided, then this method is favoured. It is also useful when on-line computing power is limited, since all the large matrix inversions to create the harmonics are done off-line and only once. The Fourier analysis is then the major part of the on-line calculation. The method is also 'resistant' to random errors in the position monitor readings. If, however, there are problems with reliability when powering all corrector magnets, or with the ripple control at low current levels, it is better to leave this method and pass to the next category of correction strategies which specifically search for only a small number of 'effective' correctors.

(iii) *Finding the 'best' corrector*

In colliders with their rather inhomogeneous lattices, it is more useful to search for a small group of the most effective correctors. The very simplest approach is to scan through the correctors one by one to find the 'best' by whatever criterion is most easily applied. The following applies the least squares minimisation method to finding the corrector which has the largest correlation product with the observed distortion. Let Δy_n be the deflection seen in the nth monitor due to a kick δ_m from the mth corrector, so that,

$$\Delta y_n = t_{nm}\delta_m \qquad (11.8)$$

where t_{nm} is the matrix element that relates the kick δ_m to the orbit shift Δy_n and is found by applying the closed-orbit distortion formulae (5.3) or (5.4).

If y_n is the observed orbit distortion, then the corrected orbit (using the mth corrector) would be $(y_n - \Delta y_n)$. The correction criterion requires that the sum of the squares of the residual distortion,

$$S_m = \sum_n (y_n - t_{nm}\delta_m)^2 \qquad (11.9)$$

be a minimum, i.e.

$$0 = \frac{\partial S_m}{\partial \delta_m} = -2\sum_n (y_n - t_{nm}\delta_m)t_{nm},$$

which yields,

$$\delta_m = \sum_n y_n t_{nm} \bigg/ \sum_n (t_{nm})^2. \qquad (11.10)$$

By performing these simple sums, it is possible to find a 'best' corrector and the optimum current level to set.

The main advantages of the above method are simplicity and a modest need for on-line computing power. Since modern accelerators are not usually limited by the latter more sophisticated algorithms are preferred; however, the above can still be useful in certain specific applications. For example, in the ISR a physics detector magnet known as the Split Field Magnet (SFM) required two dipole correctors to compensate the otherwise catastrophic orbit distortion. Occasions arose when this magnet was used at unmapped field levels and the compensator settings could only be estimated. The above method was then used to correlate the residual distortion with the two compensators in order to optimise their settings.

(iv) *Finding the 'best' group of correctors*

One sure way of finding the best group of correctors is to solve the least squares minimisation of the orbit distortion for all possible sub-sets of the correctors. This would be a long, if not impossible, task and unpopular with machine operators who like to see the numerical prediction of a correction within tens of seconds. It is therefore necessary to find a way of selecting sub-sets of correctors that are likely to give a good correction.

The method in the previous section selects the 'best' corrector. Once this corrector is activated (either numerically or in practice), the residual orbit can be re-analysed and the 'next best' can be found. Each iteration will add a new member to the corrector set. This has the disadvantage of 'freezing-in' the strength of a corrector at the time it is selected. It is not difficult to find scenarios in which this strategy will lead to conflict between the correctors set and the emergence of orbit bumps. Since orbit bumps require the combination of at least two correctors, an algorithm which rigorously selects one magnet at a time will continually introduce compromise correctors into the sub-set. Even with more sophisticated algorithms, it is not wise to repeatedly set up a machine with the closed-orbit correction from the previous run and then to simply add a further correction.

The above objection of 'freezing-in' correctors is partly overcome in the MICADO[10] algorithm, developed for the ISR. The algorithm considers corrector sub-sets of increasing size until halted. For each sub-set, it selects the correctors on the basis of correlation with the residual orbit and then solves the least squares minimisation, as formulated in (11.7), for all the correctors in that sub-set. This allows the algorithm to re-adjust all corrector strengths for each sub-set and even to effectively remove a corrector by setting it to zero. The least squares minimisation filters out noise that does not correlate with the optics model, i.e. it is insensitive to errors introduced via the monitors[8]. MICADO also makes a numerical prediction of the closed-orbit distortion after each iteration. The standard deviation of the orbit will be progressively reduced, but this does not mean that the peak-to-peak value will necessarily be reduced, or that the corrector strengths will remain reasonable. The operator therefore selects the iteration which in his estimation is the most satisfactory.

So far only the least squares method has been mentioned. In general, a class of norms for the vector $[y_n]$ can be defined as,

$$\| y_n \|_l = \left(\sum_n |y_n|^l \right)^{1/l}$$

where l is an integer. When $l = 2$ then $\| y_n \|$ is the Euclidean norm which is used in the least squares minimisation. At the extreme of $l = \infty$ then $\| y_n \|$ is the Tchebychev norm and the peak–peak value of the orbit is selected. The $l = 1$ norm treats the errors without a weighting for their amplitude. These alternative possibilities have been used in the CERN SPS[11].

It was noted above that the corrector strengths may not remain reasonable in magnitude. In fact, the magnet builders and survey teams have, by their skill, been remarkably successful in avoiding this problem and the MICADO algorithm has become extremely popular, but an alternative, which includes the minimisation of the corrector strengths did already exist. This method[12], which was used in the CERN Booster, starts by minimising the squares of the amplitudes and slopes of the corrected orbit summed over the monitors, plus the squares of the kick strengths summed over the correctors, rather than just the amplitudes as in (11.9). The analysis finds eigenvectors for the correction kicks and arranges them in order of importance.

11.1.4 Measurements and diagnosis

(i) 'Sawtooth'

In electron–positron colliders, the energy loss due to synchrotron radiation is frequently sufficient to make a measurable shift in the closed orbit via the dispersion. The energy loss and hence also the orbit shift progressively increases through the bending sections of the lattice to be abruptly corrected in the rf cavities. This gives rise to the popular name *energy sawtooth*. The sawtooth should be removed numerically before a closed orbit algorithm is applied. This can be a theoretical correction (see Chapter 10 for the general formulae governing radiation) or an empirical one made by measuring the difference orbit between the electrons and positrons.

(ii) Rapid measurements[8]

Modern beam position monitors are often capable of storing the values for many individual passages of the beam either consecutively, or at a sampling rate. In a situation where the beam is lost after a few tens or hundreds of turns this facility can be invaluable.

The beam position is given by the linear sum of the closed-orbit distortion and the superimposed betatron oscillation. The latter can be calculated for n turns by applying the nth power of the single-turn transfer matrix [i.e. (4.18) with $\mu_0 = 2\pi Q$ for a single turn], so that,

$$y_n = y_{\text{c.o.}} + (\cos 2n\pi Q + \alpha \sin 2n\pi Q)y_0 + \beta \sin 2n\pi Q y_0'. \tag{11.11}$$

Equation (11.11) predicts that two consecutive turns will be related according to,

$$\left. \begin{array}{l} y_1 = y_{\text{c.o.}} + (\cos 2\pi Q + \alpha \sin 2\pi Q)y_0 + \beta \sin 2\pi Q y_0') \\ y_2 = y_{\text{c.o.}} + (\cos 4\pi Q + \alpha \sin 4\pi Q)y_0 + \beta \sin 4\pi Q y_0') \end{array} \right\} \tag{11.12}$$

where y_0 is the position of the beam at the start to the first turn and y_1 and y_2 the corresponding values at the ends of the first and second turns. After a minor re-arrangement of (11.2), the two equations can be subtracted to eliminate y' and to give,

$$\frac{(y_1 - y_{\text{c.o.}})}{\sin 2\pi Q} - \frac{(y_2 - y_{\text{c.o.}})}{\sin 4\pi Q} = \left[\frac{1}{\tan 2\pi Q} - \frac{1}{\tan 4\pi Q} \right] y_0$$

which reduces to,

$$\cos 2\pi Q = \frac{1}{2} \left[\frac{y_0 + y_2 - y_{\text{c.o.}}}{y_1 - y_{\text{c.o.}}} \right]. \tag{11.13}$$

Since equation (11.13) is valid for any two consecutive turns, then a third turn would be related to the second by,

$$\cos 2\pi Q = \frac{1}{2}\left[\frac{y_1 + y_3 - y_{c.o.}}{y_2 - y_{c.o.}}\right].$$ (11.14)

Equations (11.13) and (11.14) can now be solved for Q and $y_{c.o.}$

$$\cos 2\pi Q = \frac{1}{2}\left[\frac{y_0 - y_1 + y_2 - y_3}{y_1 - y_2}\right]$$ (11.15)

$$y_{c.o.} = \left[\frac{y_1{}^2 - y_2{}^2 + y_1 y_3 - y_2 y_0}{2y_1 - 2y_2 + y_3 - y_0}\right].$$ (11.16)

Hence with four consecutive position measurements (just three turns) each position monitor will yield a value for the non-integral part of the tune and the local closed-orbit distortion. Apart from the obvious value of this technique when commissioning a machine, it also provides a unique way of measuring the tune when the beam is unstable. For example, the tune shifts due to families of sextupoles can be measured individually at all current levels, regardless of the beam stability.

(iii) *Diagnosis*

Large errors can be found with reasonable accuracy by fitting a sine wave through consecutive groups of normalised orbit readings. When the fit breaks down with a phase shift between the sine waves on either side of a problem region, this indicates the presence of a dipole error. By extrapolating towards the error region from each side, the probable position and magnitude can be found[12,13]. Once a specific magnet is suspected of causing the error, the distortion due to this magnet can be calculated and then correlated with the observed orbit. Faulty monitor readings are relatively easy to recognise especially when isolated.

11.2 TUNE MEASUREMENT

In most cases, it is the incoherent tune value that is required, whereas in all the following methods, except the use of Shottky noise, it is the coherent tune, albeit often of a small fraction of a beam, that is measured (see Chapter 8 for details of incoherent and coherent tune). The distinction is usually small enough to be ignored, but it should not be completely forgotten.

11.2.1 The 'kick method' in the time domain

Consider a single bunch circulating with a coherent betatron oscillation. This oscillation will most probably have been excited by a kicker with a pulse time less

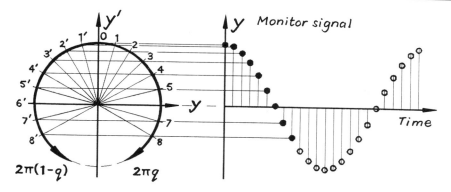

Figure 11.3. The transverse position of an oscillating bunch recorded once per turn.

than one revolution period, but may equally well be the residual oscillation from an injection error. If the beam is now observed by a position monitor a series of readings will be obtained which give the oscillation amplitude at each full turn of the machine. The fractional part q of the tune is responsible for the phase lag or lead between consecutive points (see Figure 11.3).

As Figure 11.3 shows, the integer part of the tune value is not deducible from such an observation, but fortunately it is easily found by studying a full single-turn orbit measurement of a betatron oscillation. Figure 11.3 also shows that there is a confusion between oscillations that are lagging ($0.5 < q < 1.0$) or leading ($0 < q < 0.5$) in phase on each turn. Again this can only be resolved by referring to an alternative method such as studying the form of a single-turn oscillation, or by changing a focusing element and checking whether q increases or decreases. When a machine is first commissioned, it is unlikely that all the diagnostic equipment will be available from the first day and the tune may have to be estimated from a photograph of an oscilloscope trace similar to that in Figure 11.3.

11.2.2   The 'kick method' in the frequency domain

The above method is simple and easy to understand, but automated measurements are more quickly and more accurately made by analysing the frequency spectrum of the pulse.

Consider the transverse betatron oscillation of a very short bunch or a narrow slice of a coasting beam, which has been excited by a fast kicker. The displacement of the transverse oscillation has already been given in (11.11). For the present purpose, there is no loss of generality if the harmonic terms are summed and the origin is chosen to suppress the constant phase term. In complex notation, the signal will then be,

$$y = y_{c.o.} + A \, e^{j2\pi n Q_{coh}} \qquad \text{for } n = 1, 2, 3, \ldots \qquad (11.11)'$$

263

where the subscript 'coh' denotes the coherent value for the bunch. These points lie on the curve (11.17) illustrated in Figure 11.3,

$$y = y_{\text{c.o.}} + A\, e^{j\Omega_{\text{coh}}qt}.\tag{11.17}$$

Longitudinally, it will be assumed that the bunch current appears as a delta function repeating once per turn [see Figure 11.4(a)], which can be re-interpreted in the frequency domain as an infinite series of equal amplitude harmonics of the revolution frequency [see Figure 11.4(b)]. These two interpretations are expressed in (11.18),

$$\lambda(t) = \frac{\Lambda}{C}\sum_{n=-\infty}^{n=\infty}\delta(t - nT_{\text{coh}}) = \frac{\Lambda}{C}\sum_{n=-\infty}^{n=\infty} e^{jn\Omega_{\text{coh}}t}\tag{11.18}$$

where $\lambda(t)$ is the line charge density and Λ is the charge in the oscillating bunch. The finite width of a practical bunch would cause these harmonics to diminish in amplitude as the order increases and for a long bunch, it would also become apparent that the envelope of the amplitudes is modulated periodically to zero. For a practical system, it is prudent to check that the first zero in the envelope does not happen to be at, or close to, the frequency foreseen for making the measurements.

The beam monitor will respond to the product of (11.17) and (11.18), which is the *dipole moment* of the beam,

$$y\lambda(t) = y_{\text{c.o.}}\frac{\Lambda}{C}\sum_{n=-\infty}^{n=\infty} e^{jn\Omega_{\text{coh}}t} + \frac{\Lambda}{C}A\sum_{n=-\infty}^{n=\infty} e^{j\Omega_{\text{coh}}(n+q)t}.\tag{11.19}$$

The first term in (11.19) is known as the *common mode signal*. It depends on the

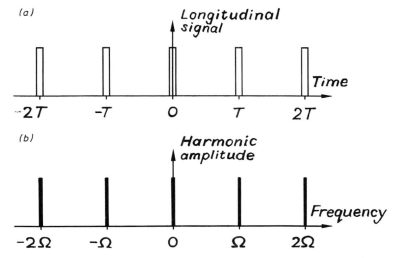

Figure 11.4. Signal from a delta pulse in time and frequency domains. (*a*) Time domain, (*b*) frequency domain.

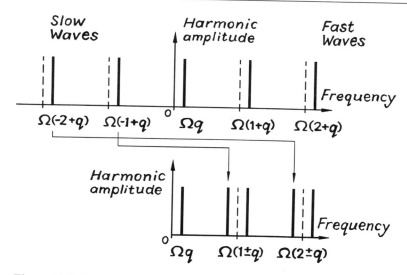

Figure 11.5. Formation of the frequency spectrum. The lower part of the diagram shows reflection of negative frequency slow waves onto positive frequency side of spectrum.

closed orbit and is of no direct interest here, except to say that it spoils the signal-to-noise ratio for the second term and in some applications it is worthwhile making orbit corrections or even designing the monitor so that it can be centred on the beam. It should be noted that q could be replaced by Q or any finite integer plus q and the expression (11.19) remains unchanged. Thus the integer part of Q cannot be extracted, as was already seen in Section 11.2.1.

The second term in (11.19) contains both negative and positive frequencies, which corresponds to the fast and slow waves. The beam monitor, however, can only sense transverse displacements and not the direction of energy flow in the wave, so this distinction is lost and all frequencies appear as positive in the spectrum (see Figure 11.5). The second term expressed with only positive frequencies takes the form given in (11.20), from which a spectrum analyser could extract the value of q from either the low frequency line $q\Omega$ or from the higher frequency harmonics. Since the fast and slow waves cannot be distinguished, it is not possible to detect from the spectrum whether $0 < q < 0.5$ or $0.5 < q < 1.0$ as was already found in Section 11.2.1.

$$\text{2nd term L.H.S. of (11.19)} = A\frac{\Lambda}{C}e^{j\Omega_{\mathrm{coh}}qt} + A\frac{\Lambda}{C}\sum_{n=1}^{n=\infty}[e^{j\Omega_{\mathrm{coh}}(n+q)t} + e^{j\Omega_{\mathrm{coh}}(n-q)t}] \quad (11.20)$$

11.2.3   RF knockout, swept-frequency and Q-diagram meter

In contrast to the 'kick method' described above, the beam can be continuously excited. This method was already in use at the Michigan synchrotron in

1954[14]*. If the excitation matches one of the frequencies in (11.20), then the particle amplitudes will grow in resonance and the particles will eventually be lost. This is known as *RF knockout*. By placing a collimator close to the beam, sweeping the frequency and monitoring for loss, the tune of a beam pulse or the tune spread in a beam can be measured. In this simple form the method is semi-destructive, but it is possible to be more subtle.

One common setup is to use a swept frequency spectrum analyser and generator. The generator, which is used to excite the beam, is made to track the spectrum analyser, which is watching the transverse signal from a beam position monitor. The response of the beam will change from non-resonance to resonance once the generator frequency matches the tune value. This *swept-frequency tune measurement* can be made at very low excitation levels and can in practice be regarded as non-destructive. An extremely useful refinement of this method is the measurement of the *beam transfer function*, which is described separately in Section 11.3, since it is better known for the information it yields on beam stability.

Another variant is the *Q-diagram meter*[15], which was specially adapted to coasting beams with large momentum spreads. This method has the advantage of relating the tune values to the radial position, or momentum deviation. An empty rf bucket is scanned through the stack and at a series of radial positions the swept frequency tune measurement is made, as described above, except that the position monitor is made sensitive to the inverse signal coming from the empty bucket. This makes it possible to detect the radial position of the resonant particles. The results can be directly plotted as a working line in the tune diagram, which gives the method its name.

With modern electronics, it is increasingly easy to apply sophisticated signal treatment techniques. Reference 16 describes the tune measurement systems implemented in LEP and includes *resonant excitation* with a *Phase-Locked Loop* (*PLL*) and *random noise excitation* with *Fast Fourier Transform* (*FFT*).

11.2.4   Schottky noise

The finite number of discrete charges in an accelerator beam and their random distribution gives rise to an electromagnetic noise called *Schottky noise*, named after the same phenomenon seen much earlier in electron tubes[17]. This noise, which can be detected by the longitudinal and transverse beam monitors, was first observed in accelerators in the ISR[18], where the stable beams afforded sufficient integration time for the signals to become apparent. The mathematical description of Schottky noise is a simple extension of the signal theory developed in the previous sections for tune measurements on bunched beams. Schottky measurements are totally passive and

---

* In Ref. 14, the rf excitation was applied to a bunched beam and both the betatron and synchrotron frequencies were observed. This is similar to the synchrotron frequency measurement to be described in Section 11.2.4(iii).

can be used to find betatron tune values, longitudinal beam densities and emittances in coasting beams and the synchrotron and betatron tune values in bunched beams[19,20].

(i) *Longitudinal spectra in coasting beams*

Equation (11.18) is directly applicable to a single particle once the charge Λ is reduced to e and the coherent bunch parameters C_{coh}, T_{coh} and Ω_{coh} are replaced by C, T and Ω for the specific particle.

$$\lambda(t) = \frac{e}{C} \sum_{n=-\infty}^{n=\infty} \delta(t - nT) = \frac{e}{C} \sum_{n=-\infty}^{n=\infty} \mathrm{e}^{\mathrm{j}n\Omega t} = \frac{e}{C} + 2 \sum_{n=1}^{n=\infty} \mathrm{e}^{\mathrm{j}n\Omega t}. \tag{11.21}$$

The longitudinal spectrum of a coasting beam will simply be the sum of the signals, of the form (11.21), from all of the particles at different phases. The full spread $\Delta\Omega$ in the angular revolution frequencies, due to the full spread $\Delta p/p_0$ in momentum according to (7.28), will transform the lines in Figure 11.4 into bands known as *Schottky bands*. As n increases these bands become wider, until they start to overlap and finally merge to form a continuous spectrum (see Figure 11.6).

Band width:
For no overlap:

$$\left. \begin{aligned} \Delta\Omega_{\mathrm{band}} &= n\,\Delta\Omega = n\Omega_0|\eta|\,\Delta p/p_0 \\ n &< (|\eta|\,\Delta p/p_0)^{-1}. \end{aligned} \right\} \tag{11.22}$$

Within a band the Schottky signal (voltage) is proportional to the square root of the number of particles at that point in momentum space. Thus each band contains basically the same information: an image of the square root of the current density in the beam versus momentum (or revolution frequency). The choice of the band is a compromise between too high frequencies that overlap and too low frequencies that require longer integration times. This is already a valuable diagnostic tool, but it becomes even more useful once combined with the transverse scan.

(ii) *Transverse spectra in coasting beams*

The transverse spectrum of a coasting beam can be found in an analogous way to the longitudinal one by applying (11.20) to each unit charge circulating with a transverse betatron oscillation. The spectrum will then simply be the sum of the slow wave and the fast wave signals from all of the particles. The individual lines will be spaced out into bands by the spread in the angular revolution frequencies due to the momentum spread (as before) and by the spread in the betatron tune, due to chromaticity (see Figure 11.7). Any amplitude–tune dependence will also contribute to widening the bands. The width of the bands can be found by introducing the angular revolution frequency spread, $\Delta\Omega$, and the betatron tune spread, Δq, into the exponent of (11.20) to give,

$$\Delta\Omega_{\mathrm{fast,\,slow}} = (n \pm q)\,\Delta\Omega \pm \Omega_0\,\Delta q$$

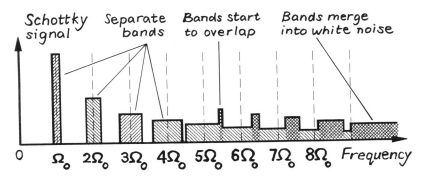

Figure 11.6. Longitudinal Schottky bands. (This figure is schematic, e.g. for the ISR ~ 3000 bands occurred before overlapping started at ~ 1 GHz.)

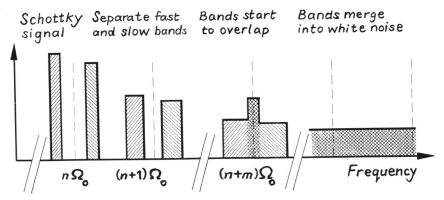

Figure 11.7. Transverse Schottky bands (schematic).

where $\Delta\Omega = \Omega_0 \eta\, \Delta p/p_0$ and $\Delta q = Q'\, \Delta p/p_0$ (assuming no amplitude–tune dependence). This reduces to,

$$\Delta\Omega_{\text{fast, slow}} = \Omega_0[\eta(n \pm q) \pm Q']\, \Delta p/p_0. \tag{11.23}$$

Thus the fast and slow wave bands will in general be of different widths according to the signs and values of η and Q'. As in the longitudinal case, each band has basically the same information and as the order increases the bands widen and finally overlap.

In the case of the transverse scan the monitor is sensitive to the dipole moment of each beam particle, so that within a band the Schottky signal is proportional to the product of the square root of the particle density at that frequency and the rms amplitude of the betatron oscillations, assuming that there is no coherence. If no beam loss occurs between two scans, they can be compared to reveal changes in the beam emittance.

268

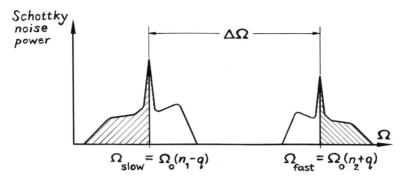

Figure 11.8. An identifiable feature in the n_1th slow-wave band and the n_2th fast-wave band.

Identifiable features such as the stack edges, peaks due to particles trapped in high-order non-linear resonances, troughs cut by strong low-order resonances, etc., can be located in tune space from the fast and slow wave bands (see Figure 11.8) by a simple manipulation of the measured frequencies.

$$q = \frac{(\Omega_{\text{fast}} + \Omega_{\text{slow}})(n_1 + n_2)}{2(\Omega_{\text{fast}} - \Omega_{\text{slow}})} - \tfrac{1}{2}(n_2 - n_1). \tag{11.24}$$

Equation (11.24) has been generalised to treat bands of different orders since this may be convenient from a practical point of view, as was the case in the ISR.

The above technique can be generalised to yield the fractional tune value at all points through the scan without the need for identifiable features*. This refinement relies on the fact that the noise power, which is proportional to the Schottky signal squared, is equal in all bands. Thus by integrating the area under the noise power curve and noting the frequencies at regular fractional steps of this area the corresponding points in two bands can be found (see Figure 11.8 for the relevant areas). Equation (11.24) can then be used to find the tune.

A yet more sophisticated technique is to use the longitudinal scan to establish the particle density in momentum space in order to extract the information on emittance from the transverse scan[21]†.

(iii) *Spectra in bunched beams*[20]

In a bunched beam, the revolution period of a given particle will be modulated about the synchronous period imposed by the rf system. The modulated period will have the form $[T_1 + \tau \sin(\Omega_s t + \phi)]$, where T_1 is the synchronous period, Ω_s is the

---

* In *ISR Performance Report, Run 593, ISR-RF/LT/ps*, 25th April 1975, this technique is attributed to a suggestion by H. G. Hereward.
† This technique is also attributed to H. G. Hereward in Ref. 21.

synchrotron angular frequency and τ is the maximum deviation. All particles are therefore centred on the same synchronous frequency and if the synchrotron motion were of negligibly small amplitude the Schottky bands described in the previous sections would shrink into lines. In practice, the modulation splits each of the lines into an infinite number of satellite sidebands spaced at the synchrotron frequency and since the modulation is sinusoidal it introduces a Bessel function dependence into the amplitudes of these satellites.

In the longitudinal case, at higher orders the modulation envelope of the satellites has about the same width as the band for an unbunched beam of the same momentum spread. The separation of the satellites can be used to measure the synchrotron frequency, although the detection of the satellites, which are incoherent signals, is rather difficult in the presence of the central line, which is a coherent signal.

In the transverse case, the signals are easier to observe, since the central line is again incoherent due to the random betatron motion. At high orders, the satellites form bands which resemble those of the unbunched beam, but at low orders the signal is concentrated in the central line and well-defined tune values for the synchronous particle can be measured.

11.3   BEAM TRANSFER FUNCTION

The study of beam stability, using an external sinusoidal excitation, was first tried by the MURA group[22]. Later the theoretical link between the beam response, or *beam transfer function*, and the theory of coherent instabilities was established[23] and practical demonstrations were made[24,25].

The very large momentum spread in the ISR made this technique of particular interest and it was developed into an operational instrument for optimising the physics beams[26,27]. Either a network analyser can be used to excite the beam with a swept frequency and to analyse its response, or the beam can be excited with white noise and the response found by an FFT spectrum analyser (see Figure 11.9). The latter is an application of the techniques developed for Schottky noise measurements and is the faster and often preferred method. Both methods are, in practice, non-destructive. The beam transfer function technique can be used to optimise the beam stability[28], to measure the wall impedance and the betatron tune values (working line) across the stack[29], to monitor feedback systems, to observe resonances and to minimise beam–beam coupling.

11.4   PROFILES

11.4.1   Emittance

The definition of emittance was given in Sections 2.5 and 4.4, and one method for measuring in a coasting beam by using Schottky scans has already been mentioned

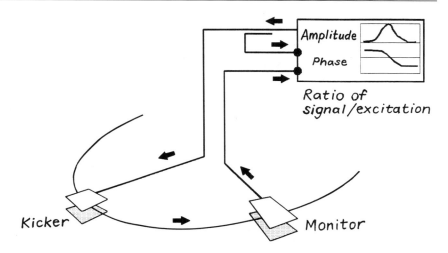

Figure 11.9. Measurement of the beam transfer function. Method 1, swept frequency excitation with network analyser. Method 2, white noise excitation with FFT.

in Section 11.2. This method, however, is better suited to relative measurements rather than absolute ones. The absolute methods are either semi-destructive and measure the projected beam density directly by the signals from a grid of parallel wires, a flying wire, or by digitising a luminescent screen or gas curtain monitor, or they are totally destructive and use a scraper, or a scanning slit.

The wire-grid and flying-wire monitors are frequently found in transfer lines and circular machines respectively. These monitors are known as *secondary emission monitors*, or *SEMs*, since they measure the current produced by electron emission from the wire surface when bombarded with beam. Descriptions of these and other monitors can be found in Refs 1–4.

In order to determine the emittance only one measurement at one azimuth is needed on a circulating beam, since the betatron function is uniquely determined by the cyclic symmetry. However, in a transfer line the betatron function is not uniquely determined and three measurements are needed to solve for the emittance, the betatron function and the alpha function. This basic difference has already been described more fully in Section 4.5. It may appear quicker and simpler therefore to make measurements on the circulating beam, but it is then too late to see any emittance mismatch at injection, since the beam has already fila-mented. The equations for use in a transfer line are given below and it is left to the reader to perform the exercise of verifying them (they are expressed using the 'principal trajectory' notation described in Section 4.6.1 to avoid multiple suffixes). Figure 11.10 shows the relationships between the various parameters.

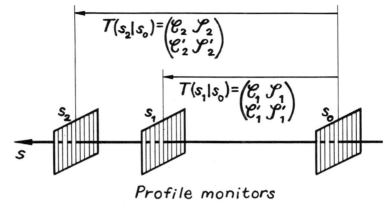

Profile monitors

Figure 11.10. Emittance and mismatch measurements in transfer lines.

$$\varepsilon = (\pi\sigma_0^2)\sqrt{[(\sigma_{1_2}/\sigma_0)^2/\mathscr{S}_{1_2}^2 - (\mathscr{C}_{1_2}/\mathscr{S}_{1_2})^2 + (\mathscr{C}_{1_2}/\mathscr{S}_{1_2})\Gamma - \Gamma^2/4]}$$

$$\beta_0 = (\pi\sigma_0^2)/\varepsilon, \qquad \alpha_0 = \tfrac{1}{2}\beta_0\Gamma$$

where

$$\Gamma = \frac{(\sigma_2/\sigma_0)^2/\mathscr{S}_2^2 - (\sigma_1/\sigma_0)^2/\mathscr{S}_1^2 - (\mathscr{C}_2/\mathscr{S}_2)^2 + (\mathscr{C}_1/\mathscr{S}_1)^2}{(\mathscr{C}_1/\mathscr{S}_1) - (\mathscr{C}_2/\mathscr{S}_2)}.$$

(11.25)

The parameters σ_0, σ_1 and σ_2 are the standard deviations of the beam profiles measured on the three monitors. The emittance is defined here for a 'beam size' of one standard deviation as in Section 4.4.

Equations (11.25) also make it possible to measure β and α and to find the mismatch between the existing and the design values.

11.4.2 Scraper scans

The scraper scan is one measurement, which requires some interpretation in order to arrive at the density profile needed to determine the emittance[30] [see Figure 11.11(a)]. The starting point is the betatron motion of a single particle as derived in Section 4.1,

$$y(s) = A\sqrt{\beta}\, \sin[\mu(s) + B].$$

(11.26)

The differentiation of (11.26) with respect to s gives,

$$\frac{dy}{ds} = \frac{A}{\sqrt{\beta}}\cos[\mu(s) + B] - \frac{A\alpha}{\sqrt{\beta}}\sin[\mu(s) + B].$$

(11.27)

Equations (11.26) and (11.27) can be re-written with the use of the normalised

272

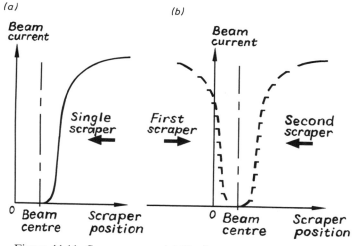

Figure 11.11. Scraper scans. (*a*) Single scraper scan, (*b*) double scraper scan.

betatron amplitude defined in (4.3), to give,

$$Y = y(s)/\sqrt{\beta} \qquad = A \sin[\mu(s) + B] \tag{11.28}$$

$$\frac{\mathrm{d}Y}{\mathrm{d}\mu} = y\alpha/\sqrt{\beta} + y'\sqrt{\beta} = A \cos[\mu(s) + B]. \tag{11.29}$$

Consider now a beam in a dispersion-free region, so that the momentum spread does not affect the beam size. The particles will turn on circular paths filling out a circular area in the (Y, Y') normalised phase-space as shown in Figure 11.12. The profile of interest for the emittance is the projection of the particle density onto the Y-axis, i.e.

$$\frac{\mathrm{d}N}{\mathrm{d}Y} = \int_{-\infty}^{\infty} n(Y, Y') \, \mathrm{d}Y'. \tag{11.30}$$

The total number of particles in the beam, N, is the integral of (11.30) with respect to Y. A flying-wire monitor, for example, would directly evaluate $\mathrm{d}N/\mathrm{d}Y$ at a series of Y-positions across the beam, but a scraper works over many turns and current losses affect the whole distribution.

The first step in interpreting the scraper scan is to replace the variable Y' in (11.30) by means of the amplitude χ (i.e. the distance from origin in the $Y-Y'$ plane). It is also possible to write the distribution $n(Y, Y')$ in terms of the single variable χ, because filamentation ensures a uniform distribution in phase. Since

$$\chi = (Y^2 + Y'^2)^{\frac{1}{2}} \quad \text{and} \quad \frac{\mathrm{d}\chi}{\mathrm{d}Y'} = Y'(Y^2 + Y'^2)^{-\frac{1}{2}} = \pm(\chi^2 - Y^2)^{\frac{1}{2}}/\chi$$

(11.30) can be re-written as,

$$\frac{\mathrm{d}N}{\mathrm{d}Y} = 2 \int_{Y}^{\infty} \chi(\chi^2 - Y^2)^{-\frac{1}{2}} n(\chi) \, \mathrm{d}\chi. \tag{11.31}$$

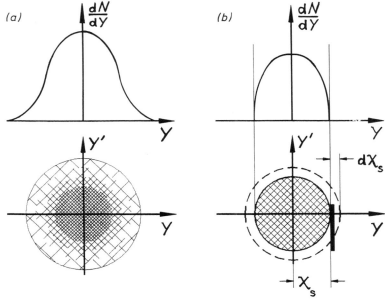

Figure 11.12. Action of a scraper in normalised phase space. (a) Before scraping, (b) during scraping. The projection of beam density on the Y-axis is shown above, with the phase space plot of the beam below.

Equation (11.31) shows that only the particles with $\chi > Y$ contribute to the particle density at position Y. If a scraper is advanced into the beam, then all particles with χ greater than the radial position of the scraper will eventually be lost and the particle density will become zero at the radius of the scraper (see Figure 11.12).

The parameters that can be measured are the scraper position, Y_s, the number of particles lost, dS, when advancing the scraper, and the number of particles remaining S. The remaining particles S can be found by integrating over the circle, whose radius χ_s is defined by the scraper position in Figure 11.12, i.e.

$$S(\chi_s = Y_s) = \int_0^{\chi_s} 2\pi\chi\, n(\chi)\, d\chi \qquad (11.32)$$

and the incremental loss is given by differentiating (11.32), i.e. by considering the annular volume defined by the scraper movement,

$$\left[\frac{dS}{d\chi}\right] = 2\pi\chi\, n(\chi). \qquad (11.33)$$

By logging the step losses as the scraper advances $[dS/d\chi]$ can be found for all values of χ. The substitution of (11.33) into (11.31) gives,

$$\frac{dN}{dY} = \frac{1}{\pi}\int_Y^\infty \left[\frac{dS}{d\chi}\right](\chi^2 - Y^2)^{-\frac{1}{2}}\, d\chi. \qquad (11.34)$$

Equation (11.34) converts the scan of the current loss versus scraper position into the particle density distribution of the original beam. In order to determine the position of the scraper with respect to the centre of the pulse, a double scraper can be used which cuts in small steps alternately from each side. The scan obtained from such a scraper is illustrated in Figure 11.11(b). This technique, which is described more fully in Ref. 31, is extremely accurate and is well suited to the measurement of emittance and luminosity (luminosity is discussed in Section 12.2), the general study of beam profiles and closed-orbit bump calibration.

The current measurements are made by a beam transformer, which can have a high absolute precision as well as an excellent differential accuracy. In the *pulsed current transformer* the bunched beam is simply the secondary winding, but because the chamber may also carry currents (including the image current) the monitor has to be isolated and the wall currents conducted around the device. The *d.c. current transformer* is more sophisticated. Basically it comprises two ferromagnetic rings placed side by side and through which the beam passes. The rings are equipped with identical windings that excite them at more than 100 Hz but with opposite polarities. Secondary windings or pickup coils are also wound on both rings and connected in series. The rings are carefully matched to have similar properties, so that the sum of the signals induced in the pickup coils would normally be zero. The presence of the magnetic field of a beam, however, introduces a bias in the excitation of the cores and a signal appears. The beam current is then measured indirectly by powering a further set of compensation windings, which are adjusted to bring this signal to zero. Further details can be found in Refs 1–4.

The effect of betatron coupling on the beam distribution when scraping is analysed in Ref. 32.

11.4.3   RF scans

Although the word 'profile' is usually applied to the transverse plane, it is also necessary to measure the longitudinal, or momentum, profile in a coasting beam. The longitudinal Schottky scan is one method that has already been mentioned in Section 11.2.4 and the *rf scan*, described below, is another.

This method determines the momentum distribution in a coasting beam by scanning with small rf buckets. By 'small', it is meant that the momentum bite of the bucket should be less than the resolution required. The bucket is created at a frequency that puts it completely outside the beam in real space. The frequency is then ramped so that the bucket approaches the beam and slowly passes through it. By 'slowly', it is meant that a distance corresponding to the bucket height is covered in a time which is long compared to the synchrotron oscillation period of a fictitious particle inside the bucket. In this way the diffusion of particles into the bucket will be slow and can be neglected. With these conditions fulfilled the 'sum' signal seen by a capacitive pickup will be proportional to the density of the beam at the

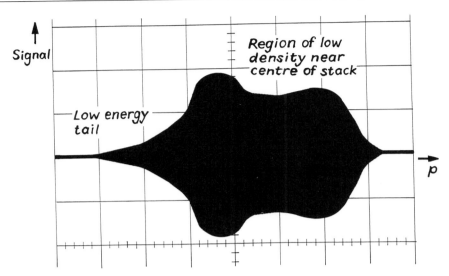

Figure 11.13. Typical rf scan in the ISR.

momentum of the empty bucket (see Figure 11.13). In order to avoid a permanent phase displacement of the beam, the scan must be made in both directions.

This method was first demonstrated in the MURA FFAG machine[33] using the phase-displacement acceleration system and later in the CERN CESAR[34] (electron storage ring), which was designed with a special rf scanning system. The method became less important with the advent of Schottky scans, since it causes an appreciable disturbance to the beam. It is, however, the natural technique to use with phase-displacement acceleration where empty rf buckets are anyway repeatedly passed through the stack.

11.5   LACK OF REPRODUCIBILITY

11.5.1   Hysteresis errors

When the closed orbit and/or the tune values are not reproducible from run to run, it may be due to hysteresis errors in the magnets. The first principles to follow are that:

- the magnets should be measured on the same current cycle as they will be operated on,
- the machine or transfer line should always be set using the same current cycle, and
- during the set-up cycle, current overshoot must be avoided.

The first principle appears to be simple, but in practice magnets are often recuperated from earlier applications and since there is no universally right or wrong way to cycle

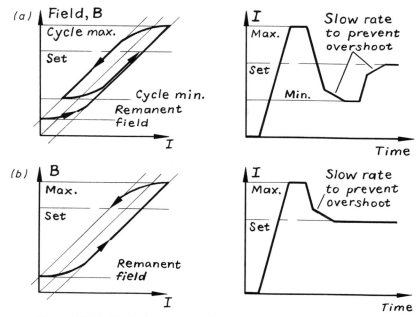

Figure 11.14. Typical magnet cycles. (a) Setting on upward branch of hysteresis curve, (b) setting on downward branch of hysteresis curve.

a magnet the measurements on different magnets may not be consistent. The current cycle for setting up a constant-field, or slow-ramping machine, would typically go from zero to maximum (in the saturation region) to a low value (usually not zero) and then to 'set'. A quicker cycle would be to go directly to 'set' on the downward branch of the hysteresis curve. In order to observe the third principle of no overshoot, the current ramp rate should be reduced just before the minimum setting in the cycle and just before 'set' (see Figure 11.14). At the maximum of the cycle the sensitivity to an overshoot is reduced as the magnet is saturated. In a pulsed machine, the operating cycle is set up and allowed to run for a few cycles. In this case, the least trouble with hysteresis is experienced if the maximum and minimum of the cycles are maintained constant and the ramp is stopped at intermediate levels for injection and the different extractions.

Hysteresis problems are a little more complicated for correction magnets and auxiliary windings. For example, a correction winding on a main dipole can have the same effect as a current overshoot when looking at the hysteresis loop (see Figure 11.15), but since it is an individual unit it can cause more disturbance than a systematic error in all units that occurs regularly in the set-up cycle. If this proves to be a problem, a solution is always possible, in principle at least, by making a sufficient number of measurements to equip the on-line computer with a hysteresis model. Since the main dipoles in many machines work at several current levels up to saturation this model could be rather complicated[35]. Steering magnets whose

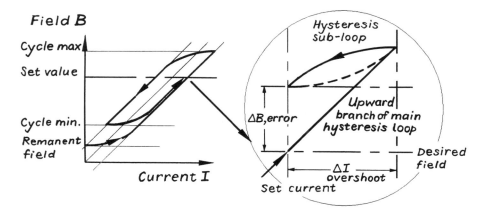

Figure 11.15. Effect of current overshoot or cycling a correction winding.

nominal field is zero and which work in a basically linear region can be equipped with far simpler hysteresis correction models[36].

11.5.2   Fringe field effects

Like hysteresis errors, fringe fields are very non-linear and appear more readily at high field levels. Fringe field effects fall into three main categories:

- direct action of a fringe field onto a beam,
- the modification of a magnet's field pattern caused by the proximity of a large mass of iron (e.g. an adjacent magnet yoke), which shunts (i.e. short-circuits) the magnet's end field,
- the two-stage action in which a fringe field is shunted into an adjacent magnet's yoke and, by altering slightly the flux level in the iron, it changes the permeability and hence the reluctance of the circuit so affecting the main field.

The direct effect is the most obvious and machine designers often specify limits of the order of a few Gauss meters for the fringe fields of detector magnets which are to be mounted temporarily in the storage rings. Shielding with several thin sheets of a high permeability alloy is an equally temporary and efficient solution. The second problem occurs, for example, when a small dipole corrector has to be placed close to a large aperture detector magnet. In such a case the fields of both magnets are affected and measurements must be made with both units in their standard configuration. The last category is a subtle version of the second. The magnet causing the stray field is not appreciably affected, but the yoke of a nearby unit shunts the stray field. If the second unit is operating close to saturation, the permeability of that unit will be at the point in its characteristic where it is changing rapidly with field.

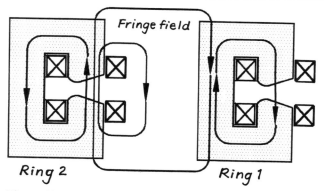

Figure 11.16. The 'magnetic amplifier' fringe-field effect.

A change in flux in the second magnet's yoke, of say 1 per mil, may change its permeability by a few per mil, which is reflected directly into its own field. This 'magnetic amplifier' effect seems unlikely but was experienced in the ISR in 1982 when the sensitive optics of the superconducting low-β insertion were combined with ramping to the highest possible field of 31.4 GeV. The effect is illustrated in Figure 11.16.

11.5.3   On-line field display system

The inclusion of spare magnet units equipped with field measurement coils in the main dipole and quadrupole circuits can be used to provide an on-line display of the guide field in the machine[37]. These reference units are best installed in an area accessible during machine operation and should be fully equipped with the same auxiliary windings as the main units. From the series magnetic measurements small numerical correction factors can be calculated in order to find the total bending and focusing power in the machine from the reference readings. To keep this numerical correction small, it is best to select the reference units during the series measurements, so that they have the average characteristics of the production. In a large system, the precaution of connecting the upper dipole coils in series with the incoming power cable and the lower coils in series with the outgoing power cable averages out the effect of any earth leakage current.

A magnet is usually set or incremented according to the current measurement system in its power converter. The field display system can be used either to check the reproducibility, or to set the magnets directly. With such a system, it is also easier to set up the machine at intermediate field levels that have not been calibrated during the series magnetic measurements. It also allows on-line measurements of the hysteresis effects when changing auxiliary windings. Usually the range of settings with poleface, correction and backleg windings is too large to have been measured in detail during the series measurements and in this case the field display system provides an invaluable service. The integrated bending field obtained from the reference reading is the only on-line measurement of the machine energy on central orbit.

11.6   RF MANIPULATIONS

The longitudinal particle motion can also be disturbed by imperfections and it is necessary to provide ways of observing the perturbations and remedies for compensating them. In some respects, the transverse and longitudinal cases are analogous. For instance, rf voltage and/or dB/dt disturbances influence the bunch as a whole, just as field errors affect the beam's transverse coherent motion. Mismatches between bunch and bucket shapes lead to dilution of the longitudinal emittance, in the same way as mismatches between phase-space ellipses of beams and lattices lead to dilution in the transverse case. The technical approaches to these problems are, of course, very different and it is worth noting that the time scales, which are set by the natural oscillation frequencies of the beam, are also very different. The betatron freqencies are usually several orders of magnitude higher than the synchrotron frequency. A few examples of these problems will be described in the following sections.

11.6.1   Beam control by phase lock and radial steering

The phase stable angle can be measured by comparing the passage of the bunch through a pickup with the phase of the rf cavity (with account taken of the various delays), but this is usually automatic, since the rf system operates under its own surveillance.

The stability of the rf phase angle is very important in order to minimise the slow dilution of longitudinal phase-space density. This is normally done by a *phase-lock system*[38,39], the principle of which is sketched in Figure 11.17. The phase of the bunch as observed on a sum pickup is compared with the phase of the rf cavity. The error signal is fed back to the rf cavity through the appropriate circuitry to correct the phase.

For this to be effective and to be stable, the response time of the system must be short compared with the period of the phase oscillation. It is this requirement that has led to the term 'phase lock' and the fact that it is often referred to as being 'fast'. However, since the phase oscillations are rather slow, typically ~ 100 Hz, the loop is still, electronically speaking, rather slow.

With only the phase-lock loop, the beam may still drift in or out radially in the vacuum chamber. In order to avoid this, it is common to combine the phase-lock system with a *radial feedback*. A signal taken from a different pickup is also fed back

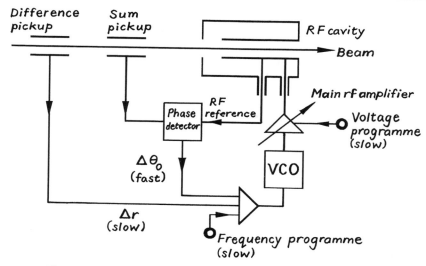

Figure 11.17. Phase-lock system for bunched beams.

into that part of the phase-lock system that determines the phase angle. It can alternatively act on the rf voltage, but the phase is more commonly used. Thus the beam is steered back to its nominal orbit. In order to provide stability, the radial feedback must be 'slow', again compared with the phase oscillation.

With these two feedback systems, the beam control is stable. In principle, the system can provide its own frequency, but it is more common to use a frequency programme to which relatively small corrections are then applied. A more detailed description can be found in Ref. 40.

11.6.2   RF matching

When a bunch is incorrectly matched to its rf bucket, it will tumble over and over inside the bucket at the synchrotron frequency. The non-linearities in the motion will cause the particles with larger amplitudes to rotate more slowly leading to filamentation and dilution of the longitudinal emittance in an exactly analogous way to the transverse filamentation described in Section 4.5.1. The mismatch can be observed by monitoring the bunch length oscillations on an intensity-sensitive pickup. The signal will appear at twice the synchrotron frequency. The rf engineer has a number of techniques at his disposal to remove a mismatch. RF matching can be carried out with stationary or accelerating buckets. The physical principle is unchanged, but the theoretical algorithms are more complicated for the latter. Only stationary buckets will be considered here, which is the more common approach in practice[41].

(i) *Simple bucket shaping*

For small mismatches, it is sufficient to make minor adjustments to the voltage in order to shape the bucket for minimum bunch length oscillation.

(ii) *Voltage-step matching*

For much larger mismatches, a *voltage-step* or '*quarter-wave transformer*' technique is used. For example, the bunch in Figure 11.18(*a*) is allowed to rotate for a quarter of a synchrotron period at an intermediate voltage, after which the voltage is reduced to match the new bunch shape with its increased phase extent and reduced momentum spread. In principle, any desired bunch shape change can be achieved by this simple method. However, in practice it is occasionally necessary to perform the matching in several steps, for example when the system cannot provide the voltage needed for the single-step matching.

(iii) *Phase jump*

A bunch can be compressed or expanded in momentum spread by shifting it to the unstable phase point for a short period. For example, in Figure 11.18(*b*) the bunch has too great a momentum spread, so after $\frac{3}{8}$ of a synchrotron period the rf phase is shifted by π so placing the bunch on the unstable phase point. Once the desired degree of compression is obtained the phase is switched back. In a different situation, the momentum spread could be expanded by waiting only $\frac{1}{8}$ of a synchrotron period before jumping the rf phase.

(iv) *Rebunching*

This is a relatively simple operation that requires only adiabatic changes in the rf voltages. It could be used, for example, when changing to an rf system with a different harmonic number.

(v) *Closed-loop matching*

This technique is similar to voltage-step matching except that the changes in the rf voltage are controlled via a feedback loop from a pickup monitoring the bunch length oscillations. The matching is carried out over several synchrotron oscillation periods[42].

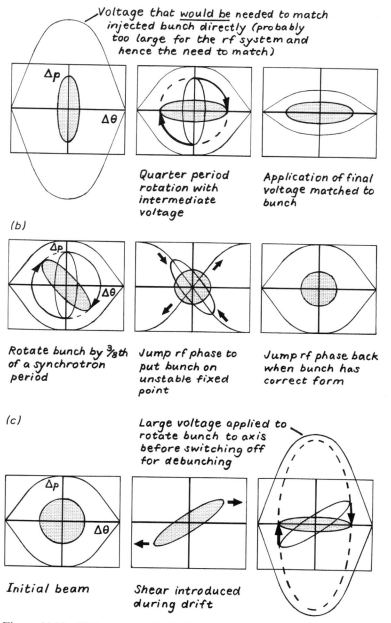

Figure 11.18. RF 'gymnastics'. (*a*) Voltage step, (*b*) phase jump, (*c*) drift.

(vi) *Drift*

When a bunch is left without rf, the dispersion in the lattice will cause a shear to appear in the phase space diagram as shown in Figure 11.18(*c*). This has been used for one rather special matching problem, which required an extracted beam to be debunched with minimum momentum spread[43]. Once the shear had developed the beam was passed through a cavity which rotated the bunch back to the axis, so giving the minimum momentum spread.

11.6.3 Bunch length measurements

With long bunches, there is little problem in measuring the bunch length with an intensity pickup, but for short bunches this method is limited by cable response to bunch lengths around 200 ps. This can be improved for electrons to 100 ps by extracting the synchrotron light from the ring and processing with a photo-diode. According to the sophistication of the equipment this can be reduced to a few tens of ps, but ultimately the best resolution of about 5 ps is obtained with a streak camera.

The so-called *mountain-range display* is frequently used when observing bunch length oscillations and more complex behaviour. The sweep of the monitor is stepped down the screen so that the bunch signal appears repetitively on a grid. The evolution of the bunch can then be followed visually by looking down what is very reminiscent of a mountain range. For short bunches the main limitations are due to the signal cables and the oscilloscope bandwidth. A resolution approaching 50 ps can be achieved[44].

CHAPTER 12

Special aspects of circular colliders

In the preceding chapters, much of the basic accelerator theory has been introduced and the ideas were so fundamental that they were often equally applicable to transfer lines and linear accelerators, as well as to circular machines, the central theme of this book. The present chapter will depart from this pattern and look briefly at some special aspects of circular colliders. This review will be rather superficial, but its aim is to make the reader aware of the problems and to provide references rather than to lay out detailed derivations of formulae.

From the introductory chapter, it is evident that high-energy physics was profoundly influenced by the inventions that made colliding beams feasible experimental tools. This did not happen suddenly. The main ideas on how to accumulate sufficiently intense beams to achieve significant interaction rates originated in the MURA Group in 1956, but considerable scepticism on the part of the physics community had to be overcome through practical demonstrations before the consensus of opinion turned in favour of colliders.

For electrons, many facilities were built and operated, but the highlights of this development were SPEAR and DORIS. Following their outstanding physics discoveries in the 1970s, all later proposals for electron machines were for colliders. LEP at CERN is the most recent and will probably remain the largest circular collider to be built for electrons. Fixed-target electron accelerators have become obsolete for high-energy physics, although existing ones still continue to do valuable research.

For protons, the ISR served the purpose of demonstrating the performance potential of hadron colliders and that their very high energies could be exploited by appropriately designed detectors. This led to a remarkable change in the attitudes of experimentalists, which became prominent with the spectacular successes of the conversions of the large synchrotrons at CERN and at Fermilab into proton–antiproton colliders. Since then no high-energy, fixed-target hadron facilities have been built, or proposed, and all future plans are for colliders.

12.1   ENERGY RELATIONS

The energy available in a collision between two particles (denoted by subscripts 1 and 2), which are moving along the same straight line is given by,

$$E_{cm} = c^2 \sqrt{[m_1{}^2 + m_2{}^2 + 2m_1 m_2 \gamma_1 \gamma_2 (1 - \beta_1 \beta_2)]}$$ (12.1)

where the signs of β_1 and β_2 indicate the direction of motion and 'cm' denotes the centre of mass system.

In the fixed-target facility β_2 will be zero, so that the available energy becomes,

$$m_1 \bullet \rightarrow \bullet m_2 \qquad E_{cm} = c^2 \sqrt{(m_1{}^2 + m_2{}^2 + 2m_1 m_2 \gamma_1)}.$$ (12.2)

If the two colliding particles have the same rest mass m, a simpler formula applies,

$$E_{cm} = mc^2 \sqrt{(2 + 2\gamma_1)}$$ (12.3)

$$\simeq mc^2 \sqrt{(2\gamma_1)} \qquad (\text{for } \gamma \gg 1).$$ (12.4)

The available energy therefore goes as the square root of the energy of the accelerator, making it necessary to have rather large steps to get significant increases E_{cm}.

For colliding beams the situation is different. In the event of a head-on collision the sign of β_2 is inverted, so that (12.1) becomes,

$$m_1 \bullet \rightarrow \leftarrow \bullet m_2 \qquad E_{cm} = c^2 \sqrt{[m_1{}^2 + m_2{}^2 + 2m_1 m_2 \gamma_1 \gamma_2 (1 + \beta_1 \beta_2)]}.$$ (12.5)

The importance of this expression is best seen by making the simplifying assumption that both particles are relativistic, in which case,

$$E_{cm} \simeq 2\sqrt{(E_1 E_2)}.$$ (12.6)

The advantage of a circular collider over a fixed-target installation, in terms of the energy available for particle production, is therefore very pronounced and becomes more so as the particle energy increases. This is illustrated by the equivalent particle energy needed on a fixed-target, E_{eq}, to match a given collider beam energy,

$$E_{eq}/m_1 c^2 \simeq 2\gamma_1 \gamma_2 - \tfrac{1}{2}(m_1/m_2 + m_2/m_1).$$ (12.7)

These formulae need to be modified when collider beams cross at a small angle, α, but in order to have simple expressions, this will only be done for the case of identical particles with the same energy. The expression (12.6) can then be replaced by,

$$E_{cm} \simeq 2E\sqrt{[1 - \beta^2 \sin^2(\alpha/2)]} \simeq 2E \cos(\alpha/2) \qquad (\text{for } \gamma \gg 1),$$ (12.8)

which is equivalent to bombarding a stationary target with an energy of,

$$E_{eq}/mc^2 \simeq 2\gamma_{coll}{}^2 \cos^2(\alpha/2) - 1.$$ (12.9)

This is the expression used to incorporate the hadron colliders into the extended Livingston Chart (see Figure 1.1). The electron–proton collider HERA at DESY, has been included by the use of (12.7) with the choice of the electron as the

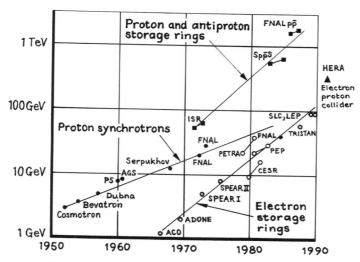

Figure 12.1. Transformed Livingston Chart for fixed-target synchrotrons and colliders showing centre-of-mass energy versus time.

bombarding particle for the equivalent energy. For electrons against positrons, the direct application of these formulae does not give results that fit meaningfully into the Livingston Chart. For this, and other reasons, it is more useful to transform the Livingston Chart and to plot centre-of-mass energy against time, rather than equivalent fixed-target energy. This has been done in Figure 12.1. In this transformed chart, the various accelerator facilities fall neatly into three groups: fixed-target machines, lepton colliders and hadron colliders. Not only is the energy advantage of colliders clearly illustrated, but also that their development is on a much steeper slope. It appears that the hadron colliders have about an order-of-magnitude advantage over the lepton colliders, but this is not quite true. Electrons are already basic constituents of matter, whereas in protons the energy is divided between the basic constituents of quarks and gluons. Very approximately each constituent has about one tenth of the full proton energy. Thus the hadron and lepton colliders are about equivalent, where energy is concerned.

In the light of the energy argument, the collider may seem an obvious choice now, but in the beginning there were many voices proclaiming they would not work. There were two enormous extrapolations to be made:

- *firstly* the beams were so tenuous that it seemed impossible to reach a useful interaction rate, especially when all experimental expertise was centred on the analysis of the vast floods of particles emerging from fixed-target experiments;
- *secondly* the beams would need to circulate stably many orders of magnitude longer than in an ordinary accelerator.

The following sections will be devoted to these problematic aspects of colliders.

12.2   LUMINOSITY

Although the energy relations given in Section 12.1 are the driving argument for colliders, they are basically trivial and the people working with accelerators were aware of their potential from very early on. The first obstacle was how to obtain high enough beam densities, so that useful event rates could be achieved. A break-through in this field came from the MURA Group[1,2]. Nevertheless, the issue of the event rate has stayed as a major design problem for most colliders and will therefore be addressed further in this section.

   The measure of the intensity performance of a collider is the *luminosity*, which is defined as the ratio of the event rate of a chosen interaction at a beam crossing to the cross-section of that interaction. The luminosity depends on the geometry of the interaction region and the characteristics of the beams. The definition can be given a general mathematical form, but with integrals that can only be solved for special geometries and particle distributions. For a physical understanding, it is preferable to consider specific types of machine and find and discuss the corresponding expressions for the luminosity. The aim of this analysis will be to gain insight into how to obtain the highest possible luminosity and it will become clear that different types of machine require different approaches to their luminosity optimisation.

12.2.1   Two ribbon-shaped beams crossing at an angle

Colliding beam machines with two separate rings and a crossing angle can operate with coasting, i.e. unbunched, beams. The interaction region length is then defined by the beam dimension in the plane of the crossing angle as illustrated in Figure 12.2. A well-known example of this type of beam crossing was the CERN ISR proton–proton collider.

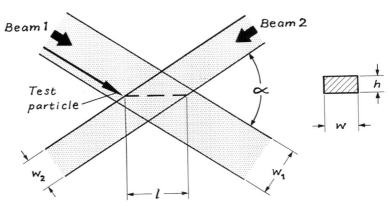

Figure 12.2. Schematic of a horizontal crossing of coasting beams.

In the first instance, a simple luminosity formula will be derived for two uniform ribbon beams of widths w_1 and w_2 and equal height h crossing horizontally with an angle α as shown in Figure 12.2. The derivation will follow that of Middelkoop and Schoch in Ref. 3 for the ISR machine. The particles of beam 2 that are crossed by a test particle in beam 1 will lie along a locus drawn by the intersection point on beam 2. This locus will stretch across beam 2 diagonally (see dotted line in Figure 12.2) and will be of length

$$l = w_2/\sin(\alpha/2). \tag{12.10}$$

Because of the crossing angle, the centre-of-mass (c.m.) is moving inwards (or outwards), but since the locus of the crossing is perpendicular to the c.m. motion, it has the same length in the c.m. and laboratory reference frames. The number of particles encountered by the test particle must be the same for all observers, and when calculated in the c.m. system will be,

$$\frac{w_2}{\sin(\alpha/2)}\, n_2{}^* \Sigma = \frac{w_2}{\sin(\alpha/2)} \frac{n_2}{\gamma_{c.m.}} \Sigma \tag{12.11}$$

where

$$\gamma_{c.m.}{}^{-1} = \sqrt{[1 - (v_{c.m.}/c)^2]} = \sqrt{\{1 - [v\sin(\alpha/2)/c]^2\}} \tag{12.12}$$

and $v_{c.m.}$ is the velocity of the c.m. system, Σ is the cross-section of the interaction and n_2 is the particle volume density in the laboratory system and $n_2{}^*$ in the c.m. system, which is assumed to be uniform for the time being. Since $n_1 w_1 h v$ beam 1 particles cross the interaction region per second, the luminosity is,

$$L = w_1 w_2 h n_1 n_2 \beta c \sqrt{[1 - \beta^2 \sin^2(\alpha/2)]}/\sin(\alpha/2), \tag{12.13}$$

which in terms of the beam currents and after some trigonometrical manipulation becomes,

$$L = \frac{I_1 I_2}{\beta c e^2 h} [\operatorname{ctg}^2(\alpha/2) + \gamma^{-2}]^{\frac{1}{2}}. \tag{12.14}$$

For relativistic particles this simplifies to,

$$L = \frac{I_1 I_2}{c e^2 h \tan(\alpha/2)} \tag{12.15}$$

and for non-relativistic particles to,

$$L = \frac{I_1 I_2}{\beta c e^2 h \sin(\alpha/2)}. \tag{12.16}$$

A good approximation for any energy will therefore be,

$$L = \frac{I_1 I_2}{\beta c e^2 h(\alpha/2)}, \tag{12.17}$$

289

since $\alpha/2$ will always be a relatively small angle, e.g. in the CERN ISR $\alpha/2 \simeq \pi/24$, but since in all practical cases the beam particles are relativistic, (12.15) will be used in what follows. A more detailed and rigorous derivation of the above expressions is given in Ref. 3.

The important assumption of uniform particle densities was made in the above derivation of the luminosity, but it is noticeable that the result (12.15) does not contain the widths of the beams. It can be concluded from this that, even if the densities had been made dependent on the horizontal position, it would not have changed the final result. The beam height, however, does enter into (12.15), which means that vertical density distributions would affect the result. Such cases can easily be calculated as an extension of the uniform beam case.

Consider two infinitesimally thin ribbons of height dz crossing each other. The currents in the beam ribbons can be expressed as,

$$dI_{1,2} = i_{1,2}\,dz. \tag{12.18}$$

After substitution into (12.15), this gives a luminosity of,

$$dL = \frac{i_1 i_2\,dz}{ce^2 \tan(\alpha/2)}, \tag{12.19}$$

which when integrated over z gives,

$$L = \frac{1}{ce^2 \tan(\alpha/2)} \int_{-\infty}^{\infty} i_1 i_2\,dz = \frac{I_1 I_2 \int_{-\infty}^{\infty} i_1 i_2\,dz}{ce^2 \tan(\alpha/2) \int_{-\infty}^{\infty} i_1\,dz \int_{-\infty}^{\infty} i_2\,dz}. \tag{12.20}$$

Comparison with the simpler formula (12.15), shows that the beam height h has been replaced by an *effective height*[4] h_{eff} given by,

$$h_{eff} = \int_{-\infty}^{\infty} i_1\,dz \int_{-\infty}^{\infty} i_2\,dz \Big/ \int_{-\infty}^{\infty} i_1 i_2\,dz. \tag{12.21}$$

This equation takes on a simpler form if normalised distribution functions $\psi_{1,2}(z)$ are introduced with the definitions,

$$i_{1,2} = I_{1,2}\psi_{1,2}(z). \tag{12.22}$$

This gives,

$$h_{eff} = \frac{1}{\int_{-\infty}^{\infty} \psi_1 \psi_2\,dz}. \tag{12.23}$$

Both (12.21) and (12.23) not only take care of non-uniform current distributions, but also of beams which are vertically displaced.

It is now instructive to consider the example of two centred beams with Gaussian distributions of the form,

$$\psi(z) = \frac{1}{\sqrt{(2\pi)}\,\sigma_z}\,e^{-z^2/(2\sigma_z^2)}. \tag{12.24}$$

The substitution of (12.24) into (12.23) for each beam then yields,

$$h_{\text{eff}}^{-1} = \frac{1}{2\pi\sigma_{z_1}\sigma_{z_2}} \int_{-\infty}^{\infty} e^{-\frac{1}{2}z^2(\sigma_{z_1}^{-2}+\sigma_{z_2}^{-2})} \, dz \tag{12.25}$$

which yields,

$$h_{\text{eff}} = \sqrt{[2\pi(\sigma_{z_1}^2 + \sigma_{z_2}^2)]} \quad \text{(Gaussian).} \tag{12.26A}$$

For the simple case of equal distributions this simplifies to

$$h_{\text{eff}} = 2\sqrt{\pi}\,\sigma_z \quad \text{(Gaussian).} \tag{12.26B}$$

It is now interesting to return to the case of the uniform distribution, where $h_{\text{eff}} = h$. The standard deviation of a uniform distribution is equal to,

$$\sigma_z = \frac{h}{2\sqrt{3}} \quad \text{(uniform)} \tag{12.27}$$

which can be re-expressed as,

$$h_{\text{eff}} = 2\sqrt{3}\,\sigma_z \quad \text{(uniform).} \tag{12.28}$$

The fact that the effective heights of these two very different distributions are so close when expressed in terms of their standard deviations may appear accidental, but it is in fact difficult to construct distributions for which this is not true. Hereward[5] has shown that the effective height of any realistic distribution can be written to a good approximation as,

$$h_{\text{eff}} = 3.59\sigma_z \pm 3.7\%. \tag{12.29}$$

Thus h_{eff} is largely distribution-independent.

Van der Meer invented an elegant method of measuring the effective height, as given by (12.21), for the geometry in Figure 12.2[6]. This method was used with great success in the CERN ISR, where an accurate knowledge of the luminosity was crucial for many experiments, such as the discovery of the unexpected rise of the total proton–proton cross-section with energy.

Sometimes it is convenient to express the luminosity in terms of beam emittance instead of the effective beam sizes. This will only be done for a Gaussian distribution, by introducing the beam emittance definition (4.31) into (12.26) to give,

$$h_{\text{eff}} = 2\sqrt{(\beta_z \mathcal{E}_z)} \tag{12.30}$$

so that the luminosity becomes,

$$L = \frac{I_1 I_2}{ce^2 2\sqrt{(\beta_z \mathcal{E}_z)} \tan(\alpha/2)}. \tag{12.31}$$

It should be noted that the β_z introduced in (12.30) and (12.31) is the betatron amplitude function. The significance of this presentation of the luminosity formula will become clearer later in this chapter.

So far only coasting beams have been considered. In order to get some feeling for the effect of bunching with the same geometry, the simple model of a rectangular, longitudinal bunch distribution will be assumed that is long with respect to the interaction region. It will also be assumed that both beams have the same number of bunches, of the same length and that the beams are properly synchronised. The bunching factor is defined as,

$$\Lambda = \frac{\text{Average beam current}}{\text{Peak beam current}}. \qquad (12.32)$$

These changes lead to an increase in the instantaneous luminosity of Λ^{-2}, but the active time is reduced by Λ. The net result is a factor Λ^{-1} in the luminosity formula, which is now for the average luminosity,

$$\bar{L} = \frac{\bar{I}_1 \bar{I}_2}{\Lambda c e^2 h_{\text{eff}} \tan(\alpha/2)}. \qquad (12.33)$$

It is reasonable to conclude from this that bunching gives an improvement in luminosity, but this is only true if the bunching is achieved without a reduction in the average circulating currents. For instance, in the CERN ISR the highest luminosities were obtained with coasting beams (max. 1.4×10^{32} cm^{-2} s^{-1}). There were two reasons for this:

- *firstly* the momentum width of the stacked beam was so large that unrealistic rf voltages would have been required for bunching,
- *secondly*, even disregarding this practical limitation, the ISR was filled to its full momentum acceptance and bunching would therefore have led to a current loss of Λ in each beam and hence by (12.33) a net reduction in luminosity of Λ.

This situation could be different for other filling methods. If, for example, filling were carried out at a relatively low energy, then the acceleration would liberate momentum space that (in principle) could be exploited for bunching.

12.2.2 Bunched beams in head-on collision

Electron–positron and proton–antiproton colliders, which have opposing beams in a single ring, and certain double-ring machines, such as HERA at DESY, operate with head-on collisions between bunched beams. This is also the case, to a good approximation, for the planned colliders LHC at CERN and SSC in Texas, where the crossing angle is small compared with the ratio of the width to the half-length of the bunch. In this section the interaction rate in a single crossing region with bunched beams in the head-on geometry will be considered.

If there are k bunches in each ring and the revolution frequency is f_{rev}, there will be $k f_{\text{rev}}$ bunch crossings per second in each crossing region. With this geometry, the longitudinal distribution of the particles is unimportant and it is the projection of this distribution onto the cross-sectional area that enters the calculation. Let this projected density in each bunch be n, which is then a function of the transverse coordinates only. The total number of particles in beam 2 encountered by a single particle of beam 1 during a single bunch crossing will be $n_2 \Sigma$ and the contribution to the luminosity of the particles within a cross-sectional element dS will therefore be,

$$dL = (\Delta t)^{-1} n_1 n_2 \, dS \tag{12.34}$$

where $\Delta t = (k f_{rev})^{-1}$, the time between two bunch crossings. It follows from (12.34) that the total luminosity at a crossing point will be,

$$L = k f_{rev} \int_S n_1 n_2 \, dS. \tag{12.35}$$

The integration must be made over the total overlapping area S of the two beams. If the population varies from bunch to bunch then n_1 and n_2 can be taken as the average values.

The number of particles in the full circumference is,

$$N_{1,2} = k \int_S n_{1,2} \, dS. \tag{12.36}$$

Equation (12.35) for the luminosity can now be rewritten using (12.36) in the form,

$$L = \frac{N_1 N_2 f_{rev}}{k} \frac{\int n_1 n_2 \, dS}{\int n_1 \, dS \int n_2 \, dS} \tag{12.37}$$

or in terms of the average currents $I_{1,2}$,

$$L = \frac{I_1 I_2}{e^2 f_b} \frac{\int n_1 n_2 \, dS}{\int n_1 \, dS \int n_2 \, dS}, \tag{12.38}$$

where $f_b = k f_{rev}$, the bunch frequency. When written in this form, the distribution functions $n_{1,2}$ can be scaled by any convenient value. The simplest formulation is obtained by scaling each distribution function, so that its integral becomes unity. Such a normalised distribution will be denoted by ψ and

$$\int_S \psi_{1,2} \, dS = 1. \tag{12.39}$$

With this inserted, (12.38) takes the simplified form,

$$L = \frac{I_1 I_2}{e^2 f_b} \int_S \psi_1 \psi_2 \, dS. \tag{12.40}$$

Two illustrative examples will be worked out, both with circular symmetry of the two beams and a common axis of symmetry. With this geometry (12.38) can be rewritten as,

$$L = \frac{I_1 I_2}{e^2 f_b} \frac{\int_0^\infty n_1 n_2 r \, dr}{2\pi \int_0^\infty n_1 r \, dr \int_0^\infty n_2 r \, dr}. \tag{12.41}$$

In the first example, the beams are assumed to have uniform distributions out to radii $r_{1,2}$, so that,

$$\int_0^{r_{1,2}} n_{1,2} r \, dr = n_{1,2} \int_0^{r_{1,2}} r \, dr = \tfrac{1}{2} n_{1,2} r_{1,2}^2 \tag{12.42}$$

and,

$$\int_0^{r_{\min}} n_1 n_2 r \, dr = \tfrac{1}{2} n_1 n_2 r_{\min}^2 = \tfrac{1}{2} n_1 n_2 r_1^2 \tag{12.43}$$

where $r_{\min}$ is the smaller radius, which has been assumed to be r_1. The luminosity then becomes,

$$L = \frac{I_1 I_2}{e^2 f_b \pi r_2^2} = \frac{I_1 I_2}{e^2 f_b S_2}, \tag{12.44}$$

where S_2 is the cross-section of the larger beam. For estimates, it is important to notice (what is intuitively obvious) that for two beams of unequal cross-section, it is the larger beam that determines the luminosity.

In the second example the beams are attributed Gaussian distributions,

$$\psi_{1,2}(r) = \frac{1}{2\pi \sigma_{1,2}^2} e^{-\frac{1}{2} r^2 / \sigma_{1,2}^2} \tag{12.45}$$

and since (12.45) is normalised the simplified expression in (12.40) can be used for the luminosity.

$$L = \frac{I_1 I_2}{e^2 f_b} \int_0^\infty \frac{1}{4\pi^2 \sigma_1^2 \sigma_2^2} e^{-\frac{1}{2} r^2 (1/\sigma_1^2 + 1/\sigma_2^2)} 2\pi r \, dr$$

$$= \frac{I_1 I_2}{e^2 f_b} \frac{1}{2\pi(\sigma_1^2 + \sigma_2^2)}. \tag{12.46}$$

An expression similar to (12.44) can be obtained by defining an effective area,

$$S_{\text{eff}} = 2\pi(\sigma_1^2 + \sigma_2^2) \tag{12.47}$$

to give,

$$L = \frac{I_1 I_2}{e^2 f_b S_{\text{eff}}}. \tag{12.48}$$

The result now depends on both beams, but if the sizes differ substantially the luminosity is again dominated by the larger beam.

On some occasions it is convenient to write the luminosity in terms of the beam emittances. With the assumptions already made, the vertical and horizontal emittances are equal and are given by (4.31). When the emittances are introduced into (12.46), this gives,

$$L = \frac{I_1 I_2}{2e^2 f_b (\beta_1 \varepsilon_1 + \beta_2 \varepsilon_2)}. \tag{12.49}$$

A Gaussian distribution may not always be a good representation of an actual beam. Other distributions require similar, but normally more complicated, calculations and the reader should refer to more specialised literature[7].

12.2.3   Some general remarks on luminosity formulae

All luminosity formulae exhibit some common traits. It is clear that the beam currents are important for obtaining high luminosity and that the requirement is simply for maximum possible currents. Secondly, it is important to reduce to the minimum the beam size in the crossing region. With reference to the formulae expressed with beam emittances, it can be seen that this can be achieved by not only keeping the emittances of the beams as small as possible, but also by acting on the values of the betatron amplitude functions in the crossing regions. Robinson and Voss[8] first showed that special insertions, now known as low-β insertions, could be used to strongly focus the beams down to very small cross-sections compared to those in other parts of the beam, without affecting the rest of the machine. Such insertions typically require stronger quadrupoles with larger apertures than those in the normal lattice. If the insertion is in a dispersion-free region, then the minimum number of quadrupoles needed per beam to both focus and match to the lattice is four (i.e. a doublet on either side of the low-β, see Section 4.7.3).

A balance must be sought between these three basic approaches and there is no universal way of finding the optimum, since this depends strongly on the type of collider under consideration. For example, the radiation in electron machines, which is almost non-existent for hadron machines, makes the optimisation process very different for these two categories.

12.2.4   The ISR approach[9]

The CERN ISR was an extreme case where unusually high beam currents were accumulated by the so-called *stacking in momentum space*, as invented by the MURA Group[1,2]. Once account had been taken of the transverse beam emittance and the space needed for injection, the momentum phase space available for stacking in the aperture of the ISR was typically of the order of a thousand times the momentum phase space occupied by a single injected pulse from the CERN-PS. In the limit of no phase dilution, Liouville's Theorem would allow that number of pulses to be stacked side by side across the aperture.

The accumulation cycle started by trapping the injected beam in the buckets of the rf system on the injection orbit close to the inner wall of the vacuum chamber and then accelerating them so that their closed orbit moved outwards to a point closer to the outer wall of the vacuum chamber where the rf voltage was switched off. This cycle was repeated many times to build up the stack. This works because each time a bunched beam is accelerated to the top of the stack the unbunched beam from the earlier pulses is phase displaced inwards, which is equivalent to putting the bunches side by side. The inward phase displacement depends on the phase-space area of the buckets, so by the time the buckets had entered the stack it was important to have reduced the rf voltage to the point where the buckets were completely full in order to minimise the phase-space dilution of the stack. There were also a number of technical refinements applied, such as suppressed buckets[10], and operational techniques such as the use of a variable step-back of the stacking orbit in order to control the longitudinal density, but the essential points are described above.

The technique was very successful and operational physics beams were typically 30–40 A of protons for runs of 50–60 h. The record single-beam current was 57 A. In order to reach such high currents the vacuum system had to be improved beyond the original specification. The working lines in the tune diagram were carefully tailored to ensure maximum and uniform Landau damping across the aperture and a third-harmonic cavity was added to enhance the longitudinal damping. At very high beam currents, transverse feedback was applied for extra stability.

The ISR machine was designed and constructed before low-β insertions had been developed, although the vertical β-values did have local minima of 14 m in the crossing regions. Towards the end of the ISR's life two low-β insertions were constructed. The first used conventional quadrupoles, but the second and stronger insertion with a β-minimum of 0.3 m consisted of eight superconducting quadrupoles. These lenses were built in industry and were the first superconducting lattice elements in an operational storage ring, which at that time was routinely operating with 40 A beams.

The measured normalised emittances delivered by the CERN PS were $\sim 13\pi \times 10^{-6}$ m.rad. horizontally and $\sim 8\pi \times 10^{-6}$ m.rad. vertically and the ISR had only to avoid any increase in these emittances during the injection process. However, it became standard practice to shave the vertical dimension of the beams. This decreased the currents, but for given current levels the luminosity was higher. For a while the maximum current was vacuum-limited and wide, strongly shaved stacks were more efficient. Shaved stacks also had lower loss rates, which meant less background in the experiments. It then became standard practice to shave physics beams whenever the background deteriorated.

Near the end of its life, the ISR set a luminosity record of 1.4×10^{32} cm^{-2} s^{-1} in its superconducting low-β insertion at the start of a physics run (Dec. 1982); a record that has proven hard to beat*.

12.2.5   Proton–antiproton colliders

During the planning stages of the ISR, the possibility of colliding protons with antiprotons was studied[11]. These studies were rather discouraging and further investigations along the same lines[12] did not raise any hopes for several years, although some interesting ideas, such as injection via antihyperon decays[13], were envisaged.

In 1966, Budker and Skrinsky[14] at Novosibirsk suggested that high-luminosity proton–antiproton collisions in a single ring might become feasible by means of a technique called *electron cooling*, which could be used to obtain a high-density antiproton beam. The technique consists of passing a high-current electron beam together with the antiproton beam along a common straight section of the ring. By precisely matching the beam velocities, the Coulomb interaction is able to act towards equalizing the beam 'temperatures' so that an electron beam with very little transverse motion will reduce that of the antiprotons, thus increasing the antiproton density.

Two years later, in 1968, van der Meer[15] at CERN proposed a very different cooling method called *stochastic cooling*. He suggested that it might be possible to detect the statistical fluctuations of the centre-of-gravity of a beam (due to the finite number of particles) and then to amplify this signal and apply it with the appropriate phase to a corrector some way downstream in the ring. After some time, the feedback loop between the pickup and corrector should reduce the statistical fluctuations and so increase the beam density.

In subsequent years experimental and theoretical studies were pursued on electron cooling at Novosibirsk and on stochastic cooling at the ISR. These studies showed increasingly promising results and interest was revived in colliding proton and antiproton beams. In 1976, Rubbia[16] proposed a variety of schemes using electron and stochastic cooling for collecting an antiproton beam, which could then be injected into the CERN SPS and accelerated together with a proton beam up to 270 GeV to give collisions at a centre-of-mass energy of 540 GeV; almost a factor of ten higher than at any other existing machine.

These proposals were studied and a decision was taken to construct the scheme illustrated in Figure 12.3, which became operational in Oct. 1982 with 2×270 GeV[17] (subsequently raised to 2×315 GeV). Fermilab built a similar facility, which started operation in Oct. 1985 with the considerably higher beam energies of 2×800 GeV[18] (subsequently raised to 2×900 GeV). The two facilities are, however, basically the same so that only the former will be used to illustrate the techniques.

---

* CESR at Cornell reached 1.7×10^{32} cm^{-2} s^{-1} early in 1991.

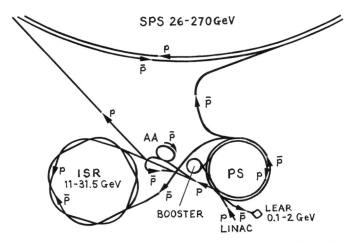

Figure 12.3. Layout of the CERN antiproton complex (1984).

At the start, the protons are accelerated to 26 GeV/c in the PS in the normal way, but before ejection they are concentrated into one quarter of the PS circumference. The protons are then ejected along the line towards the Antiproton Accumulator (AA) (see Figure 12.3), where they are focused onto an antiproton-producing target. The yield is maximum at 3.5 GeV/c and the antiprotons at this energy are collected and injected into the AA near to the outer wall of the vacuum chamber (see Figure 12.4). On this orbit the antiprotons are subjected to fast stochastic cooling. They are then trapped by an rf system and, after the removal of the shutter system that electromagnetically isolates the injection region from the accumulation region, they are moved sideways into the stacking region. The whole process takes place within a single pulsing period of the PS, so that the injection is free to accept the next batch of antiprotons. The antiprotons in the stack are subjected to continuous stochastic cooling by systems independent of those operating on the injection orbit. A dense core of antiprotons is slowly built up near to the inner wall of the vacuum chamber. The accumulation process typically takes about one day after which a part of the dense core can be rebunched, ejected from the AA and returned to the PS for acceleration to 26 GeV/c before being transferred to the SPS.

By the time the antiprotons are ready for transfer, the SPS has already been filled with three proton bunches at 26 GeV/c, to which are now added three bunches of antiprotons circulating in the opposite direction. Finally, both beams are accelerated to 315 GeV/c, the highest energy the SPS is capable of in d.c. operation. At top energy, the low-β insertions are turned on ready for physics.

At the end of 1987, the above system was upgraded by the commissioning of a second ring (ACOL) around the AA. The ring is dedicated to the initial fast cooling of the injected pulse, after which it is transferred to the AA for slow cooling. This separation of functions improved the performance considerably. It has also become

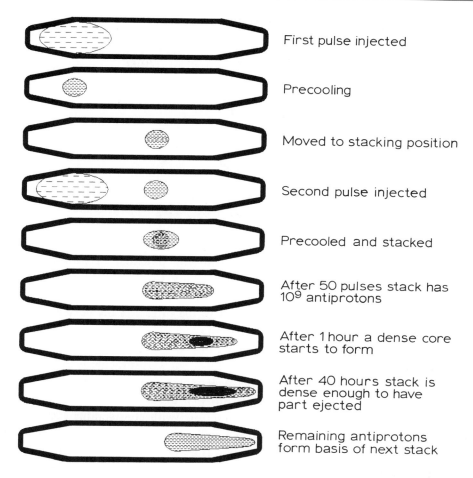

First pulse injected

Precooling

Moved to stacking position

Second pulse injected

Precooled and stacked

After 50 pulses stack has 10^9 antiprotons

After 1 hour a dense core starts to form

After 40 hours stack is dense enough to have part ejected

Remaining antiprotons form basis of next stack

Figure 12.4. Antiproton accumulation in the CERN AA (based on CERN PS/88-16 (AR) and SPS/88-10 (DI)).

possible to work with six bunches in each beam in the SPS. The performance improved steadily up to 1991 (see below), when the facility stopped regular operation.

- Peak luminosity 5.9×10^{30} cm^{-2} s^{-1} at 2×315 GeV/c.
- Low-β values 2 m × 1 m.
- Longest coasting beam time 45.5 h.
- Daily integrated luminosity exceeded 200 nb^{-1}.

These results depend on the antiproton supply being reliable and efficient. At best the AA has functioned without interruption for nearly 48 days, the peak stack accumulated was more than 10^{12} antiprotons and the peak stacking rate was 5.8×10^{10} particles per hour. Comparable performance is available at the Fermilab collider at the much higher energy of 2×900 GeV/c.

In the early days of the AA, antiprotons were also transferred to the ISR[19], where the first proton–antiproton collisions occurred (4 April 1981). The ISR was equipped with its own stochastic cooling systems for maintaining the antiproton beam quality, which has not yet been done in the SPS since it is more difficult to stochastically cool bunched beams at high energies.

The other use of antiprotons at CERN has been in LEAR, a low energy storage ring (indicated on Figure 12.3), where the particles can be decelerated to as low as 2 MeV or 61 MeV/c. There are also plans to extract the antiproton beam via an RFQ and to decelerate to 10 keV.

Several other projects are planned around the world, often with combinations of stochastic and electron cooling, for nuclear physics experimentation.

The main advantage of the 'antiproton approach' in high energy colliders has been that existing rings could be used to increase the available centre-of-mass energy substantially at modest cost. Luminosities have been obtained that few people believed possible. Nevertheless, compared with the potential of proton–proton colliders, they suffer from relatively modest luminosity. Consequently, in spite of the success of the two proton–antiproton colliders, future plans for hadron colliders are based on intersecting proton rings, which will be the subject of the next section.

12.2.6   Future hadron colliders

The stacking methods used in the ISR gave very high beam currents, but suffered from the drawback of making coasting beams that are difficult to accelerate. It is true that phase displacement acceleration was used with considerable success, but only from 26.5 GeV/c to 31.5 GeV/c. When a collider needs to work with a relatively low injection energy, an accumulation method is needed that preserves the bunches. This means lower average currents and for high luminosities emphasis must be put on reducing beam sizes and increasing the number of bunches and the bunching factor. Some reduction in beam size comes automatically with the adiabatic damping due to the increase in energy, but the reduction of the β-values at the crossing point is the most important approach.

In principle, the accumulation is fairly simple and is similar in the various projects under consideration such as UNK* (2×3 TeV)[20], LHC (2×8 TeV)[21], SSC (2×20 TeV)[22] and ELOISATRON (2×100 TeV)[23]. As a representative example a short description of the CERN LHC project will be given here. Not all parameters are determined by accelerator physics. For instance, particle detectors have a resolution time of around 15 ns, which determines the bunch-separation time in LHC and hence the bunch frequency at ~ 67 MHz. Injection into LHC is via the existing CERN accelerator complex (see Figure 12.5). The LHC bunch spacing is already formed at the top energy in the PS by a dedicated rf system and the bunches are compressed

---

* UNK also keeps open the option of proton–antiproton collisions.

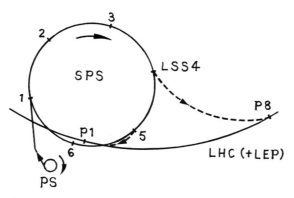

Figure 12.5. Injector chain for the LHC.

so as to fit the SPS buckets. The PS pulses are then transferred one after the other into the buckets around the SPS circumference with exact synchronisation until the machine is full. This is called *box-car stacking*. The SPS accelerates the beam to the LHC injection energy (450 GeV) and again box-car stacking is applied. According to the beam intensity required, four to twelve SPS pulses are stacked. The second LHC ring is filled in the same way and both beams are accelerated to the top energy. Finally the low-β insertions are powered and the beams prepared for physics.

It is estimated that beam currents of nearly one ampere can be accumulated by this method, which is an impressive current at that energy, but, even so, strong beam focusing in the crossings is still needed to achieve the desired luminosity. Low-β insertions with β-values down to 0.25 m have been designed.

The emittances of the final beams usually depend on the various instabilities that may occur in the accelerators of the injector chain. One of the more serious limitations comes from the non-linear forces in the beam–beam interactions in the LHC itself. Although some aspects of this effect are still not well understood, experimental data coupled with analysis play a governing role in the determination of the optimum balance between beam currents and emittances and hence luminosity. Once all elements are included, it is estimated that the LHC should be capable of reaching a luminosity of 1.65×10^{34} cm^{-2} s^{-1} in each of the three interaction regions. At more than two orders of magnitude above what has been so far demonstrated in practice, this is not only a challenge for the accelerator builders, but perhaps even more so for the detector designers.

12.2.7 Electron–positron colliders

Nature provides a very simple way of accumulating electrons in a ring by means of synchrotron radiation damping (see Sections 10.5 and 10.7). By the right choice of lattice parameters all three degrees of freedom of the electrons' oscillations can be damped and, to some extent, the damping rates can be set by the lattice design. A more flexible control of the damping can be obtained by installing special *dipole wiggler magnets* (see Section 10.9).

Because damping compresses the beam in phase space, it is possible to repeatedly inject with a small, intentional injection error. Providing the damping time is shorter than the repetition period, this injection error is damped away, compressing the pulse onto the equilibrium orbit and thus liberating the phase space for injection, before the next pulse arrives.

At first sight, very high beam intensities would be possible by this method, but many different phenomena limit both the intensity and the emittance. The most fundamental is the quantum excitation (see Sections 10.6 and 10.8) and the equilibrium between this excitation and the damping is the dominant factor in the final emittances. The beam currents are further limited by the direct space-charge fields and the wake fields acting within a bunch and between bunches. These effects eventually need active feedback systems to stabilise the beams[24]. The beam–beam interaction also limits the beam sizes, although, by virtue of the damping, lepton colliders are more tolerant than hadron ones. Full separation of the beams at injection and in the unused crossing regions at top energy are important remedies for this effect. Low-β insertions are essential features in these colliders, as they have little negative effect on the above limitations, while dramatically improving the luminosity. Values down to $\beta \simeq 1$ cm have been tried*. The insertion is less efficient if the β-value is made less than the bunch length, which sets a practical limitation.

The severity and complexity of these limitations can be judged by the fact that electron–positron colliders often have difficulty in reaching their design luminosities, which makes the highest value reached so far of 1.7×10^{32} cm^{-2} s^{-1} in CESR at Cornell (beg. 1991) a great achievement. Although synchrotron radiation is useful for injection and beneficial for stability, it is also a limitation on the maximum energy. For this reason LEP[25] will probably be the largest collider of this type to be built. Once the present machine has been upgraded from 2×55 GeV to 2×100 GeV with a superconducting rf system, it will be time for the linear electron colliders to take over the frontiers of lepton, particle physics research.

12.2.8 Electron–proton colliders

This type of collider has frequently been studied, usually as an option to be added to an existing or planned accelerator, but the first major facility is the HERA machine at DESY with an 820 GeV proton ring against a 30 GeV electron (or positron) ring[26]. HERA became operational for physics in 1992. Basically the accumulation process in the electron ring is as described in Section 12.2.7 and for the proton ring as described in Section 12.2.6. However, the different beams make the optimisation of

---

* Linear colliders are presently considered with sub-millimeter β-values in the final focus.

the luminosity different in detail. The LHC design study also has an electron–proton option for colliding the LEP beam with one of the proton rings.

The design luminosity of HERA is 1.5×10^{31} cm^{-2} s^{-1} and for the electron–proton option in LHC 2.8×10^{32} cm^{-2} s^{-1} with an electron energy of about 60 GeV. The luminosity depends strongly on the electron energy.

12.3   SUMMARY OF SOME OTHER EFFECTS

12.3.1   Current and luminosity lifetime

For a luminosity of the order of 10^{32} cm^{-2} s^{-1} the particle loss from the beam is appreciable and the useful luminosity lifetime can be reduced to a few hours. This makes it necessary to optimise the run time as a function of the filling time and luminosity lifetime. The particle loss due to luminosity is given by,

$$\frac{1}{I}\frac{dI}{dt} = -ef_{\mathrm{rev}}\Sigma_{\mathrm{tot}}L/I \qquad \text{per intersection} \qquad (12.50)$$

where, Σ_{tot} is the total cross-section.

For example, in the LHC proposal one of the three interaction regions has the exceptionally high luminosity of 3.8×10^{34} cm^{-2} s^{-1} dominating the expected current losses from the beams. The time constant for the current loss is estimated at 35 h, assuming 4.8×10^{14} beam particles (I/ef_{rev}) and 10^{-25} cm^2 for the total proton–proton cross-section. Since $L \propto I^2$, the time constant for the decay of the luminosity will be half this value, i.e. 17.5 h.

12.3.2   Vacuum

In the previous sections there were references to beams in colliders circulating, not just for hours, but for days or even weeks. This imposes extremely strict requirements on the vacuum and introduces a whole range of new problems.

(i) *Nuclear scattering on the residual gas*[27]

Nuclear scattering causes beam loss, but long before this is serious from the point of view of beam lifetime, it will cause intolerable background in the physics experiments. For example, the pressure in the main arcs of the ISR was maintained at better than 10^{-11} torr, which corresponded to a beam loss rate from nuclear scattering of 0.5×10^{-5} fractional part per minute, but, because of the very stringent background requirement, the pressure was kept at less than 10^{-12} torr in the interaction regions.

The particle loss on the residual gas is given by,

$$\frac{1}{I}\frac{dI}{dt} = -\beta cn\Sigma_{n,N} \tag{12.51}$$

where $\Sigma_{n,N}$ is the cross-section for nitrogen [m²], n is the equivalent density of nitrogen atoms for scattering [m⁻³].

(ii) *Multiple scattering on the residual gas*[27]

This will cause a slow growth in beam dimensions given by,

$$\frac{1}{\sigma_y^2}\frac{d}{dt}\sigma_y^2 = \frac{e^4}{4\varepsilon_0^2 c}\frac{\bar{\beta}_y}{\varepsilon_y p^2}\frac{Z_b^2 Z_g^2}{A_b^2}n_g\ln\left[\frac{3.84\times 10^4}{(A_g Z_g)^{1/3}}\right] \tag{12.52}$$

where $\sigma_y = \sqrt{(\varepsilon_y\beta_y/\pi)}$ as defined in (4.31); $\bar{\beta}_y$ is the average of the betatron amplitude function; p is the momentum of the beam particle; ε_0 is the permittivity of vaccum; Z_b, Z_g are the atomic charges of the beam particles and residual gas atoms respectively; A_b, A_g are the atomic numbers of the beam particles and residual gas atoms respectively; and n_g is the density of the atoms in the residual gas.

In the ISR example mentioned earlier, the average pressure was better than 10^{-11} torr, which corresponded to a growth in beam size due to multiple scattering of around 1% per hour for large stacks (i.e. ≥ 30 A).

(iii) *Gas desorption in proton machines*

This is caused by ions that are created in scattering interactions with the residual gas and are then accelerated by the beam potential onto the chamber walls. This bombardment releases more gas, which is then ionised by the beam so intensifying the bombardment. At a critical current level this chain reaction cause a catastrophic breakdown of the vacuum ending in total beam loss[28]. The problem can be partly solved by higher pumping speeds, but ultimately it is necessary to reduce the desorption coefficient to less than unity by higher bakeout temperatures and glow-discharge cleaning of the chambers.

(iv) *Gas desorption in electron machines*

This is driven by the synchrotron radiation which bombards the chamber wall[29-31]. The beam will eventually clean its own chamber bringing the desorption coefficient below 10^{-3}, but it should be noted that the synchrotron light from the electrons and positrons falls on different parts of the chamber, so that both beams are needed to clean a collider.

(v) *Charged particle trapping*

In the potential well of the beam this can be a serious problem. In Chapter 8 the effect of partial neutralisation has already been discussed and the resultant tune shift evaluated. If the tune shift moves the beam onto a resonance the resultant beam loss will bombard the chamber wall releasing more gas and causing further neutralisation. The beam current will be seen to decay in a matter of tens of seconds as the resonance cuts deeper and deeper. There are many aspects of neutralisation and a good pedagogic account is to be found in Ref. 32, which itself contains many references. To a large extent, bunched beams are self-clearing, but the reader may be interested by such elusive effects as the 'ion-ladder'[33]. Some of the latest observations in antiproton, electron and positron beams and developments in beam clearing by shaking can be found in Ref. 34.

(vi) *Pumping systems*

These are very varied. The old weak-focusing machines had large apertures and vacuum chambers into which a person, in some cases, could crawl. Localised pumps were the obvious solution for these machines. Alternating-gradient focusing reduces the aperture requirement, the magnet cost and the power bill, but creates a problem by also reducing the vacuum conductance. At first more lumped pumping was added, but as designers built longer dipoles with ever smaller apertures this became impractical. *Distributed pumping* is the only way to combat the poor conductance, but the technologies for very high energy proton and electron machines are totally different. All TeV proton machines will use high-field superconducting magnets and by adopting a *cold-bore* design the vacuum chamber will be one enormous cryo-pump. This additional use of the very low temperature of liquid helium is an argument against the new warm superconductors being employed in accelerator magnets, except perhaps for specialised applications such as connection leads. Electron–positron machines, like LEP, use a pumping chamber with a getter or non-evaporable getter (NEG) running along the side of the main chamber with regular communication ports[25].

12.3.3   Intra-beam scattering

By virtue of their transverse motion the particles in a beam can be scattered by their neighbours. This scattering is classified into two broad categories,

- *Touschek effect* or single scattering events, which cause direct loss of particles
- *Intra-beam scattering*, which causes beam growth.

Touschek was the first to observe and understand beam loss by single scattering events, while observing the lifetime of the bunched beams in the first electron–positron storage ring AdA (1963). The theoretical model was published the same year[35]. Small discrepancies between this model and the observations led Bruck to suggest that the anomalous beam dimensions were the result of weaker scattering events that did not immediately cause particle loss[36]. This multiple scattering effect was generalised for proton storage rings in 1974 by Piwinski[37] and became renamed as intra-beam scattering.

This internal scattering mechanism is a basic limitation on beam size. There is a possibility that beams with extremely low transverse temperatures finally settle into what is effectively a crystal lattice[38], which prohibits scattering and removes Schottky noise, but for all practical accelerators the only counter action is to apply some form of cooling.

The mechanism of intra-beam scattering couples all three degrees of freedom. The sum of the growth rates $(\tau_x^{-1} + \tau_z^{-1} + \tau_p^{-1})$ is always positive, although individually the growth rates can be positive or negative according to the conditions. This indicates that the beam does not tend towards a stable configuration, but instead continues to grow. The scattering is basically the same phenomenon as occurs between the beam and the residual gas. It is proportional to the atomic charge to the fourth power and inversely proportional to the atomic number squared. This strong dependence on the particle charge makes heavy-ion colliders especially sensitive to this effect. The mathematical expressions are complicated[37], but many computer programs have been written to perform the calculations (for example see Ref. 39). In Ref. 40 a quasi-universal curve is given which once calculated for a given machine should be accurate to within 50% for all conditions in that machine.

In some machines, intra-beam scattering is an important design factor since it constitutes a substantial limitation on the lifetime and average luminosity. For instance, in the relativistic heavy-ion collider RHIC[41], planned at BNL, the aperture has been chosen larger than one would otherwise expect, just in order to accommodate this beam growth. Nevertheless, at the lower operational energies, this growth is so rapid as to be a significant limitation on the average luminosity and background.

12.3.4 Beam–beam effect

The lens formed by the space-charge distribution in a beam is the most nonlinear element in the machine. Each beam sees the space-charge lens of the other giving rise to the infamous beam–beam effect. This mutual interaction can drive resonances of all orders and is infinitely complex in all its aspects. A very complete survey of the beam–beam interaction can be found in Refs 42 and 43.

Since the tune shift arising from the linear part of the beam–beam effect appears as a factor in the driving terms of all the higher order non-linear resonances, the linear tune shift is considered as an indicator of the strength of the beam–beam effect.

As a design guide, maximum values for the beam–beam tune shift are frequently quoted. These values differ for lepton and hadron machines and, since they are mainly phenomenological, they also tend to vary according to their source. The examples quoted below are for the total tune shift from all intersections and are considered as operational values in the quoted references,

LEP[25] electron–positron collider $\Delta Q_{\text{b–b,max}} = 0.03$

SPS[44] proton–antiproton collider $\Delta Q_{\text{b–b,max}} = 0.02$

LHC[21] proton–proton collider $\Delta Q_{\text{b–b,max}} = 0.015.$

The linear beam–beam tune shift provides a simple criterion for designers, but should be applied with care. In the proton case, it appears that the spacing between strong resonances is the factor of paramount importance and that the strength of the individual resonances in the beam correlates well with observed conditions[45]. In electron colliders, the tune shift per crossing is often quoted as being more important than the total and individual resonances are not able to explain observations. Reference 44 gives experimental data showing that the relative sizes of the two beams in hadron colliders can have a large effect and Ref. 46 draws on recent linear collider studies and suggests that the disruption parameter is more fundamental than the tune shift in lepton colliders.

There appears to be no sure way of beating the beam–beam effect. The best that can normally be done is to adjust the ratio of the horizontal and vertical emittances so as to equalise the tune shifts in both planes in order that the beams are not limited prematurely in one plane. The beams are also separated, usually by electrostatic separators, in all of the unused crossings, so that the most efficient use of the limit is made.

12.3.5 Aperture

The tails of the particle distribution are cut by the chamber walls into which the beam slowly diffuses. In the ISR it was found necessary to leave $\pm 10\sigma_y$ apertures in the interaction regions and $\pm 7\sigma_y$ in the arcs for low background conditions. In addition the beam was positively limited at one place by a dump block, which was followed by collimators to trap the scattered halo. This system also absorbed accidental beam losses, so protecting the vacuum chamber and experiments.

12.3.6 Stored beam energy

Although the designers of the first colliders were worried about lack of luminosity, their subsequent success in making intense beams can perhaps be said to have been too successful, since the stored energy in such beams can become so great as to pose its own problems.

$$E_{\text{stored}} = N(E - m_0 c^2) = I f_{\text{rev}}{}^{-1}(E - m_0 c^2)/e. \qquad (12.53)$$

A 30 A beam at 26.5 GeV/c in the ISR had an energy of 2.4 MJ, which required a substantial dump block and abort system. The beam was quite capable of piercing the vacuum chamber, as it did on a small number of occasions. However, this stored energy is relatively modest compared to the proposed TeV colliders. In the future electron linear colliders, the risk of beam loss will also pose a very serious threat. In very high energy accelerators a definite effort is therefore made to keep the stored energy down. This makes it necessary to progressively squeeze the emittances and low-β insertions in order to maintain the luminosity with fewer particles.

12.4 CONCLUSION

Particle physics research remains the driving force behind the development of accelerators and at present accelerators are in a phase dominated by circular storage rings and colliders. It appears certain that there will be yet bigger hadron colliders before a new technology is born out of need or chance discovery, but for leptons it seems very likely that the largest circular machine, LEP, is already built and working. Once LEP has been upgraded to its final energy of 2×100 GeV, the alternative linear-collider technology must be developed, if research is to reach into yet higher energy domains. However, many of the earlier accelerator technologies, with the passing of their supremacy as the highest energy machines available for research, diversified into other applications. The betatron and cyclotron became far more numerous as medical machines than they had been as research machines and linacs appeared in their thousands as ion implanters and radiotherapy machines. The lepton colliders are already performing the same feat. LEP is sometimes referred to as a Z_0 factory, but now one also hears of Beauty Factories and Tau Charm Factories, which are very high luminosity colliders with less than 10 GeV per beam. These machines promise a long and interesting future still within the realm of particle physics research. The electron storage ring has already diversified to become the ubiquitous synchrotron light source. These machines are also poised to enter the medical field for angiography and the semi-conductor industry for lithography. The compact synchrotron light source, presently foreseen for industrial lithography, could become as widespread as electron microscopes in research departments, or X-ray machines in hospitals. This promises a rich future for circular accelerators, which it is hoped the reader will enjoy and enrich by his own effort.

APPENDIX A

Transverse particle motion in an accelerator

The derivations given in Chapters 2 and 3 for the transverse motion appear very simple and easy to understand, which was the motivation for doing them in the way shown. The final results are valid and are universally applied, but behind this economy of the truth there are some pitfalls.

As often happens in applied physics, the final expressions are relatively simple, but only as a result of making a number of approximations, which are justified as having very small effects. Usually this is fully satisfactory compared to the desired accuracy, but in some cases, after very many oscillations of the system*, the consequences may no longer be negligible. In such cases the approximations are violating, however slightly, some fundamental principle of physics, which for accelerators will be seen later to be the *conservation of phase space*. The same problem reappears in celestial mechanics where again expressions are required that will accurately represent the motion of a planetary system through extremely large numbers of oscillations. In both cases the conservation of phase space must be rigorously observed. At present there is a lot of research activity motivated by the design of new accelerators to determine the limit between stability and instability for very large numbers of oscillations in the presence of non-linear fields. It is therefore imperative to be sure that at least the equations for the simple linear motion are stable and to find a safe method for treating this and more complicated cases. This method is the *Hamiltonian formalism*, but since it is not possible to give a full mechanics course here, the reader is recommended to consult a standard work such as Ref. 1 or Ref. 2 in order to supplement the following summary.

(i) General formulation

The motion of a single particle under an external force can be described by a Hamiltonian $H(\mathbf{q}, \mathbf{P}, t)$, where $\mathbf{q}$ is the position vector, $\mathbf{P}$ the canonically conjugate

---

* In the CERN ISR, it was possible to have stable beams for 50 h, or more, without cooling, which corresponds to 4.6×10^{11} betatron-oscillations.

momentum (P distinguishes the conjugate momentum from the more usual kinetic momentum p) and t is an independent variable such as time. H is chosen in such a way that,

$$\frac{dq_i}{dt} = \frac{\partial H}{\partial P_i} \tag{1A}$$

and

$$\frac{dP_i}{dt} = -\frac{\partial H}{\partial q_i}. \tag{2A}$$

Thus in a system with n degrees of freedom the dynamics will be described by $2n$ first-order equations. The Hamiltonian can be found by expressing the system in whatever generalised set of co-ordinates is convenient and then constructing the Lagrangian. The Lagrangian is then the key to finding the conjugate momenta and the Hamiltonian via,

$$P_i = \frac{\partial L}{\partial \dot{q}_i} \tag{3A}$$

and

$$H = \sum_i P_i \dot{q}_i - L. \tag{4A}$$

In practice the construction of the Lagrangian may not be easy, but in this case the general expression for the Lagrangian of a relativistic charged particle in an electromagnetic field will be simply quoted and taken as the starting point, i.e.

$$L = -m_0 c^2 \gamma^{-1} - e(\phi - A \cdot v) \tag{5A}$$

where $A(q)$ is the vector potential of the external magnetic field such that,

$$B = \nabla \times A \tag{6A}$$

and $\phi(q)$ is the scalar potential of the electric field so that,

$$E = -\nabla \phi - \frac{\partial A}{\partial t} \tag{7A}$$

remembering $\gamma = (1 - v^2/c^2)^{-\frac{1}{2}}$.

The second term in (5A), $(\phi - A \cdot v)$, is a generalised potential, which has the somewhat unusual feature of a velocity-dependent term. It is left to the reader to verify that differentiation of this generalised potential according to,

$$F = e\left\{ -\nabla U + \frac{d}{dt}\left[\frac{\partial U}{\partial v} \right] \right\} \quad \text{where } U = \phi - A \cdot v \tag{8A}$$

gives the Lorentz force on a moving charge*

$$F = e(E + v \times B). \tag{9A}$$

---

* Make use of $d/dt = \partial/\partial t + v \cdot \nabla$, the standard expansion for a triple vector product written as $v \times (\nabla \times A) = \nabla(A \cdot v) - (v \cdot \nabla)A$, and finally the expressions $E = -\nabla \phi - \partial A/\partial t$ and $B = \nabla \times A$.

Attention is drawn to this generalised potential, since it is this term that makes the conjugate momentum different from the kinetic momentum in the presence of a magnetic field.

The application of (3A) to (5A) to find the components of the conjugate momentum in terms of the kinetic momentum, p, gives

$$P_i = (m_0\gamma v_i + eA_i) = (p_i + eA_i). \qquad (10A)$$

By the use of (4A), the Hamiltonian can be expressed in the kinetic variables used in the Lagrangian,

$$H = m_0\gamma c^2 + e\phi. \qquad (11A)$$

The first term in (11A) is the expression for the total energy of a free particle. This can be re-expressed in terms of the kinetic momentum as

$$(m_0\gamma c^2)^2 = c^2 p^2 + m_0^2 c^4. \qquad (12A)$$

The substitution of (10A) and (12A) into (11A) yields a well-known expression (13A), which is often used as a starting point for relativistic charged-particle mechanics. The important point to note is that P is the momentum which is canonically conjugate to the chosen position coordinate q and that it is related to the kinetic momentum by (10A).

$$H = c[(P - eA)^2 + m_0^2 c^2]^{\frac{1}{2}} + e\phi. \qquad (13A)$$

The strict mathematical approach to the Hamiltonian is via the Lagrangian as outlined above, but it turns out to have more familiar associations. The Hamiltonian is related to the total energy of the system and the conjugate variables to the kinetic (T) and potential (V) energies. Furthermore if the potential function is conservative and the kinetic energy does not contain time explicitly then the Hamiltonian is identical to the total energy,

$$H = T + V.$$

In the relativistic case T also contains the rest mass energy.

(ii) Tailoring the Hamiltonian to a circular accelerator

Once a designer has established the principal parameters for an accelerator, he will determine an ideal central orbit for the beam with a reference kinetic momentum p_0. He then needs to know whether the orbits close to this central orbit will oscillate stably about it and whether the central orbit itself is stable against small field errors. In order to answer these questions it is necessary to re-express (13A) in a curvilinear coordinate system and to tailor it to suit this particular problem. For those readers interested in more detail, Ref. 3 by Bell is a description of Hamiltonian mechanics especially for accelerator physicists and specifically dealing with this problem. This is also well described in Ref. 4 by Montague. The reader might also find it useful to read about the Frenet–Serret formulae for the curvature and torsion of surfaces in, for example, Ref. 5 by Massey and Kestelmann.

Equation (13A) is re-expressed in a curvilinear coordinate system (x, s, z), following the ideal central orbit, which should be planar and closed, but its radius of curvature $\rho_0(s)$ may vary. The coordinates x and z measure transverse displacements and s measures the distance along the trajectory, so that,

$$H = c[(P_s - eA_s)^2(1 + x/\rho_0)^{-2} + (P_x - eA_x)^2 + (P_z - eA_z)^2 + m_0^2c^2]^{\frac{1}{2}}. \qquad (14A)$$

The term $e\phi$ has been dropped since only magnetic fields are of interest for the moment. The coordinate transformation is made according to,

$$P_x = \mathbf{P} \cdot \mathbf{n} \qquad P_z = \mathbf{P} \cdot \mathbf{b} \qquad P_s = \mathbf{P} \cdot \mathbf{t}(1 + x/\rho)$$

$$A_x = \mathbf{A} \cdot \mathbf{n} \qquad A_z = \mathbf{A} \cdot \mathbf{b} \qquad A_s = \mathbf{A} \cdot \mathbf{t}(1 + x/\rho)$$

where $\mathbf{n}$, $\mathbf{b}$ and $\mathbf{t}$ are the unit normal, bi-normal and tangential Frenet vectors. One point to note is that the components P_s and A_s are not generally parallel to the reference orbit.

Next the independent variable is changed from t to s, which is more convenient in a cyclic accelerator, since the lattice fields are periodic in s. This change of the independent variable is a standard canonical transformation[3], which creates a new Hamiltonian H_1, equal to $-P_s$, with $(x, P_x; t, -H; z, P_z)$ as the new canonically-conjugate variables, i.e.

$$H_1 = -eA_s \pm (1 + x/\rho_0)[H^2/c^2 - m_0^2c^2 - (P_x - eA_x)^2 - (P_z - eA_z)^2]^{\frac{1}{2}}.$$

For a non-dissipative system H is the total energy $(m_0\gamma c^2)$, so with the use of (12A) this equation becomes,

$$H_1 = -eA_s - (1 + x/\rho_0)[p^2 - (P_x - eA_x)^2 - (P_z - eA_z)^2]^{\frac{1}{2}}. \qquad (15A)$$

The choice of the negative sign, from the first equation for H_1, implies that the particle is travelling in the positive s-direction with positive momentum. Equation (15A) is the general form used for the study of the long-term stability of the betatron oscillations.

Equation (15A) can be greatly simplified by considering only 2-dimensional fields, so that $A_x = A_z = 0$, and by expanding the square root and neglecting high-order terms on the assumption that $(P_x^2 + P_z^2) \ll p^2$.

$$H_1 \simeq -eA_s - \left[1 + \frac{x}{\rho_0}\right]p\left\{1 - \frac{1}{2}\left[\frac{P_x}{p}\right]^2 - \frac{1}{2}\left[\frac{P_z}{p}\right]^2\right\}. \qquad (16A)$$

The component A_s can be expanded up to second order as,

$$eA_s = -p_0 \left\{ \frac{x}{\rho_0} + \left[\frac{1}{\rho_0^2} - k \right] \frac{x^2}{2} + k \frac{z^2}{2} \right\}. \tag{17A}$$

This takes into account the curvature term and the definition of the normalised gradient,

$$k = \frac{e}{p_0} \left[\frac{\partial B_z}{\partial x} \right]_0 = -\frac{1}{B_0 \rho_0} \left[\frac{\partial B_z}{\partial x} \right]_0.$$

It should be noted in (17A), that p_0 is the momentum of a fictitious reference particle on the central orbit and therefore depends only on the strength of the dipole fields. The final Hamiltonian is obtained by substituting (17A) into (16A) and again dropping the high-order terms to give,

$$H_1 = -p - \frac{(p - p_0)}{\rho_0} x + \frac{p_0}{2} \left[\frac{1}{\rho_0^2} - k \right] x^2 + \frac{p_0}{2} k z^2 + \frac{p}{2} \left[\frac{P_x}{p} \right]^2 + \frac{p}{2} \left[\frac{P_z}{p} \right]^2. \tag{18A}$$

The first order equations for the transverse motion are obtained by applying the equations of Hamiltonian (1A) and (2A) to (18A), with the independent variable s replacing t, to give:

$$\frac{d}{ds} x = \frac{P_x}{p}; \qquad \frac{d}{ds} P_x = \frac{(p - p_0)}{\rho_0} - p_0 \left[\frac{1}{\rho_0^2} - k \right] x;$$

$$\frac{d}{ds} z = \frac{P_z}{p} \qquad \text{and} \qquad \frac{d}{ds} P_z = -kz,$$

which in turn yield the second order equations of motion in a slightly more general form than derived earlier in Chapters 2 and 3:

$$\frac{1}{p_0} \frac{d}{ds} \left[p \frac{d}{ds} x \right] + \left[\frac{1}{\rho_0^2} - k \right] x = \frac{\Delta p}{p_0} \cdot \frac{1}{\rho_0} \tag{19A}$$

$$\frac{1}{p_0} \frac{d}{ds} \left[p \frac{d}{ds} z \right] + \qquad kz = 0. \tag{20A}$$

Note that p is the momentum of the particle, p_0 is related to the dipole field strength and $\Delta p = p - p_0$. Both p and p_0 can be functions of time.

(iii) Conservation of phase space

Accelerator physicists should be aware of *Liouville's theorem* which can be stated as 'The area of an element of phase space $(\delta x \, \delta P)$ along a phase-space trajectory is invariant provided only pure Hamiltonian forces are acting'. The essence of a proof of this theorem using the properties of the Hamiltonian will be seen in the next section. For discerning people two points should be noted:

- the individual particles of a beam are not in detail the same as the continuous 'fluid' of phase space, and
- not all forces in accelerators are Hamiltonian.

Thus when a beam filaments, its particles will appear to the user as being uniformly distributed over an enlarged phase-space area. Knowing that on an infinitesimally small scale the strands of coiled phase space still conserve their original area, will give the user little consolation. Stochastic cooling also relies on this difference between individual particles and the continuum of phase space by performing the seemingly impossible task of arranging the empty space on the outside and the particles at the centre of a phase-space area.

An example of the second point above is the phase-space damping that occurs when an electron beam radiates its energy in a bending field and is then re-accelerated by the rf cavities.

(iv) Effects of small terms and approximations on the motion[6]

Consider the equation for some oscillation in the form,

$$\frac{d}{dt}\left[m\frac{d}{dt}x\right] + kx = 0. \tag{21A}$$

Adiabatic damping and such phenomena as the transition energy arise from variations with time of the coefficients m or k or both in an equation of this type.

It is well-known that a term of the form $(b\, dx/dt)$, small compared with the existing terms, if added to the equation will produce a small damping effect, which can make a large change in the amplitude if one waits long enough. This is the most obvious case where the smallness of a term is not sufficient to justify neglecting it. From this it follows that if m varies with time, even slowly, one cannot move it to the front of the d/dt operator and neglect the (dm/dt) term that then arises.

To see where other small terms may affect the damping, it is convenient to work with the simultaneous first-order equations, obtained from (21A) by choosing $p \equiv m\, dx/dt$.

$$\frac{d}{dt}p = -k(t)x \tag{22A}$$

and

$$\frac{d}{dt}x = [m(t)]^{-1}p. \tag{23A}$$

It can be shown that (22A) and (23A) are the canonical equations of a Hamiltonian,

$$H = k(t)x^2/2 + [m(t)]^{-1}p^2/2$$

and therefore Liouville's Theorem will hold in the (x, p) space, x and p are the canonically conjugate variables in this example. The only additional requirements for the adiabatic damping law to be true is that k and m^{-1} should vary slowly and not contain Fourier components that excite resonances.

It is convenient at this point to extend the analysis to include the longitudinal phase oscillations of Chapter 7. Since these equations are more complicated, more general forms for the righthand sides of (22A) and (23A) are needed, i.e.

$$\frac{\mathrm{d}}{\mathrm{d}t} p = -F(t, x, p) \tag{24A}$$

and

$$\frac{\mathrm{d}}{\mathrm{d}t} x = G(t, x, p). \tag{25A}$$

Equations (24A) and (25A) can be used to verify in what circumstances these equations can be the canonical equations belonging to a Hamiltonian H. If they are, then

$$\frac{\mathrm{d}}{\mathrm{d}t} p = -\frac{\partial H}{\partial x} \tag{26A}$$

and

$$\frac{\mathrm{d}}{\mathrm{d}t} x = \frac{\partial H}{\partial p}. \tag{27A}$$

Therefore

$$\frac{\partial H}{\partial x} = F \tag{28A}$$

and

$$\frac{\partial H}{\partial p} = G. \tag{29A}$$

For H to be a single-valued continuous function of (t, x, p), then

$$\frac{\partial^2 H}{\partial x\, \partial p} = \frac{\partial^2 H}{\partial p\, \partial x} \quad \text{and therefore} \quad \frac{\partial F}{\partial p} = \frac{\partial G}{\partial x}. \tag{30A}$$

Clearly (22A) and (23A) satisfy this; in their case both sides of (30A) are zero. However any p-dependence of F or x-dependence of G will spoil the Hamiltonian formalism unless both occur to the same extent.

It is possible, whether the system is Hamiltonian or not, to consider what happens to areas in phase space. In fact a small area $(\delta x\, \delta p)$ will change according to,

$$\frac{1}{\delta x\, \delta p} \frac{\mathrm{d}}{\mathrm{d}t} (\delta x\, \delta p) = \frac{\mathrm{d}}{\mathrm{d}t} [\log(\delta x\, \delta p)] = \frac{\partial \dot{p}}{\partial p} + \frac{\partial \dot{x}}{\partial x} = -\frac{\partial F}{\partial p} + \frac{\partial G}{\partial x} \tag{31A}$$

where use has been made of (24A) and (25A). This relation is often found as one step in the deduction of *Liouville's theorem*. Thus (31A) shows that the criterion for constant phase-space area $(\delta x\, \delta p)$ is just the same as the criterion (30A) that a Hamiltonian exists for the variables.

In a particular problem it may be desirable to make approximations in deducing the functions F and G. For most purposes it is sufficient if the terms neglected are small (in relation to the wanted accuracy) compared with the main terms, but it can be seen from (30A) that it is not permissible to neglect a small p-dependence of F or a small x-dependence of G when damping is being studied; such terms can have a large effect if they are given time to do so.

It should be noted in particular that non-linearities, either higher order x-terms in F or higher order p-terms in G, have no effect on (31A). This sort of term may therefore be neglected, without affecting the damping, but it should be checked that the linearised approximation so obtained is good enough from other points of view.

Those terms that affect damping, i.e. an x-term added to (21A) or a p-dependence of F or an x-dependence of G, are usually associated with dissipation forces, but they can also arise from an imprudent change of variables. As a very simple example, suppose that in (22A) and (23A) it was decided to use the variable Λ defined by,

$$\Lambda = \gamma p \tag{32A}$$

instead of p. In this case γ is some function of time so that,

$$\dot{\Lambda} = \gamma \dot{p} + p\dot{\gamma} \tag{33A}$$

and the pair of equations would become

$$\frac{\mathrm{d}}{\mathrm{d}t}\Lambda = -k\gamma x + p\,\frac{\mathrm{d}}{\mathrm{d}t}\gamma \tag{34A}$$

and

$$\frac{\mathrm{d}}{\mathrm{d}t}x = m^{-1}\gamma^{-1}\Lambda. \tag{35A}$$

The last term in (34A) will be small if γ varies slowly, but it is just the type of term that must *not* be neglected in damping. By using (31A) the reader may verify that this term results in $(\delta x\, \delta p)$ varying in proportion with γ, which is just the variation required by the known $(\delta x\, \delta p)$ conservation.

APPENDIX B

Accelerator magnets

In Chapters 2 and 3 the analysis of the particle motion is developed in the environment of sharply defined regions of constant dipole and quadrupole fields. Real-world magnets are more complicated.

(i) Multipole expansion of a 2-dimensional magnetic field

First consider a purely 2-dimensional field, which is a rather reasonable approximation since the majority of accelerator magnets are long compared with their aperture. In the source-free region of the magnet gap, the field can be derived from a scalar potential ϕ. In the local cylindrical coordinate system for the magnet, the general Fourier expansion of this scalar potential will be[1],

$$\phi = \sum_{m=0}^{\infty} r^m \{A_m \cos m\theta + B_m \sin m\theta\} \tag{1B}$$

where, A_m and B_m are constants. The zero order term is a constant and can be disregarded, since it will make no contribution to the fields derived later. Equation (1B) contains two orthogonal sets of multipoles;

- the *sine terms* are designated *normal or right multipoles**, and
- the *cosine terms* are designated *skew multipoles*.

An alternating-gradient lattice has two normal modes for particle oscillations and by using only *normal lenses* these modes are made *horizontal* and *vertical*. As discussed in Chapter 5 the inclusion of skew lenses in a normal lattice will cause coupling between these modes. It is clear from (1B) that the skew multipoles can be made by simply rotating the normal multipoles by $\pi/(2m)$.

Equation (1B) is a natural starting point in a mathematical sense, but is slightly

---

* The terms *normal* or *right* are generally omitted and only the term *skew* serves to distinguish between the two possibilities.

inconvenient for certain applications. It will therefore be used for the basic derivations and then, at the appropriate places, correspondences will be established with the Taylor expansion of the field, which is used in optics, and a normalised field expansion, which is used by magnet builders.

A quick way of transforming (1B) into Cartesian coordinates is to use complex notation in which the *skew terms* are the *real parts* and the *normal terms* are the *imaginary parts*, so that,

$$\phi = \sum_{m=1}^{\infty} r^m \{A_m \, \text{Re}[\cos\theta + j\sin\theta]^m + B_m \, \text{Im}[\cos\theta + j\sin\theta]^m\} \qquad (2B)$$

where $[\cos m\theta + j\sin m\theta] = [\cos\theta + j\sin\theta]^m$ (De Moivre's theorem). Equation (2B) can now be written directly in Cartesian coordinates as,

$$\phi = \sum_{m=1}^{\infty} \{A_m \, \text{Re}[X + jZ]^m + B_m \, \text{Im}[X + jZ]^m\}. \qquad (3B)$$

The field components are obtained by differentiation and are listed in Table 1B up as far as decapole.

$$B_x = -\frac{\partial\phi}{\partial X} \qquad \text{and} \qquad B_z = -\frac{\partial\phi}{\partial Z} \qquad (4B)$$

In the above definitions, the summation index m has the useful feature of being simply related to the number of poles that a lens would need to excite the given harmonic, i.e.

$$2m = \text{number of poles needed to excite } m\text{th multipole}. \qquad (5B)$$

In Chapter 2 and Appendix A, the magnetic guide field was expanded around the median plane in a Taylor expansion for the derivation of the particle equations. The relationship between the series coefficients in (1B) and Table 1B with those in the Taylor expansion is by inspection,

$$B_m = \frac{-1}{m!}\left[\frac{\partial^{m-1}B_z}{\partial X^{m-1}}\right]_{z=0} \qquad \text{for } m \geq 1. \qquad (6B)$$

(ii) Dealing with a 3-dimensional magnet

It is not possible to apply the 2-dimensional expansion (1B) directly to the 3-dimensional fringe fields at the ends of accelerator magnets, but fortunately, when these fields are included in a field integral through the magnet, this integral behaves 2-dimensionally, providing one condition is satisfied.

If $\phi(X, Y, Z)$ is the general scalar potential, which satisfies the Laplace equation $\nabla^2\phi = 0$ inside the magnet aperture and if the integral I is defined as,

$$I = \int_{-Y_0}^{Y_0} \phi(X, Y, Z)\, dY, \qquad (7B)$$

Table 1B. *Potential and field components in 2-dimensional multipole lenses up to decapole*

Multipole of order m	Skew lenses $\phi = A_m \, \mathrm{Re}(X + jZ)^m$	Normal lenses $\phi = B_m \, \mathrm{Im}(X + jZ)^m$
Dipole*, $m = 1$	$\phi = A_1 X$ $B_x = -A_1$ $B_z = 0$	$\phi = B_1 Z$ $B_x = 0$ $B_z = -B_1$
Quadrupole, $m = 2$	$\phi = A_2(X^2 - Z^2)$ $B_x = -2A_2 X$ $B_z = 2A_2 Z$	$\phi = 2B_2 ZX$ $B_x = -2B_2 Z$ $B_z = -2B_2 X$
Sextupole, $m = 3$	$\phi = A_3(X^3 - 3Z^2 X)$ $B_x = -3A_3(X^2 - Z^2)$ $B_z = 6A_3 ZX$	$\phi = B_3(3ZX^2 - Z^3)$ $B_x = -6B_3 ZX$ $B_z = -3B_3(X^2 - Z^2)$
Octupole, $m = 4$	$\phi = A_4(X^4 - 6Z^2 X^2 + Z^4)$ $B_x = -4A_4(X^3 - 3Z^2 X)$ $B_z = 4A_4(3ZX^2 - Z^3)$	$\phi = 4B_4(ZX^3 - Z^3 X)$ $B_x = -4B_4(3ZX^2 - Z^3)$ $B_z = -4B_4(X^3 - 3Z^2 X)$
Decapole, $m = 5$	$\phi = A_5(X^5 - 10Z^2 X^3 + 5Z^4 X)$ $B_x = -5A_5(X^4 - 6Z^2 X^2 + Z^4)$ $B_z = 20A_5(ZX^3 - Z^3 X)$	$\phi = 5B_5(ZX^4 - 2Z^3 X^2)$ $B_x = -20B_5(ZX^3 - Z^3 X)$ $B_z = -5B_5(X^4 - 6Z^2 X^2 + Z^4)$

* The horizontal field dipole falls naturally into the skew category but it is never referred to as a skew lens.

then the above statement can be formulated as,

$$\frac{\partial^2 I}{\partial X^2} + \frac{\partial^2 I}{\partial Z^2} = 0, \tag{8B}$$

in an exactly analogous way to the original potential ϕ in the 2-dimensional central region of the magnet, providing that,

$$\left[\frac{\partial \phi}{\partial Y}\right]_{Y_0} = \left[\frac{\partial \phi}{\partial Y}\right]_{-Y_0}. \tag{9B}$$

The condition (9B) can always be satisfied by going far enough away from the magnet for the field to drop to zero. The proof of the above can be found in Ref. 2.

Since most magnets are short compared with the betatron wavelength, the field integral through the magnet is all that it is necessary to know and to correct. Not only does this justify the neglect of the 3-dimensional nature of the fringe fields in many cases, but it also justifies the use of localised shims for correcting field errors.

It follows from (8B) that the integral I can be expanded in a Fourier series in an analogous way to ϕ in (1B), i.e.

$$I = \sum_{m=1}^{\infty} r^m \{A_{I,m} \cos m\theta + B_{I,m} \sin m\theta\}. \tag{10B}$$

The multipole coefficients $A_{I,m}$ and $B_{I,m}$ in the expansion (10B) will be composed of two contributions[3]:

$$\left.\begin{array}{l} A_{I,m} = A_m l_{a,m} + E_{a,m} \\ B_{I,m} = B_m l_{b,m} + E_{b,m} \end{array}\right\} \tag{11B}$$

where $l_{a,m}$ and $l_{b,m}$ are the effective lengths of the 2-dimensional, mth order components with amplitudes A_m and B_m and $E_{a,m}$ and $E_{b,m}$ are the fringe-field contributions for the same components. A perfect vertical field pole would have, for example,

$$B_{I,1} \neq 0, \qquad B_{I,m} = 0 \text{ for } m > 1 \text{ and } A_{I,m} = 0 \text{ for all } m.$$

Since the end effects will always be present, unless excluded by symmetry, (11B) implies that the magnet designer should build into the 2-dimensional part of the magnet compensating multipole components according to,

$$A_m l_{a,m} = -E_{a,m} \qquad \text{and} \qquad B_m l_{b,m} = -E_{b,m} \tag{12B}$$

for all the multipole components that should be zero. Calculation, however, is not always perfect and the designer may need to fine tune these corrections by adding shims during the magnetic measurements.

(iii) Rotating-coil measurements

Equation (1B) is ideally suited to the analysis of what is perhaps the most elegant method for measuring accelerator magnets, i.e. the *rotating coil* or *harmonic coil* method[2,1]*. Consider first a short length of wire of length l mounted parallel to and at a distance r from the magnetic axis of a lens in its 2-dimensional field region and imagine that this wire rotates with a constant angular velocity ω about the axis [see Figure 1B(a)]. As the wire rotates, it will cut the radial lines of magnetic flux and and an e.m.f. will be induced, given by,

$$\text{e.m.f.} = -\omega r B_r l = \omega r l \left[\frac{\partial \phi}{\partial r}\right]$$

$$\text{e.m.f.} = \omega l \sum_{m=1}^{\infty} m r^m \{A_m \cos m\theta + B_m \sin m\theta\}. \tag{13B}$$

* The measurement of static magnetic fields has always been carried out by rotating, flipping, or shifting coils, but originally an emphasis was put on small coils sensitive to only the main field component with the aim of making point measurements. In the sixties, this philosophy changed and large coils filling the whole aperture were used to analyse the full multipole content of the field. The mathematical treatment of a rotating coil was already known and it was probably the availability of the necessary technology that held this development back.

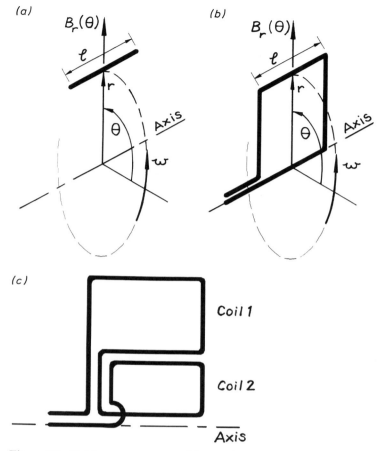

Figure 1B. Field measurements with rotating coils. (*a*) Rotating wire segment, (*b*) single rotating coil, (*c*) two rotating coils connected in opposition.

Thus the induced e.m.f. has a form that is similar to the field potential and can be Fourier analysed to yield the coefficients A_m and B_m. The simplest practical arrangement would be to complete the coil circuit with radii leading to a return wire on the axis, so that the additional wires cut no flux and (13B) is still directly applicable [see Figure 1B(*b*)]. If the return wire is at a finite radius the e.m.f. induced in the return wire opposes that in the outer wire. Often it is more important to measure the errors in a magnet than to measure the main field component. In order to increase the sensitivity to the error harmonics the signal from the main harmonic can be 'backed off' by winding a second coil with more turns but reaching to a smaller radius. The radii and numbers of turns in the two coils are carefully adjusted so that for the principal harmonic of the field their outputs are equal. By opposing the coils [see Figure 1B(*c*)], only the error harmonics will have a net output. In Ref. 1 an alternative scheme is proposed in which the return wires from a single coil are split into two groups set at different azimuthal angles. The main harmonic can again be cancelled by a suitable choice of parameters.

The short coils described above would be used to explore the central 2-dimensional region of a magnet. Long coils spanning the whole magnet and reaching beyond the fringe fields would be used to investigate the multipole content of the total integral through the magnet as described earlier.

Rotating coils are not always the most convenient or best way of measuring magnetic fields. For a general, but brief, survey of the available methods and their applicability see Ref. 4.

(iv) Nomenclature for magnet measurements

Magnet-measurement experts prefer to work with a specially normalised expansion of the field. This can be derived by re-expressing (1B) in a slightly different, but equivalent, form as,

$$\phi = -\sum_{m=1}^{\infty} \frac{B_{ref} r_{ref}}{m} \left[\frac{r}{r_{ref}} \right]^m \{ -a_m \cos m\theta + b_m \sin m\theta \} \tag{14B}$$

where $(B_{ref} r_{ref})$ is a reference potential, whose value is chosen according to convenience, r_{ref} is a reference radius and is often taken equal to the estimated limit to the good-field region, B_{ref} is a reference field, the choice of which will be discussed later.

The relationship between the coefficients in (14B) and (1B) is given by,

$$a_m = \frac{m}{B_{ref}} r_{ref}^{m-1} A_m \quad \text{and} \quad b_m = -\frac{m}{B_{ref}} r_{ref}^{m-1} B_m \quad \text{for } m \geq 1. \tag{15B}$$

With the potential in the form (14B), the radial and azimuthal field components become,

$$B_r = \sum_{m=1}^{\infty} B_{ref} \left[\frac{r}{r_{ref}} \right]^{m-1} (-a_m \cos m\theta + b_m \sin m\theta) \tag{16B}$$

$$B_\theta = \sum_{m=1}^{\infty} B_{ref} \left[\frac{r}{r_{ref}} \right]^{m-1} (a_m \sin m\theta + b_m \cos m\theta) \tag{17B}$$

so that the new coefficients a_m and b_m give directly the field contributed by each multiple at the reference radius as a fraction of the reference field.

Readers should be aware that there is some controversy over the form of the above equations with references to the summation index m. Some people prefer to consider the field, rather than the potential, as the fundamental quantity and therefore re-write (16B) and (17B) with the index m running from 0 to infinity. This removes the inelegant powers of $(m - 1)$, but it destroys the simple relationship (5B) between m and the number of magnet poles. Both systems are in regular use. The appearance, or absence, of b_0 is a quick guide to which system is in use.

It is usual to choose the reference field to be equal to the field given by the main multipole of the magnet element in question at the reference radius. Thus for a dipole, the reference field would be the nominal dipole field of the magnet and the coefficients would then be equal to the field errors of each multipole at the reference radius expressed as a fraction of the nominal field. For a quadrupole, the reference field would be made equal to the gradient times the reference radius and so on for the higher order lenses. This system provides a convenient picture of the quality of a magnet with respect to its main field component.

The choice of the reference field is free and it is interesting to consider an alternative to the above. When designing a machine the tolerances are set globally according to criteria arising from the machine optics. It would be sensible therefore to define a global reference field for all elements. The natural choice would be the main dipole field. The errors in all units would then be directly comparable. The b_1 component of a vertical field steering magnet, for example, would directly give its strength as a fraction of the main field and it would be clear that it would not need the same intrinsic field quality as the main magnet.

(v) Practical lenses

(a) Conventional accelerator magnets

In a conventional (iron-yoke, copper-coil) magnet, the shape of the iron pole is far more important than the position of the coil for determining the field quality. The ideal pole profile is simply the equipotential surface for the desired multipole, but in practical magnets the poles have to be truncated both axially and transversally and the iron is subject to non-uniform saturation effects, which makes the detailed design complicated. Very often a magnet has to be shimmed during measurements in order to satisfy the field tolerances.

Conventional accelerator magnets with iron yokes are limited to dipole fields of ~ 2 T and quadrupole gradients of ~ 20 T/m. The schematic cross-sections of the dipole and quadrupole types are shown in Figure 2B with the approximate expressions for the ampere-turns needed to excite them.

(b) Superconducting accelerator magnets

The technology of superconducting magnets relies on relatively recent developments in materials and designs of conductors and cables. There is also a strong element of experience entering into the mechanical design. Readers will find Ref. 5 a good introduction to these points. Current technology for accelerator magnets can reliably give 6 T for dipole fields and 100 T/m for quadrupole fields and there is every hope that in the next few years dipole fields of 10 T will also be achieved with the required quality and reliability.

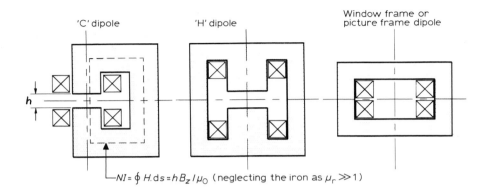

'C' dipole   'H' dipole   Window frame or picture frame dipole

$$NI = \oint H.ds \simeq hB_z/\mu_0 \; (\text{neglecting the iron as } \mu_r \gg 1)$$

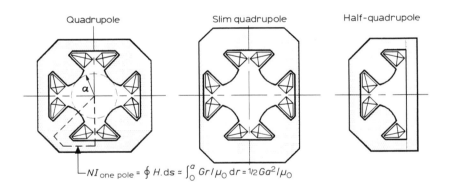

Quadrupole   Slim quadrupole   Half-quadrupole

$$NI_{\text{one pole}} = \oint H.ds = \int_0^a Gr/\mu_0 \, dr = \tfrac{1}{2} Ga^2/\mu_0$$

Combined-function magnet (dipole + quadrupole)

where
NI = number of ampère-turns [A]
h = gap height in dipoles [m]
a = radius of inscribed circle [m]
s = integration path [m]
H = magnetic field intensity [A.m^{-1}]
G = gradient, $dB_z/dx = (dB_r/dr)_{45°}$

Figure 2B. Schematic cross-sections of conventional magnets.

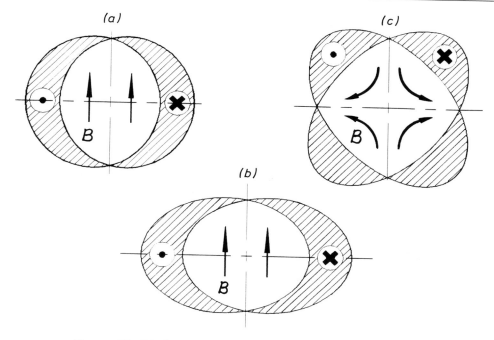

Figure 3B. Ideal current distributions used in superconducting magnets. (*a*) Dipole, (*b*) dipole, (*c*) quadrupole.

The iron yoke in superconducting magnets plays only a limited role and it is the positioning of the conductors which is of paramount importance for the field quality. The coil design in the majority of these magnets is based on some interesting properties of uniformly-distributed, counter-flowing currents in the following geometries[6].

- *Overlapping conductors of circular cross-section* [see Figure 3B(*a*)].

    The overlap region forms a current-free aperture in which there is an exactly uniform dipole field.
- *Overlapping conductors of elliptical cross-section* [see Figure 3B(*b*)].

    This similar arrangement also gives an exactly uniform dipole field in the current-free overlap region. This result also hints at a useful property of these configurations. The magnetic forces in Figure 3B(*a*) tend to flatten the circular conductors into an approximation of the elliptical geometry of Figure 3B(*b*), so it is not surprising to learn that the field quality in practical magnets is insensitive to this type of distortion.
- *Overlapping elliptical conductors set at 90°* [see Figure 3B(*c*)].

    This arrangement gives a perfect quadrupole field.
- *Cylindrical current sheets.*

    The same results can also be obtained using cylindrical current sheets,

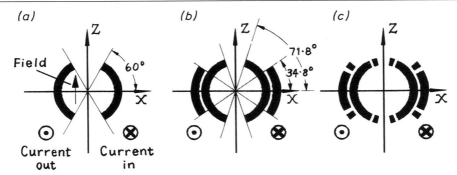

Figure 4B. Coil configurations leading up to the HERA dipole design. (a) Single-layer coil, $b_3 = 0$; (b) double-layer coil, $b_3 = 0$, $b_5 = 0$; (c) double-layer coil with spacers near edges, $b_3 = b_5 = b_7 = 0$, $b_9 \simeq 0$.

in which the current density around the cylinder varies as cos $n\theta$, where n is the order of the multipole to be generated inside the cylinder[6].

In practice the designer has to replace these ideal current distributions with conductors in blocks and layers. Careful positioning of the various blocks will yield the field required. Mess and Schmüser give some interesting details of this process for the HERA dipole in Ref. 7. In their example, it was found that a single-layer winding with two shells subtending 60° each would generate no sextupole field and the higher multipoles would only contribute an error of a few percent at the edge of the aperture. The sextupole (b_3) is the first error allowed by the dipole symmetry. This result is surprisingly good for its simplicity, but for HERA it was necessary to improve on this, so a second layer was added and the inner shells were adjusted to subtend angles of 71.8° and the outer shells 34.8°. This suppressed both the sextupole (b_3) and the decapole (b_5) components. Finally the 14- and 18-pole terms (b_7, b_9) were reduced by adding wedges near the edges of the shells, so that in the final magnet all high order multipoles are less than 10^{-4} (see Figure 4B). Higher-order correction coils are frequently needed in such magnets and these are usually a single layer directly wound on the beam pipe with a cos $n\theta$ distribution of the turns.

The whole coil assembly with its clamps is mounted inside a cylindrical iron yoke. To the first approximation this yoke does not introduce any new multipoles and by virtue of the image currents it re-inforces the main component. The proximity of the iron determines the gain in field. Warm-iron designs (i.e. the yoke is outside the cryostat) add about 10% to the field, while cold-iron designs add up to 40%. The iron yoke also reduces the stray field.

(c) Twin-bore superconducting magnets

The twin-bore magnet (see Figure 5B) was suggested by Blewett[8], as a cost-saving design, with the provisos that both asymmetric beam energies and proton–antiproton

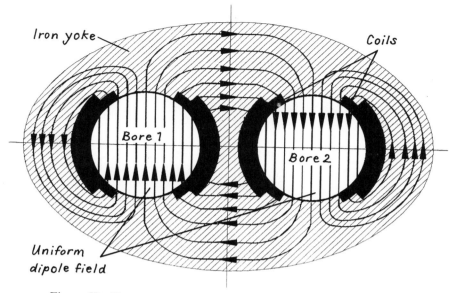

Figure 5B. Flux pattern in a twin-bore magnet.

beams would be excluded. Figure 5B shows how the two apertures use the same flux. Blewett estimated that there would be a 40% reduction in the ampere-turns required and of course there would only be one cryostat in the place of two.

This design was developed to the prototype stage by R. Palmer for the BNL machine ISABELLE, which was later called CBA[9,10], and was also considered for SSC[11], but in both cases the final decisions were made in favour of single-bore magnets in independent rings. It was felt that any reduction in cost would be offset by the operational limitations and the greater complexity of the design. The boundary conditions for LHC, however, were very different, since the machine had to fit inside the existing LEP tunnel, so the problem of available space was solved by adopting the twin-bore design[12].

APPENDIX C

Closed orbits[1]

Hamilton's principle[2] describes the motion of a mechanical system between times t_1 and t_2 in configuration space by specifying that the line integral,

$$I = \int_{t_1}^{t_2} L \, dt,$$ (1C)

where L is the Lagrangian, will have a stationary value for the path of the motion. This principle can be applied to one turn of a closed orbit at constant energy in an accelerator. The Lagrangian of a relativistic charged particle has already been introduced in Appendix A, so that over one revolution period T, on the closed orbit C,

$$I = \int_0^T -m_0 c^2 \gamma^{-1} \, dt + \oint_C eA \cdot ds.$$ (2C)

Hamilton's principle requires $\delta I = 0$ if the orbit is to exist. The first term in (2C) depends only on the particle's total energy, which will be constant in this case. This leaves,

$$\delta I = \oint_C e \, \delta A \cdot ds = 0.$$ (3C)

Stoke's theorem can be used to re-express the line integral of (3C) by a surface integral,

$$\delta I = \int_S e \, \delta \nabla \times A \cdot dS = \int_S e \, \delta B \cdot dS = e \, \delta \Phi = 0$$ (4C)

where Φ is the flux enclosed by the orbit C with area S. Equation (4C) can be interpreted that the closed orbit positions itself so that the flux it links has a stationary value, i.e. a minimum (or maximum) with respect to nearby orbits. The option of a minimum is applicable since this will lead to the closed orbit being stable with respect to small deflections.

APPENDIX D

Phase equation

(i) General derivation of the phase equation[1]

The general Lagrangian for relativistic particles has already been quoted in Appendix A, but it is repeated below since it is the starting point for determining the variables for the present derivation.

$$L = -m_0 c^2 \gamma^{-1} - e(\phi - \boldsymbol{A} \cdot \boldsymbol{v}). \tag{5A}$$

The prescription given in Appendix A specifies that the position coordinates q_i can simply be chosen according to the needs of the problem and the canonically conjugate momenta are then defined by,

$$P_i = \frac{\partial L}{\partial \dot{q}_i}. \tag{3A}$$

For the phase equation, it is convenient to use the cylindrical co-ordinates (R, Θ, z), where the azimuthal angle is defined as,

$$\Theta = s/R = 2\pi s/C. \tag{1D}$$

Differentiation of (5A) with respect to $\dot{\Theta}$ gives the momentum which is canonically conjugate to Θ. This is known as the *general angular momentum S*,

$$S = m_0 \gamma R^2 \dot{\Theta} + eRA_\Theta$$

$$S = \frac{C}{2\pi} (p_\Theta + eA_\Theta) \tag{2D}$$

where p_Θ is the tangential component of the kinetic momentum. Thus (Θ, S) are the canonically conjugate variables for the azimuthal (longitudinal) motion. The simplest route to the motion equation is via the well-known Lagrangian equations of motion.

$$\frac{\mathrm{d}}{\mathrm{d}t} \left[\frac{\partial L}{\partial \dot{q}_i} \right] - \frac{\partial L}{\partial q_i} = 0. \tag{3D}$$

By virtue of (3A), (3D) can be written as,

$$\frac{\mathrm{d}}{\mathrm{d}t} P_i = \frac{\partial}{\partial q_i} [-m_0 c^2 \gamma^{-1} - e(\phi - \boldsymbol{A} \cdot \boldsymbol{v})].$$

The first term on the righthand side is not an explicit function of position, so that,

$$\frac{\mathrm{d}}{\mathrm{d}t} P_i = -e \frac{\partial}{\partial q_i} (\phi - \boldsymbol{A} \cdot \boldsymbol{v}). \tag{4D}$$

Thus when a magnetic field is present, *the time rate of change of the conjugate momentum can be equated to the position derivative of the generalised potential* [see (i) in Appendix A]. This is a more generalised version of the better-known equation which equates the rate of change of kinetic momentum to the force, when the latter can be derived from a scalar potential.

The first-order equation for the azimuthal (longitudinal) motion can now be written directly by substitution of the conjugate variables (Θ, S),

$$\frac{\mathrm{d}S}{\mathrm{d}t} = -e \frac{\partial}{\partial \Theta} (\phi - \boldsymbol{A} \cdot \boldsymbol{v}). \tag{4D}'$$

Certain approximations are now required to develop (4D)':

- It will be assumed that the guide field of the synchrotron is 2-dimensional and can therefore be derived from A_Θ only. This assumption has already been made in Appendix A for the transverse motion. It neglects the longitudinal components in the end fields of the magnets.
- It will be assumed that $\partial A_\Theta / \partial \Theta = 0$. This implies a constant guide field, but since transitions between magnets will be short compared with the synchrotron wavelength, this is still a good approximation for a practical machine.
- Only the para-axial region in the cavity will be considered where the magnetic field is essentially zero. The electric field can then be represented by a time-dependent scalar potential as in Chapter 7.

With these approximations, and since $\boldsymbol{v}$ does not explicitly depend on Θ, (4D)' reduces to,

$$\frac{\mathrm{d}S}{\mathrm{d}t} = -e \frac{\partial \phi}{\partial \Theta},$$

which can then be rewritten with the help of the active component of the electric field given in (7.19) derived from (7.13) to yield,

$$\frac{\mathrm{d}}{\mathrm{d}t} S = \frac{e\hat{u}}{2\pi} \cos\left(h\Theta - \int \omega \, \mathrm{d}t\right). \tag{5D}$$

The potential ϕ could possibly contain electric bending and focusing forces, but for

(5D) it is assumed that such forces would only act radially and therefore do not appear.

Equation (5D) describes the motion of a single particle using the conjugate variables (Θ, S). In Chapter 7, it is the relative particle motion with respect to an equilibrium particle, called the *synchronous particle*, that is of importance. This is achieved by applying (5D) to an arbitrary particle and a synchronous particle and forming the difference to give,

$$\frac{\mathrm{d}}{\mathrm{d}t}\Delta S = \frac{e\hat{u}}{2\pi}(\cos\theta - \cos\theta_0) \qquad (6D)$$

where the argument of the cosine in (5D) has been replaced by θ as used in (7.20), the synchronous particle is denoted by the subscript 0 and

$$\Delta S = S - S_0 = [(p_\Theta + eA_\Theta)C - (p_0 + eA_0)C_0]/(2\pi). \qquad (7D)$$

Further, $A_\Theta C$ can be expanded in a series of ΔC in the median plane. The expression (7D) is then linearised by retaining only the first order terms to give,

$$\Delta S = \left\{1 + \left(1 + \frac{e}{p_0}\left[\frac{\partial(A_\Theta C)}{\partial C}\right]_0\right)\alpha\right\}\frac{C_0}{2\pi}\Delta p \qquad (8D)$$

where α is the momentum compaction defined in (2.41) and Δp is the *longitudinal momentum deviation*. Since it has already been assumed that A_Θ is uniform, or close enough that average values can be used,

$$\left[\frac{\partial(A_\Theta C)}{\partial C}\right]_0 = \frac{C_0 B_0}{2\pi}. \qquad (9D)$$

The substitution of (9D) into (8D) gives,

$$\Delta S = \left\{1 + \left(1 + \frac{eC_0 B_0}{2\pi p_0}\right)\alpha\right\}\frac{C_0}{2\pi}\Delta p. \qquad (10D)$$

The expression linking $\Delta\dot{\theta}$ and Δp, which introduces the optical properties of the lattice, has already been derived in Chapter 7.

$$\Delta p = \frac{m_0\gamma C_0}{2\pi h\eta}\frac{\mathrm{d}}{\mathrm{d}t}\Delta\theta. \qquad (7.31)$$

The substitution of (7.31) into (10D) yields,

$$\frac{\mathrm{d}}{\mathrm{d}t}\Delta\theta = \frac{h\eta}{m_0\gamma}\left[\frac{2\pi}{C_0}\right]^2\left\{1 + \left[1 + \frac{eC_0 B_0}{2\pi p_0}\right]\alpha\right\}^{-1}\Delta S. \qquad (11D)$$

Since $\Delta\theta$ differs from $\Delta\Theta$ by a constant, $\Delta\theta = h\,\Delta\Theta$, the variables $(\Delta\theta, \Delta S)$ will also preserve phase-space area and can be considered as canonically conjugate. Equations (6D) and (11D) are therefore the first-order canonical equations for the relative particle motion. They can be combined to give a single, second-order equation describing the *phase oscillations*,

$$\frac{d}{dt}\left(\frac{m_0\gamma}{h\eta}\left[\frac{C_0}{2\pi}\right]^2\left\{1+\left[1+\frac{eC_0B_0}{2\pi p_0}\right]\alpha\right\}\frac{d}{dt}\Delta\theta\right)=\frac{e\hat{u}}{2\pi}(\cos\theta-\cos\theta_0).\tag{12D}$$

With pure magnetic bending, the cyclotron relation (2.9) is valid, i.e.

$$\frac{eC_0B_0}{2\pi p_0}=-1,\tag{13D}$$

which reduces (12D) to the form (7.32) already found in Chapter 7,

Magnetic guide field, $$\quad\frac{d}{dt}\left(\frac{m_0\gamma C_0}{2\pi h\eta}\frac{d}{dt}\Delta\theta\right)=\frac{e\hat{u}}{C_0}(\cos\theta-\cos\theta_0).\tag{14D}$$

When the general expression for ΔS in (10D) is re-written for a machine with pure magnetic bending by the use of (13D), the result is,

$$\Delta S=\frac{C_0}{2\pi}\Delta p=R_0\,\Delta p.\tag{10D$'$}$$

Thus ΔS and Δp differ by only a constant in this case and $(\Delta\theta,\Delta p)$ like $(\Delta\theta,\Delta S)$ will conserve phase-space area and be conjugate.

The reservation made above, that the guide field should be magnetic, may seem a little unnecessary, but suppose for a moment that the guiding force was provided by a radial electric field. This would not appear in the above azimuthal motion equations, but B_0 would be zero and the phase equation would become,

Electric guide field, $$\quad\frac{d}{dt}\left((1+\alpha)\frac{m_0\gamma C_0}{2\pi h\eta}\frac{d}{dt}\Delta\theta\right)=\frac{e\hat{u}}{C_0}(\cos\theta-\cos\theta_0).\tag{15D}$$

This equation differs from (14D) because the betatron acceleration force has been removed. The factor $(1+\alpha)$, between the two equations, is independent of time, so that the damping law will be the same whichever equation is used. However, the amplitudes and frequencies of the synchrotron oscillations are different in the two cases, although these differences are likely to be small, especially in large strong focusing machines (remember, $\alpha\simeq Q_x^{-2}$).

The exact form of the phase equation is a subject that reappears from time to time in the literature. Those readers interested in further details could try the papers in Refs 2–4. For a combined Hamiltonian treatment of the longitudinal and transverse motions see Refs 5 and 6.

(ii) Choice of alternative variables (τ, E)

In Chapter 7 the canonically conjugate variables $(\Delta\theta,\Delta p)$ were used for the phase equation because the authors feel that the derivation is simplified and the physics is made intuitively easier to understand. In this Appendix, the variables $(\Delta\Theta, \Delta S)$ have been introduced, but other sets of canonically conjugate variables may seem equally natural. In particular, many people prefer to use time and energy and, since this pair of variables $(\Delta\tau, \Delta E)$ appears frequently in the literature, it is worth giving the relations between these choices:

$$\Delta\tau = \frac{C}{2\pi ch\sqrt{(1 - \gamma^{-2})}} \Delta\theta = \frac{C}{2\pi ch\beta} \Delta\theta \tag{16D}$$

$$\Delta E = \sqrt{(1 - \gamma^{-2})}c \, \Delta p \quad = \beta c \, \Delta p. \tag{17D}$$

From the above, it is easy to see that the relationship between the longitudinal acceptances (or emittances) expressed in the two systems is,

$$A_l(\Delta\tau, \Delta E) \, [\text{eV.s}] = \frac{C}{2\pi h} \, [\text{m}] A_l(\Delta\theta, \Delta p) \, [\text{eV}/c]. \tag{18D}$$

The variables $(\Delta\tau, \Delta E)$ are better adapted than $(\Delta\theta, \Delta p)$ to the description of transfers between machines.

(iii) Choice of alternative variables (Θ, W)

In much of the literature, the canonically conjugate variables (Θ, W) are used. The angle variable Θ has already been defined in (1D) and the action variable W is defined as[7],

$$W = \int_{E_0}^{E} \frac{\mathrm{d}E}{\Omega(E)}. \tag{19D}$$

The expression (12A), from Appendix A, for the total energy of the particle, provides the link to the variables used earlier. Equation (12A) is re-expressed in terms of energy to give,

$$E^2 = c^2 p^2 + E_0^2, \tag{20D}$$

which when differentiated becomes,

$$2E \frac{\mathrm{d}E}{\mathrm{d}p} = 2c^2 p \tag{21D}$$

$$\frac{\mathrm{d}E}{\mathrm{d}p} = v.$$

The substitution of (21D) into (19D) gives the relationship between W and p, i.e.

$$W = \int_{p_0}^{p} R(p) \, \mathrm{d}p, \tag{22D}$$

which shows more clearly why W was referred to as an action variable earlier. The equivalent form of (22D) for a small change in W about the central orbit is,

$$\Delta W = \Delta E/\Omega_0 = R_0 \, \Delta p. \tag{23D}$$

If the general expression for ΔS in (10D) is re-written for a machine with purely magnetic bending by the use of (13D), the result is

$$\Delta S = \frac{C_0}{2\pi} \Delta p = R_0 \, \Delta p. \tag{10D$'$}$$

It can now be seen that W is equivalent to S, for the case of pure magnetic bending. In Section 9.4.3, the longitudinal stability of a coasting beam was considered using the variables (Θ, p), but again this could have been equally done with (Θ, S) or (Θ, W).

Vlasov equation

The Vlasov equation* is a powerful tool for finding a self-consistent solution for the behaviour of an assembly of many particles (10^{10}–10^{20}), in which each particle feels the sum of the external forces and the collective force of all particles[1]. The collective forces should not be affected by the interactions between neighbouring particles, which is often expressed as: collisions can be neglected. The equation is an expression of Liouville's Theorem, which states that phase space is conserved along a dynamical trajectory when only Hamiltonian forces are acting, and a simple, but non-rigorous, derivation will be based on this[2]. As with the Hamiltonian formalism described in Appendix A, the Vlasov equation provides a standard approach to problems and ensures the physics is sound leaving only mathematical difficulties.

First, however, it is interesting to make a small digression to discuss *phase-space area* and *phase-space density*. In Appendix A and elsewhere care has been taken to use the former term in order to imply a sort of continuous incompressible fluid flowing through phase space. In contrast to this, a beam is an assembly of discrete particles embedded in phase space. Under the conditions specified for the Vlasov equation the forces cannot manoeuvre single particles, so the particle distribution moves as phase space moves. In a given element of phase space there will be a certain density of particles and since the area is conserved so also is the phase-space density (of the particles). This distinction is raised here, since the Vlasov equation is based on the dynamics of the phase-space density and, under certain conditions, density is not an adequate description of what happens to the particles, e.g. during stochastic cooling.

Let $f(\boldsymbol{q}, \boldsymbol{P})$ be the distribution function describing the phase-space density around the point $(\boldsymbol{q}, \boldsymbol{P})$. The variables $\boldsymbol{q}$ and $\boldsymbol{P}$ are the canonically conjugate position and momentum variables used in Appendix A. Liouville's Theorem says that this density will be constant in the vicinity of a trajectory, which is expressed below as the

---

* This equation, which is well-known to plasma physicists, is also called the *kinetic equation* or the *collisionless Boltzmann equation*.

complete derivative of $f(\boldsymbol{q}, \boldsymbol{P})$ *being zero.*

$$\frac{\mathrm{d}}{\mathrm{d}t} f(\boldsymbol{q}, \boldsymbol{P}) = 0 \tag{1E}$$

$$\frac{\partial}{\partial t} f(\boldsymbol{q}, \boldsymbol{P}) + \dot{\boldsymbol{q}} \frac{\partial}{\partial \boldsymbol{q}} f(\boldsymbol{q}, \boldsymbol{P}) + \dot{\boldsymbol{P}} \frac{\partial}{\partial \boldsymbol{P}} f(\boldsymbol{q}, \boldsymbol{P}) = 0 \tag{2E}$$

Equation (2E) is the Vlasov equation in its simplest form and the one which is employed in Chapter 9 on coherent instabilities. In plasma physics (2E) is usually rearranged by expressing $\dot{\boldsymbol{P}}$ in terms of the forces acting on the particle. This non-relativistic form may be more familiar to some readers,

$$\frac{\partial}{\partial t} f(\boldsymbol{r}, \boldsymbol{v}) + \boldsymbol{v} \cdot \nabla f(\boldsymbol{r}, \boldsymbol{v}) + \frac{e}{m} (\boldsymbol{E} + \boldsymbol{v} \times \boldsymbol{B}) \frac{\partial}{\partial \boldsymbol{v}} f(\boldsymbol{r}, \boldsymbol{v}) = 0.$$

where $\boldsymbol{r}$ is now the position vector and $\boldsymbol{v}$ the velocity.

A far more rigorous derivation can be found in many plasma physics text books, but it is usual in these cases that the derivation invokes a neutral plasma and Debye shielding. The Debye shielding is invoked to smoothe the fields from remote particles. Since an accelerator beam is far from being a neutral plasma, has a very characteristic energy distribution and often has an extreme aspect ratio with transverse dimensions smaller than the Debye shielding distance, the neutral plasma model is not a satisfactory justification for omitting short-range interactions. This problem has been dealt with by Ruggiero[3] and is discussed in terms of the granularity of the beam, the average kinetic (thermal) energy and the timescale of the observation. For instabilities the beam is observed over a finite number of turns and the Vlasov equation is a useful tool, but for intra-beam scattering the beam is observed on the timescale of hours and in this case the short-range scattering forces dominate. A third situation exists in very cold beams, where 'crystallisation'[4] is expected to occur.

REFERENCES

Chapter 1

1. L. C. Teng, *Perspectives of future accelerators and colliders*, Proc. 11th Int. Conf. on High-Energy Accelerators, CERN, Geneva, 1980 (Birkhäuser Verlag, Basel, 1980), 910–21.
2. J. D. Cockcroft and E. T. S. Walton, Experiments with high velocity ions, *Proc. R. Soc. London.*, A **136** (1932), 619–30.
3. R. J. Van de Graaff, A 1,500,000 volt electrostatic generator, *Phys. Rev.*, **38** (Nov. 1931), 1919–20.
4. G. Ising, *Arkiv. för Matematik, Astronomi och Fysik*, **18** (1924), 1–4.
5. R. Wideröe, *Arch. für Elektrotechnik*, **21** (1928), 387–406.
6. D. H. Sloan and E. O. Lawrence, The production of heavy high speed ions without the use of high voltages, *Phys. Rev.*, **38** (Dec. 1931), 2021–32.
7. E. O. Lawrence and N. E. Edlefsen, *Science*, **72** (1930), 376–7.
8. E. O. Lawrence and M. S. Livingston, The production of high speed light ions without the use of high voltages, *Phys. Rev.*, **40** (April 1932), 19–35.
9. H. A. Bethe and M. E. Rose, The maximum energy obtainable from the cyclotron, *Phys. Rev.*, Letter to the editor, **52** (Dec. 1937), 1254.
10. E. M. McMillan, The synchrotron – a proposed high energy particle accelerator, *Phys. Rev.*, Letter to the editor, **68** (Sept. 1945), 1434.
11. V. Veksler, *J. Phys. USSR*, **9** (1945), 153–8.
12. M. E. Rose, Focusing and maximum energy ions in the cyclotron, *Phys. Rev.*, **53** (March 1938), 392–408.
13. M. S. Livingston, *Particle accelerators: a brief history* (Harvard University Press, Cambridge, Massachusetts, 1969), 32.
14. L. H. Thomas, The paths of ions in the cyclotron, *Phys. Rev.*, **54** (Oct. 1938), 580–8.
15. W. Paul, *Early days in the development of accelerators*, Proc. Int. Symposium in Honour of Robert R. Wilson, Fermilab, 1979 (Sleepeck Printing Co. Bellwood, Ill., 1979), 25–68.
16. D. W. Kerst, The acceleration of electrons by magnetic induction, *Phys. Rev.*, **60** (July 1941), 47–53.
17. D. W. Kerst and R. Serber, Electronic orbits in the induction accelerator, *Phys. Rev.*, **60** (July 1941), 53–8.
18. F. K. Goward and D. E. Barnes, *Nature*, **158** (1946), 413.
19. M. L. Oliphant, J. S. Gooden and G. S. Hyde, *Proc. Phys. Soc.*, **59** (1947), 666.
20. N. C. Christofilos, Unpublished report (1950); and U.S. Patent no. 2,736,799, filed March 10, 1950, issued February 28, 1956.

21. E. D. Courant, M. S. Livingston and H. S. Snyder, The strong-focusing synchrotron – a new high energy accelerator, *Phys. Rev.*, **88** (Dec. 1952), 1190–6; and E. D. Courant and H. S. Snyder, Theory of the alternating-gradient synchrotron, *Annals of Physics*, No. 3 (1958), 1–48.

22. L. W. Alvarez, The design of a proton linear accelerator, *Phys. Rev.*, **70** (July 1946), 799–800.

23. J. C. Slater, The design of linear accelerators, *Phys. Rev.*, **70** (July 1946), 799.

24. D. W. Fry, R. B. R.-S.-Harvie, L. B. Mullet and W. Walkinshaw, *Nature*, **160** (1947), 351; and *Nature*, **162** (1948), 859.

25. I. M. Kapchinsky and V. A. Teplyakov, *Prib. Tekh. Eksp.*, No. 2 (1970) (in Russian).

26. K. R. Crandall, R. H. Stokes and T. P. Wangler, *RF quadrupole beam dynamics design studies*, Proc. 1979 Lin. Accel. Conf., Montauk, NY (BNL 51134, Upton, NY, 1979), 205–16.

27. D. W. Kerst, F. T. Cole, H. R. Crane, L. W. Jones, L. J. Laslett, T. Ohkawa, A. M. Sessler, K. R. Symon, K. M. Terwilliger and N. Vogt Nilsen, Attainment of very high energy by means of intersecting beams of particles, *Phys. Rev.*, **102** (April 1956), 590–1.

28. K. R. Symon and A. M. Sessler, *Methods of radio frequency acceleration in fixed field accelerators with applications to high current and intersecting beam accelerators*, Proc. CERN Symposium on High Energy Accelerators and Pion Physics, Geneva, 1956 (CERN, 1956), Vol. 1, 44–58.

29. S. van der Meer, *Stochastic damping of betatron oscillations in the ISR*, CERN/ISR-PO/72-31 (August 1972).

30. C. Rubbia, P. McIntyre and D. Cline, *Producing massive neutral intermediate vector bosons with existing accelerators*, Proc. Int. Neutrino Conf., Aachen, 1976 (Vieweg Verlag, Braunschweig, 1977), 683–7.

31. M. Tigner, A possible apparatus for electron clashing-beam experiments, Letter to the editor, *Nuovo Cim.*, **37** (1965), 1228–31.

32. R. Stiening, *The status of the Stanford Linear Collider*, Proc. 1987 IEEE Part. Accel. Conf., Washington, 1987 (IEEE, 1987), 1–7.

Chapter 2

1. E. D. Courant, M. S. Livingston and H. S. Snyder, The strong-focusing synchrotron – a new high energy accelerator, *Phys. Rev.*, **88** (Dec. 1952), 1190–6; and E. D. Courant and H. S. Snyder, Theory of the alternating-gradient synchrotron, *Annals of Physics*, **3** (1958), 1–48.

2. J. J. Livingood, *Principles of cyclic particle accelerators*, (D. Van Nostrand Co. Ltd., 1961), 25.

3. Editors: J. S. Day, A. D. Krisch, L. G. Ratner, *History of the ZGS*, AIP Conf. Proc. No. 60, AIP New York (1960).

Chapter 3

1. E. D. Courant, M. S. Livingston and H. S. Snyder, The strong-focusing synchrotron – a new high energy accelerator, *Phys. Rev.*, **88** (Dec. 1952), 1190–6; and E. D. Courant and H. S. Snyder, Theory of the alternating-gradient synchrotron, *Annals of Physics*, No. 3 (1958), 1–48.

2. N. C. Christofilos, Unpublished report (1950); and N. C. Christofilos, U.S. Patent No. 2,736,799, filed March 10, 1950, issued February 28, 1956.

3. K. G. Steffen, *High energy optics* (Interscience Publishers, Division of J. Wiley & Sons, 1965), 17–20.

338

4. H. Margenau and G. M. Murphy, *The mathematics of physics and chemistry* (D. Van Nostrand Co., Inc., 1952), 303–5.

5. B. Autin, *Influence des tolérances magnétiques sur la fonction β, application à un schema à haute luminosité dans les ISR*, CERN ISR-MA/73-29 (June 1973).

6. B. Hedin, *Effective length for quadrupole lenses*, CERN/PS/MU/EP/Note 80-11 (Sept. 1980).

7. G. Guignard, *The general theory of all sum and difference resonances in a three-dimensional magnetic field in a sychrotron*, CERN 76-06 (March 1976).

8. R. Brinkmann, *Layout and optics of the proton transfer line for HERA*, Part. Accel. Conf., Vancouver, 1965, *IEEE Trans. on Nucl. Sci.*, Vol. **NS-32**, No. 5, Part II (Oct. 1985), 3069–71.

Chapter 4

1. E. D. Courant and H. S. Snyder, Theory of the alternating-gradient synchrotron, *Annals of Physics*, **3** (1958), 1–48.

2. H. B. Dwight, *Tables of integrals and other mathematical data*, 4th edition (The Macmillan Co., New York, 1961), 85.

3. K. G. Steffen, *High energy optics* (Interscience Publishers, Division of J. Wiley & Sons, 1965), 17–20.

4. B. Autin and A. Verdier, *Focusing perturbations in alternating gradient structures*, CERN ISR-LTD/76-14 (March 1976).

5. A. H. Sørensen, *Liouville's theorem and emittance*, Proc. CERN Accelerator School 3rd General Accelerator Physics Course, Salamanca, 1988, CERN 89-05 (April 1989), 718–36.

6. T. Sigurgeirsson, *Betatron oscillations in the strong focusing synchrotron*, CERN/T/TS-2 (Dec. 1952).

7. E. Keil, *Single-particle dynamics – linear machine lattices*, Proc. Int. School of Part. Accel. Ettore Majorana, Centre for Scientific Culture, Erice, 1976, CERN 77-13 (July 1977), 22–36.

8. K. Steffen, *Periodic dispersion suppressors*, Parts I and II, Int. Rep. DESY-HERA 81/19 (Dec. 1981), and DESY-HERA 83/02 (Jan. 1983).

9. F. Willeke, '*Verbotene' Q – Werte bei PETRA*, Int. Rep. DESY-HERA 81/28 (Nov. 1981).

10. R. Brinkmann, *Insertions*, Proc. CERN Accelerator School 2nd General Accelerator Physics Course, Aarhus, 1986, CERN 87-10 (July 1987), 45–61.

11. H. B. Dwight, *Tables of integrals and other mathematical data*, 4th edition (The Macmillan Co., New York, 1961), 92.

12. E. Regenstreif, *Phase-space transformations by means of quadrupole multiplets*, CERN 67-6 (March 1967).

13. E. Regenstreif, *Possible and impossible phase-space transformations by means of alternating-gradient doublets and triplets*, CERN 67-8 (March 1967).

14. G. Guignard, *Low-β insertion for proton–antiproton collisions in the Fermilab Doubler*, 1977 Summer Study, Vol. 1, 351–65.

15. G. Wüstefeld, R. Maier and B. Simon, *The analytical lattice approach for the Ring Design of BESSY II*, Proc. 1st European Part. Accel. Conf. (EPAC), Rome, 1988 (World Scientific, Singapore, 1989), **2**, 914–16.

16. B. W. Zotter, *Calculation of the parameters of a quadrupole triplet from the transfer matrices*, CERN/ISR-TH/73-43 (Sept. 1973).

17. M. Rogge, *Analytical solutions to the problem of constructing an arbitrary beam transport matrix with a set of quadrupoles*, Verhandlungen der Deutschen Physikalischen Gesellschaft, Berlin (March 1988), 132–3.

18. M. H. Blewett, The Cosmotron-A review, *Rev. Sci. Instr.*, **24**, No. 9 (Sept. 1953), 725–37.

19. K. R. Symon and A. M. Sessler, *Methods of radio frequency acceleration in fixed field accelerators with applications to high current intersecting beam accelerators*, Proc. CERN Symposium on High Energy Accelerators and Pion Physics (CERN 1956), 44–58.

20. The CERN study group on new accelerators, *Report on the design study of intersecting storage rings (ISR) for the CERN proton synchrotron*, AR/Int. SG/64-9 (May 1964), 104–19.

21. G. I. Budker and G. I. Dimov, *On the charge exchange injection of protons into ring accelerators*, Proc. 5th Int. Conf. on High Energy Accelerators, Dubna, 1963 (translation Conf-114, US, AEC, Div. of Tech. Info., 1372–7).

22. C. W. Potts, *Negative hydrogen ion injection into the zero gradient synchrotron*, Proc. 1977 Part. Accel. Conf., Chicago, *IEEE Trans. Nucl. Sci.*, Vol. **NS-24**, No. 3 (June 1977), 1385–9.

23. D. S. Barton, L. A. Ahrens, E. Gill, J. W. Glenn, R. K. Reece and R. L. Witkover, *Charge exchange injection at the AGS*, 1983 Part. Accel. Conf. Santa Fe, *IEEE Trans. Nucl. Sci.*, Vol. **NS-30**, No. 4 (August 1983), 2787–9.

24. C. Hojvat, C. Ankenbrandt, B. Brown, D. Cosgrove, J. Garvey, R. P. Johnson, M. Joy, J. Lackey, K. Meisner, T. Schmitz, L. Teng and R. C. Webber, *The multiturn charge exchange injection system for the Fermilab booster accelerator*, Proc. 1979 Part. Accel. Conf. San Francisco, *IEEE Trans. Nucl. Sci.*, Vol. **NS-24**, No. 3 (June 1979), 3149–51.

25. V. C. Kempson, C. W. Planner and V. T. Pugh, *Injection dynamics and multi-turn charge-exchange injection into the fast cycling synchrotron for the SNS*, Proc. 1981 Part. Accel. Conf., Washington, *IEEE Trans. Nucl. Sci.*, Vol. **NS-28**, No. 3 (June 1981), 3085–7.

26. L. Criegee, *Emittance growth during H^- injection in DESY III*, DESY HERA 88-04 (March 1988).

27. G. K. O'Neill, *Colliding beam techniques*, Princeton–Pennsylvania accelerator report, PPAD 415 D (1961).

28. K. Hübner, G. Schröder, N. Siegel, A. Verdier and E. Weisse, *Beam transfer from SPS to LEP and LEP injection*, Proc. 2nd European Part. Accel. Conf. (EPAC), Nice 1990 (Editions Frontières, 1991), 1297–9.

29. T. K. Khoe and R. J. Lari, *Beam stacking in the radial betatron phase space*, Proc. 8th Int. Conf. on High-energy Accelerators, Geneva, 1971 (CERN, 1971), 98–101.

30. M. M. Gordon and T. A. Welton, The 8/4 resonance and beam extraction from the AVF cyclotron, *Bull. Am. Phys. Soc.*, **3**, 57 (1958).

31. G. H. Rees, 'Injection' & 'Extraction', Proc. CERN Accelerator School General Accelerator Physics, Gif-sur-Yvette, 1984, CERN 85-19 (Nov. 1985), 331–45 and 346–57.

Chapter 5

1. E. D. Courant and H. S. Snyder, Theory of the alternating-gradient synchrotron, *Annals of Physics*, **3** (1958), 1–48.

2. P. L. Morton, *Corrective elements in the proposed SLAC storage ring*, Symposium international sur les anneaux de collisions, Saclay, 1966 (Presses Universitaires de France), VIIb 1-1/5.

3. G. Leleux, *Behaviour of the beam in A.C.O. near the linear coupling resonance $v_x - v_z = k$*, Symposium international sur les anneaux de collisions, Saclay, 1966 (Presses Universitaires de France), VIIb 31/7.

4. G. Ripken, *Untersuchungen zur Strahlführung und Stabilität der Teilchenbewegung in Beschleunigern und Storage-Ringen unter strenger Berücksichtigung einer Kopplung der Betatronschwingungen*, Internal Bericht, DESY R1-70/4 (June 1970).

5. G. Guignard, *The general theory of all sum and difference resonances in a three-dimensional magnetic field in a synchrotron*, CERN 76-06 (March 1976).
6. P. L. Morton, Unpublished addendum to Internal SLAC report SR-11 (July 1966).
7. P. J. Bryant, *A simple theory for betatron coupling*, CERN ISR-MA/75-28 (May 1975).
8. P. J. Bryant, P. Galbraith, J. P. Gourber, G. Guignard and K. Takikawa, *Measurement of the excitation of the coupling resonance $Q_h - Q_v = 0$*, Proc. 1977 Part. Accel. Conf., Chicago, *IEEE Trans. Nucl. Sci.*, Vol. **NS-24**, No. 3 (June 1977), 1440–2.
9. P. J. Bryant and G. Guignard, *Methods for measuring the complex coupling coefficient for the second order difference resonance $Q_h = Q_v$*, CERN ISR-MA/75-42 (Sept. 1975).
10. J.-P. Koutchouk, *Linear betatron coupling measurement and compensation in the ISR*, Proc. 11th Int. Conf. on High-Energy Accelerators, CERN, Geneva, 1980 (Birkhäuser Verlag, Basel, 1980), 491–5.
11. J.-P. Gourber, G. Guignard, A. Hofmann, J.-P. Koutchouk and H. Moshammer, *Compensation of linear betatron coupling in LEP*, Proc. 2nd European Part. Accel. Conf. (EPAC), Nice 1990 (Editions Frontières, 1991), 1429–31; and G. Guignard, L. Hand, J.-P. Koutchouk and H. Moshammer, *Measurement and compensation of the solenoid effects in LEP*, ibid, 1432–4.
12. P. J. Bryant, *The revised skew quadrupole system for coupling compensation in the CERN Intersecting Storage Rings*, Proc. 1979 Part. Accel. Conf., San Francisco, *IEEE Trans. Nucl. Sci.*, Vol. **NS-26**, No. 3 (June 1979), 3499–501.
13. R. Hagedorn, *Stability and amplitude ranges of two dimensional non-linear oscillations with periodical Hamiltonian*, CERN 57-1 (March 1957).
14. A. Schoch, *Theory of linear and non-linear perturbations of betatron oscillations in alternating gradient synchrotrons*, CERN 57-23 (Feb. 1958).
15. G. Guignard, *A general treatment of resonances in accelerators*, CERN 78-11 (Nov. 1978).
16. Nonlinear dynamics aspects of particle accelerators, *Lecture Notes in Physics*, No. 247 (Springer-Verlag, 1986).

Chapter 6

1. J. Jäger and D. Möhl, *Comparison of methods to evaluate the chromaticity in LEAR*, PS/DL/LEAR/Note 81-7 (Sept. 1981).
2. W. Hardt, J. Jäger and D. Möhl, *A general analytical expression for the chromaticity of accelerator rings*, PS/LEA/Note 82-5 (Nov. 1982).
3. M. Bassetti, *A simplified derivation of chromaticity formulae*, Internal Report, LEP Note 504 (June 1984).
4. H. Zyngier, *Stratégie pour la correction de chromaticité*, Report LAL 77-35 (Nov. 1977).
5. B. W. Montague, *Linear optics for improving chromaticity correction*, Internal Report, LEP Note 165 (July 1979); and *Chromatic effects and their first-order correction*, Proc. CERN Accelerator School Advanced Accelerator Physics, Oxford, 1985, CERN 87-03 (April 1987), 75–90.
6. M. H. R. Donald, *User's guide to HARMON*, RAL Report, RL-76-052 (May 1976); and PEP Note 311 (July 1979).
7. B. Autin and A. Verdier, *Focusing perturbations in alternating gradient structures*, CERN ISR-LTD/76-14 (March 1976).
8. G. Guignard, *First order chromatic perturbations and sextupole strength*, CERN ISR-TH/82-14; and LEP Note 391 (July 1982).
9. H. Wiedemann, PEP Note 220 (1976) (program PATRICIA).
10. A. Wrulich, Particle tracking in accelerators with higher order multipole fields, Computing

in Accelerator Design, *Lecture Notes in Physics* 215 (Springer-Verlag, Berlin 1984) (program RACETRACK).

11. L. Schachinger and R. Talman, SSC-52 (1985) (program TEAPOT).
12. A. Dragt *et al.*, *MARYLIE, a program for charged particle beam transport systems based on Lie algebraic methods*, Draft Manual, University of Maryland (1985).
13. G. Guignard and J. Hagel, Sextupole correction and dynamic aperture: numerical and analytical tools, *Particle Accelerators*, 18 (1986), 129–65.

Chapter 7

1. W. Paul, *Early days in the development of accelerators*, Proc. Int. Symposium in Honour of Robert R. Wilson, Fermilab, 1979 (Sleepeck Printing Co. Bellwood, Ill., 1979), 25–68.
2. D. W. Kerst, The acceleration of electrons by magnetic induction, *Phys. Rev.*, **60** (July 1941), 47–53.
3. D. W. Kerst and R. Serber, Electronic orbits in the induction accelerator, *Phys. Rev.*, **60** (July 1941), 53–8.
4. H. G. Hereward, *What are the equations for the phase oscillations in a synchrotron?*, CERN 66-6 (Feb. 1966).
5. K. R. Symon and A. M. Sessler, *Methods of radio frequency acceleration in fixed field accelerators with applications to high current and intersecting beam accelerators*, Proc. CERN Symposium on High Energy Accelerators and Pion Physics (CERN, 1956), 44–58.
6. K. Johnsen, *Phase oscillations and transition-energy problems*, Proc. Conf. on the Alternating-Gradient Proton Synchrotron, Inst. of Physics of the Univ. of Geneva, 1953 (CERN, 1953), 83–93; see also, *Transition*, Proc. CERN Accelerator School General Accelerator Physics, Gif-sur-Yvette, 1984, CERN 85-19 (Nov. 1985), 178–94.
7. K. Johnsen, *Effects of non-linearities on the phase transition*, Proc. CERN Symposium on High Energy Accelerators and Pion Physics, Geneva, 1956 (CERN, 1956), 106–9.
8. A. Sørenssen, CERN/MPS/DL 73-9/Rev (March 1974); or see A. Sørenssen, Crossing the phase transition in strong-focusing proton synchrotrons, *Particle accelerators* (Gordon & Breach 1975), Vol. 6, 141–65.
9. W. Hardt, *Gamma-transition-jump scheme of the CPS*, Proc. 9th Int. Conf. on High Energy Accelerators, Stanford, 1974 (SLAC, 1974), 434–8.
10. S. van der Meer, *Measuring the p–p̄ mass difference by the transition energy method*, CERN/PS/AA 78-17 (Aug. 1978).
11. N. Angert, E. Brouzet, R. Garoby, S. Hancock, H. Haseroth, C. Hill, K. Schindl and P. Têtu, Accelerating and separating mixed beams of ions with similar charge to mass ratio in the CERN PS complex, *Proc. 1st European Part. Accel. Conf.* (EPAC), Rome, 1988 (World Scientific, Singapore, 1989), Vol. 2, 1367–9.
12. S. Hancock, *Transition revisited*, CERN/PS/90-53 (OP) (July 1990).
13. H. G. Hereward and K. Johnsen, *The effect of radio-frequency programme noise on the phase-stable acceleration process*, CERN 60-38 (Oct. 1960).
14. G. Dôme, *Diffusion due to rf noise*, Proc. CERN Accelerator School Advanced Accelerator Physics Course, Oxford, 1985, CERN 87-03 (April 1987), 370–401.

Chapter 8

1. D. W. Kerst, The acceleration of electrons by magnetic induction, *Phys. Rev.*, **60** (July 1941), 47–53.
2. J. van Bladel, *Image forces in the third MURA model*, Midwestern Universities Research Association Report, MURA-466 (Madison, Wisconsin, June 1959).

3. L. J. Laslett, *On intensity limitations imposed by transverse space-charge effects in circular particle accelerators*, Proc. 1963 Summer Study on Storage Rings, Accelerators and Experimentation at Super-high Energies, BNL-Report 7534, 324–67.

4. L. J. Laslett and L. Resegotti, *The space-charge intensity limit imposed by coherent oscillation of a bunched synchrotron beam*, Proc. 6th Int. Conf. on High Energy Accelerators, Cambridge, Mass., 1967 (Cambridge Electron Accelerator, 1967), 150–2.

5. B. W. Zotter, *Coherent Q-shift of a relativistic particle beam in a metallic vacuum chamber*, CERN ISR-TH/72-8 (March 1972).

6. B. W. Zotter, *Image fields of an off-centre particle beam in an elliptic vacuum chamber*, CERN ISR-TH/74-11 (February 1974).

7. B. W. Zotter, *Incoherent Q-shift of a flat, off-centred particle beam in an elliptical vacuum chamber*, CERN ISR-TH/74-38 (July 1974).

8. B. W. Zotter, *Tune shifts of excentric beams in elliptic vacuum chambers*, Proc. 1975 Part. Accel. Conf., Washington, *IEEE Trans. Nucl. Sci.*, Vol. **NS-22,** No. 3 (June 1975), 1451–5.

9. P. J. Bryant, *Dynamic compensation during stacking of the detuning caused by space charge effects*, Proc. 9th Int. Conf. on High Energy Accelerators, Stanford, 1974 (SLAC, 1974), 80–2.

10. P. J. Bryant, D. M. Lewis, B. Nielsen and B. W. Zotter, *On-line correction of the incoherent tune shifts due to space charge*, CERN ISR-MA/75-54 (Dec. 1975).

11. L. C. Teng, ANL Report, ANLAD-59 (Feb. 1961).

12. B. W. Montague, *Fourth-order coupling resonance excited by space-charge forces in a synchrotron*, CERN 68-38 (Oct. 1968), Appendix II.

13. B. W. Zotter, *Q-shift due to the direct space-charge field of stacked particle beams*, CERN/ISR-TH75-5 (Feb. 1975).

14. P. Bryant and J.-P. Gourber, *Experimental investigation of single-beam and beam–beam space charge effects*, Proc. 9th Int. Conf. on High Energy Accelerators, Stanford (SLAC, 1974), 87–91.

15. K. Johnsen, *Some intensity considerations for high energy machines*, CERN AR/Int. SR/61-25 (Nov. 1961).

16. G. Guignard, *Selection of formulae concerning proton storage rings*, CERN 77-10 (June 1977).

17. S. van der Meer, *Calibration of the effective beam height in the ISR*, CERN/ISR-PO/68-31 (June 1968).

18. S. Myers, Private communication of ISR Performance Report Run 904, *A new working line above Q = 9 (9125)* (January 11, 1978).

19. J. P. Gourber, H. G. Hereward and S. Myers, *Overlap knock-out effects in the CERN Intersecting Storage Rings (ISR)*, Proc. 1977 Part. Accel. Conf., Chicago, Ill., *IEEE Trans. Nucl. Sci.*, Vol. **NS-24**, No. 3 (June 1977), 1405–7.

Chapter 9

1. A. Hofmann, *Single-beam collective phenomena – longitudinal*, Proc. Int. School of Part. Accel. Ettore Majorana, Centre for Scientific Culture, Erice, 1976, CERN 77-13 (July 1977), 139–74.

2. B. W. Zotter and F. Sacherer, *Transverse instabilities of relativistic particle beams in accelerators and storage rings*, Proc. Int. School of Part. Accel. Ettore Majorana, Centre for Scientific Culture, Erice, 1976, CERN 77-13 (July 1977), 175–218.

3. J.-L. Laclare, *Introduction to coherent instabilities – coasting beam case*, Proc. CERN Accelerator School, Gif-sur-Yvette, 1984, CERN 85-19 (Nov. 1985), 377–414; and *Bunched*

beam coherent instabilities, Proc. CERN Accelerator School, Oxford 1985, CERN 87-03 (April 1987), 264–326.

4. A. Chao, *Coherent instabilities of a relativistic bunched beam*, Proc. US Summer School on High Energy Particle Accelerators, SLAC, 1982, AIP Conf. Proc. No. 105 (1983), 353–523.

5. A. H. W. Beck, *Space charge waves* (Pergamon Press, 1959).

6. J. C. Maxwell, Essay winning Adams Prize 1856, included in W. D. Niven, *The Scientific Papers of James Clerk Maxwell* (Dover Publications Inc., New York), 288–378.

7. F. Pedersen, unpublished John Adams Memorial Lecture given in CERN, Nov. 1988.

8. C. E. Nielsen and A. M. Sessler, Longitudinal space charge effects in particle accelerators, *Rev. Sci. Instr.*, **30**, No. 2 (Feb. 1959), 80–9.

9. A. A. Kolomensky and A. N. Lebedev, *Certain beam-stacking effects in fixed-field magnetic systems*, Proc. Int. Conf. on High Energy Accelerators and Instrumentation, Geneva, 1959 (CERN, 1959), 115–24.

10. K. Robinson, CEA-11 (1956); and CEAL-1000 (1964).

11. C. Pellegrini, On a new instability in electron–positron storage rings [the head-tail effect], *Il Nuovo Cimento*, **LXIV**, A, N.2 (Nov. 1969), 447–73.

12. J. R. Pierce, *Almost all about waves* (The MIT Press, 1974).

13. R. Littauer, *Beam instrumentation*, Proc. US Summer School on High Energy Particle Accelerators, SLAC, 1982, AIP Conf. Proc. No. 105 (1983), 869–953.

14. J. Gareyte, *Beam observation and the nature of instabilities*, Proc. US Summer Schools on High Energy Particle Accelerators, Fermilab 1987 and Cornell 1988, A.I.P. Conf. Proc. 184 (1989), 343–429.

15. F. Sacherer, *Methods for computing bunched-beam instabilities*, CERN/SI-BR/72-5 (Sept. 1972).

16. F. Sacherer, *A longitudinal stability criterion for bunched beams*, Proc. 1973 Part. Accel. Conf., San Francisco, 1973, *IEEE Trans. Nucl. Sci.*, Vol. **NS-20**, No. 3 (June 1973), 825–8.

17. A. Hofmann, *Physics of beam instabilities*, Frontiers of Particle Beams, Proc. Joint US-CERN School on Part. Accel., South Padre Island, 1986 (Springer-Verlag, Lecture Notes in Physics, No. 296, 1987), 99–119.

18. L. Palumbo and V. G. Vaccaro, *Wake fields, impedance and Green's function*, Proc. CERN Accelerator School, Oxford, 1985, CERN 87-03 (April 1987), 341–69.

19. C. E. Nielsen, A. M. Sessler and K. R. Symon, *Longitudinal instabilities in intense relativistic beams*, Proc. Int. Conf. on High Energy Accelerators and Instrumentation, Geneva, 1959 (CERN, 1959), 239–52.

20. J. D. Lawson, *The physics of charged-particle beams* (Clarendon Press, Oxford, 1977), 322–40.

21. V. K. Neil and R. J. Briggs, Stabilization of non-relativistic beams by means of inductive walls, *Plasma Physics*, **9** (1967), 631–9.

22. V. K. Neil and A. M. Sessler, Longitudinal resistive instabilities of intense coasting beams in particle accelerators, *Rev. Sci. Instru.*, **36**, No. 4 (April 1965), 429–36.

23. A. Sessler and V. G. Vaccaro, *Longitudinal instabilities of azimuthally uniform beams in circular vacuum chambers with walls of arbitrary electrical properties*, CERN 67-2 (Feb. 1967).

24. L. D. Landau, On the vibrations of the electronic plasma, *J. Phys. U.S.S.R.*, **10** (1946), 25; alternatively, see D. Terr Haar, *Collected papers of L. D. Landau* (Pergamon Press).

25. H. G. Hereward, *The elementary theory of Landau damping*, CERN 65-20 (May 1965).

26. R. D. Kohaupt, *What is Landau-damping? Plausibilities, fundamental thoughts, theory*, DESY M-86-02 (March 1986).

27. A. G. Ruggiero and V. G. Vaccaro, *Solution of the dispersion relation for longitudinal*

stability of an intense coasting beam in a circular accelerator, CERN-ISR-TH/68-33 (July 1968).

28. L. J. Laslett, V. K. Neil and A. M. Sessler, Coherent effects in high current particle accelerators: III. Electromagnetic instabilities in a coasting beam, *Rev. Sci. Instr.*, **32**, No. 3 (March 1961), 276–9.

29. E. Keil and W. Schnell, *Concerning longitudinal stability in the ISR*, CERN-ISR-TH-RF/69-48 (July 1969).

30. B. W. Zotter, *Longitudinal stability diagrams for some particular distribution functions*, CERN/ISR-GS/76-11 (Feb. 1976).

31. K. Hübner and V. G. Vaccaro, *Dispersion relations and stability of coasting particle beams*, CERN-ISR-TH/70-44 (Aug. 1970).

32. L. J. Laslett, V. K. Neil and A. M. Sessler, Transverse resistive instabilities of intense coasting beams in particle accelerators, *Rev. Sci. Instr.*, **36**, No. 4 (April 1965), 436–48.

33. W. Schnell and B. W. Zotter, *A simplified criterion for transverse stability of a coasting beam, and its application to the ISR*, CERN-ISR-GS-RF/76-26 (July 1976).

34. K. Hübner, P. Strolin, V. G. Vaccaro and B. W. Zotter, *Concerning the stability of the ISR beam against coherent dipole oscillations*, CERN/ISR-RF-TH/70-2 (Jan. 1970).

35. B. W. Zotter, *The influence of various beam profiles on the transverse resistive wall instability*, CERN-ISR-TH/68-51 (Nov. 1968).

Chapter 10

1. J. P. Blewett, Synchrotron radiation – 1873 to 1947, *Nucl. Instr. Methods in Phys. Research*, A**266** (1988), 1–9.

2. J. D. Jackson, *Classical electrodynamics*, 2nd edition (John Wiley & Sons, New York, 1975).

3. W. K. H. Panofsky and M. Phillips, *Classical electricity and magnetism* (Addison-Wesley Pub. Co. Inc., Massachusetts, 1956).

4. H. Bruck, *Accélérateurs circulaires de particles* (Presses Universitaires de France, Paris, 1966).

5. K. Hübner, *Synchrotron radiation* and *Radiation damping*, Proc. CERN Accelerator School General Accelerator Physics, Gif-sur-Yvette, 1984, CERN 85-19 (Nov. 1985), 226–38 and 239–52.

6. M. Sands, *The physics of electron storage rings. An introduction*, Proc. Int. School of Physics 'Enrico Fermi' (Academic Press, New York, 1971), 257–411; or, SLAC-121 UC-28 (ACC) (Nov. 1970).

7. A. Hofmann, *Characteristics of synchrotron radiation*, Proc. CERN Accelerator School Synchrotron Radiation and Free Electron Lasers, Chester, 1989, CERN 90-03 (April 1990), 115–41.

8. A. Piwinski, *Synchro-betatron resonances*, Proc. CERN Accelerator School Advanced Accelerator Physics, Oxford, 1985, CERN 87-03 (April 1987), 187–202.

9. K. W. Robinson, Radiation effects in circular electron accelerators, *Phys. Rev.*, **111**, No. 2 (July 1958), 373–80.

10. H. G. Hereward, *Means of damping radial betatron oscillations in strong focusing electron rings*, Proc. Int. Conf. on High-Energy Accelerators, Brookhaven National Laboratory (Sept. 1961), 222–6.

11. A. Hofmann, R. Little, J. M. Paterson, K. W. Robinson, G. A. Voss and H. Winick, *Design and performance of the damping system for beam storage in the Cambridge electron accelerator*, Proc. 6th Int. Conf. on High Energy Accelerators, Cambridge, Mass. (CEAL-2000, Dec. 1967), 123–5.

12. Y. Baconnier, R. Cappi, M.-P. Level, J.-P. Riunaud, M. Sommer, H. H. Umstätter and H.

Zyngier, Emittance control of the PS $e^{\pm}$ beams using a Robinson wiggler, *Nucl. Instr. Methods in Phys. Research*, *A***234** (1985), 244–52.

13. R. H. Helm, M. J. Lee, P. L. Morton and M. Sands, *Evaluation of synchrotron radiation integrals*, Proc. 1973 Part. Accel. Conf., San Francisco, *IEEE Trans. Nucl. Sci.*, Vol. **NS-20**, No. 3 (June 1973), 900–1.

14. R. Chasman, G. K. Green and E. M. Rowe, Proc. 1975 Part. Accel. Conf., Washington, 1975, *IEEE Trans. Nucl. Sci.*, Vol. **NS-22**, No. 3 (June 1975), 1765–7.

Chapter 11

1. *Frontiers of particle beams; observation, diagnosis and correction*, Lecture Notes in Physics No. 343 (Springer-Verlag, Heidelberg, 1989).
2. P. Strehl, *Beam diagnostics*, Proc. CERN Accelerator School 2nd General Accelerator Physics Course, Aarhus, 1986, CERN 87-10 (July 1987), 99–134.
3. J. Borer and R. Jung, *Diagnostics*, Proc. CERN Accelerator School Antiprotons for Colliding Beam Facilities, Geneva, 1983, CERN 84-15 (Dec. 1984), 385–467.
4. H. Koziol, *Beam diagnostics*, Proc. CERN Accelerator School 3rd General Accelerator Physics Course, Salamanca, 1988, CERN 89-05 (April 1989), 63–101.
5. J. Billan *et al.*, *Operational experience with the superconducting high-luminosity insertion in the CERN Intersecting Storage Rings (ISR)*, Proc. 1983 Part. Accel. Conf., Santa Fe, *IEEE Trans. Nucl. Sci.*, Vol. **NS-30**, No. 4 (Aug. 1983), 2036–8.
6. L. Vos, *Automated injection procedure for the CERN Intersecting Storage Rings (ISR)*, Proc. 1981 Part. Accel. Conf., Washington, *IEEE Trans. Nucl. Sci.*, Vol. **NS-28**, No. 3 (June 1981), 3031–3.
7. J. R. Maidment and C. W. Planner, An analytic method for closed orbit correction in high energy synchrotrons, *Nucl. Instr. Methods*, **98** (Jan/Feb. 1972), 279–84.
8. J.-P. Koutchouk, *Trajectory and closed orbit correction*, in *Frontiers of particle beams; observation, diagnosis and correction*, Proc. Joint US–CERN School on Part. Accel., Capri, 1988 (Lecture Notes in Physics, No. 343, Springer-Verlag, 1989), 46–64.
9. B. Autin and P. J. Bryant, *Closed orbit manipulation and correction in the CERN ISR*, Proc. 8th Int. Conf. on High-Energy Accelerators, Geneva, 1971 (CERN, 1971), 515–17.
10. B. Autin and Y. Marti, *Closed orbit correction of A.G. machines using a small number of magnets*, CERN ISR-MA/73-17 (March 1973).
11. J. Miles, *Preliminary note on closed orbit correction in the l_p, $1 \leq p \leq \infty$ norms including a generalised best kick algorithm*, CERN SPS/AOP/Note-83-3 (Feb. 1983).
12. G. Guignard, *Effets des champs magnétiques perturbateurs d'un synchrotron sur l'orbite fermée et les oscillations bétatroniques ainsi que leur compensation*, CERN 70-24 (1970).
13. G. Guignard, *The closed orbit measurements as a diagnostic tool for localisation and correction of misalignments in the ISR*, CERN ISR-BOM/80-21 (July 1980).
14. C. L. Hammer, R. W. Pidd and K. M. Terwilliger, Betatron oscillators in the synchrotron, *Rev. Sci. Instr.*, **26**, No. 4 (June 1955), 555–6.
15. D. Kemp, E. Peschardt and A. Vaughan, *On-line Q-measurement during phase displacement acceleration in the CERN ISR*, Proc. 1979 Part. Accel. Conf., San Francisco, *IEEE Trans. Nucl. Sci.*, Vol. **NS-26**, No. 3 (June 1979), 3352–4.
16. K. D. Lohmann, M. Placidi and H. Schmickler, *Q-monitoring in LEP*, CERN/LEP-BI/88-45 (Aug. 1988).
17. W. Schottky, *Ann. Physik* **57** (1918), 541.
18. J. Borer, P. Bramham, H. G. Hereward, K. Hübner, W. Schnell and L. Thorndahl, *Non-destructive diagnostics of coasting beams with Schottky noise*, Proc. 9th Int. Conf. on High Energy Accelerators, Stanford, 1974 (SLAC, 1974), 53–6.

19. H. G. Hereward, *Statistical phenomena – theory*, Proc. Int. School of Part. Accel. Ettore Majorana Centre for Scientific Culture, Erice, 1976, CERN 77-13 (July 1977), 281–9.

20. D. Boussard, *Schottky noise and beam transfer function diagnostics*, Proc. CERN Accelerator School Advanced Accelerator Physics, Oxford, 1985, CERN 87-03 (April 1987), 416–52.

21. S. van der Meer, *Diagnostics with Schottky noise*, in *Frontiers of particle beams; observation, diagnosis and correction*, Proc. Joint US–CERN School on Part. Accel., Capri, 1988 (Lecture Notes in Physics, No. 343, Springer-Verlag, 1989), 423–33.

22. C. D. Curtis, A. Galonsky, R. H. Hilden, F. E. Mills, R. A. Otte, G. Parzen, C. H. Pruett, E. M. Rowe, M. F. Shea, D. A. Swenson, W. A. Wallenmeyer and D. E. Young, *Beam experiments with the MURA 50 MeV FFAG accelerator*, Proc. 5th Int. Conf. on High Energy Accelerators, Dubna, 1963 (translation Conf-114, US, AEC, Div. of Tech Info., 815–47).

23. D. Möhl and A. M. Sessler, *The use of rf-knockout for determination of the characteristics of the transverse coherent instability of an intense beam*, Proc. 8th Int. Conf. on High-Energy Accelerators, Geneva, 1971 (CERN, 1971), 334–7.

24. H. A. Grunder and G. R. Lambertson, *Transverse beam instabilities at the Bevatron*, Proc. 8th Int. Conf. on High-Energy Accelerators, Geneva, 1971 (CERN, 1971), 308–10.

25. A. Hofmann and B. W. Zotter, Measurement of beam stability and coupling impedance by rf excitation, *IEEE Trans. Nucl. Sci.*, Vol. **NS-24**, No. 3 (June 1977), 1487–9.

26. J. Borer, G. Guignard, A. Hofmann, E. Peschardt, F. Sacherer and B. W. Zotter, *Information from beam response to longitudinal and transverse excitation*, Proc. 1979 Part. Accel. Conf., San Francisco, *IEEE Trans. Nucl. Sci.*, Vol. **NS-26**, No. 3 (June 1979), 3405–8.

27. J. Borer, J.-Y. Hemery, J.-P. Koutchouk, E. Peschardt and L. Vos, *ISR beam monitoring system using 'Schottky noise' and transfer function*, CERN ISR-RF/80-30 (July 1980).

28. J.-P. Koutchouk, *Transverse beam stability optimization based on beam transfer function measurement*, Proc. 11th Int. Conf. on High-Energy Accelerators, CERN, Geneva, 1980 (Birkhäuser Verlag, Basel 1980), 496–500.

29. J.-Y. Hemery and L. Vos, *A procedure for obtaining transverse wall impedance and working line from transfer function measurements*, Proc. 11th Int. Conf. on High-Energy Accelerators, CERN, Geneva, 1980 (Birkhäuser Verlag, Basel, 1980), 501–5.

30. G. Guignard, *Détermination par rabotage des faisceaux du profil vertical et de la luminosité avec application aux anneaux de stockage à intersections*, CERN-ISR-OP/73-28 (May 1973).

31. K. Potter, *Beam profiles*, Proc. CERN Accelerator School General Accelerator Physics, Gif-sur-Yvette, 1984, CERN 85-19 (Nov. 1985), 301–17.

32. G. Guignard, *Beam blow-up and luminosity reduction due to linear coupling*, CERN ISR-BOM/77-43 (June 1977).

33. E. M. Rowe, R. H. Hilden, R. F. Stump and D. A. Swenson, Radiofrequency acceleration system for the MURA 50-MeV electron accelerator: design and construction, *Rev. Sci. Instr.*, **35**, No. 11 (Nov. 1964), 1470–7.

34. F. A. Ferger, E. Fischer, E. Jones, P. T. Kirstein and M. J. Pentz, *The CERN electron storage ring model*, Proc. 5th Int. Conf. on High Energy Accelerators, Dubna, 1963 (translation Conf-114, US, AEC, Div. of Tech Info.), 417–25.

35. Private Communication, M. Bäckström, Internal Note, ISR/BOM/MB/ml, *A numerical model for the hysteresis effect in the ISR main magnets* (June 1979).

36. Private communication, T. Taylor, Internal Note, ISR-MA/TMT/rh, *More on the radial field magnet system (hysteresis model)* (Nov. 1972).

37. K. N. Henrichsen and J.-P. Gourber, *The computer controlled system for the forecast of beam parameters in the CERN Intersecting Storage Rings*, Proc. 8th Int. Conf. on High-Energy Accelerators, Geneva, 1971 (CERN, 1971), 521–2.

38. K. Johnsen and C. Schmelzer, *Beam controlled acceleration in synchrotons*, Proc. CERN

Symposium on High Energy Accelerators and Pion Physics, Geneva, 1956 (CERN, 1956), 395–403.

39. W. Schnell, *Remarks on the phase lock system of the CERN proton synchrotron*, Proc. Int. Conf. on High-Energy Accelerators and Instrumentation, Geneva, 1959 (CERN, 1959), 485–90.

40. W. Schnell, *Equivalent circuit analysis of phase lock beam control systems*, CERN 68-27 (July 1968).

41. E. Ciapala, *Stacking and phase displacement acceleration*, Proc. CERN Accelerator School General Accelerator Physics, Gif-sur-Yvette, 1984, CERN 85-19 (Nov. 1985), 195–225.

42. E. Peschardt, *Switch-tuned active filters and their application in the closed loop matching system of the ISR*, CERN ISR-RF/75-48 (Sept. 1975).

43. K. Johnsen, *The 'debuncher': a device for reducing the energy spread of a linac*, CERN-PS/KJ-29 (July 1955).

44. Private Communication, L. K. de Jonge and J. P. H. Sladen, *LEP mountain range display*, LEP Performance Note 36, 14 Sept. 1990.

Chapter 12

1. D. W. Kerst, F. T. Cole, H. R. Crane, L. W. Jones, L. J. Laslett, T. Ohkawa, A. M. Sessler, K. R. Symon, K. M. Terwilliger and N. Vogt Nilsen, Attainment of very high energy by means of intersecting beams of particles, *Phys. Rev.*, **102** (April 1956), 590–1.

2. K. R. Symon and A. M. Sessler, *Methods of radio frequency acceleration in fixed field accelerators with applications to high current and intersecting beams of particles*, Proc. CERN Symposium on High Energy Accelerators and Pion Physics, Geneva, 1956 (CERN, 1956), Vol. 1, 44–58.

3. W. C. Middelkoop and A. Schoch, *Interaction rate in colliding beam systems*, CERN/AR/Int. SG/63-40 (Nov. 1963).

4. Private Communication, P. Darriulat and C. Rubbia, Working Note, *On beam monitoring for ISR experiments*, 68/340/5 SIS/si (28 Feb. 1968).

5. H. G. Hereward, *How good is the r.m.s. as a measure of beam size?*, CERN/MPS/DL 69-15 (Nov. 1969).

6. S. van der Meer, *Calibration of the effective beam height in the ISR*, CERN/ISR-PO/68-31 (June 1968).

7. G. Guignard, *Selection of formulae concerning proton storage rings*, CERN 77-10 (June 1977), 25–30.

8. K. Robinson and G. A. Voss, *Operation of a synchrotron as a storage ring*, Symposium international sur les anneaux de collisions, Saclay, 1966 (Presses Universitaires de France), III-4-1.

9. M. Jacob and K. Johnsen, *A review of accelerator and particle physics at the CERN Intersecting Storage Rings*, CERN 84-13 (Nov. 1984).

10. W. Schnell, *Stacking in proton storage rings with missing buckets*, Proc. 6th Int. Conf. on High Energy Accel., Cambridge, Mass., 1967, CEAL-2000, 84–5.

11. K. Johnsen, *The possibility of p-p̄ colliding beam experiments in a set of storage rings*, CERN Study Group on High Energy Projects, AR/Int. SG/62-11 (Sept. 1962) 15.1–15.2.

12. K. Hübner, K. Johnsen and G. Kantardjian, *The feasibility of antiprotons in the ISR*, CERN/ISR-LTD/75-45 (Sept. 1975).

13. S. van der Meer, *Filling the ISR with antiprotons from antihyperon decay*, CERN-ISR-PO/70-5 (Feb. 1970).

14. G. I. Budker, *Status report of works on storage rings at Novosibirsk*, Symposium

international sur les anneaux de collisions, Saclay, 1966 (Presses Universitaires de France), I-1-1.

15. S. van der Meer, *Stochastic damping of betatron oscillations in the ISR*, CERN/ISR-PO/72-31 (August 1972).

16. C. Rubbia, P. McIntyre and D. Cline, *Producing massive neutral intermediate vector bosons with existing accelerators*, Proc. Int. Neutrino Conf., Aachen, 1976 (Vieweg Verlag, Braunschweig, 1977), 683–7.

17. G. Brianti, Experience with the CERN p̄p complex, *IEEE Trans. Nucl. Sci.*, Vol. **NS-30**, No. 4 (Aug. 1983), 1950–6.

18. H. Edwards, *The Fermilab Tevatron and pbar source status report* Proc. 13th Int. Conf. on High Energy Accel., Novosibirsk, 1986 (Novosibirsk, Publishing House Nauka, Siberian Division, 1987), 20–7.

19. P. J. Bryant, Antiprotons in the ISR, *IEEE Trans. Nucl. Sci.*, Vol. **NS-30**, No. 4 (Aug. 1983), 2047–9.

20. A. P. Vorobiev and K. P. Myznikov, *The status of the UNK project*, ECFA Study Week on Instr. Tech. for High Luminosity Hadron Colliders, Barcelona, 1989, CERN 89-10, ECFA 89-124, Vol. 1 (Nov. 1989), 4–11.

21. The LHC Study Group, *Design study of the large hadron collider (LHC)*, CERN 91-03 (May 1991).

22. J. Peoples, *The SSC Project*, Proc. 1st European Part. Accel. Conf. (EPAC), Rome 1988 (World Scientific, Singapore 1989), 237–41; for design details, see *Conceptual design of the SSC*, SSC-SR-2020 (Univ. Research Assoc., DOE contract) (March, 1986).

23. K. Johnsen, *Long term future*, ECFA Study Week on Instr. Tech. for High Luminosity Hadron Colliders, Barcelona, 1989, CERN 89-10, ECFA 89-124, Vol. 1 (Nov. 1989), 25–34.

24. R. D. Kohaupt, *Cures for instabilities*, in *Frontiers of particle beams: observation, diagnosis and correction*, Proc. Joint US-CERN School on Part. Accel., Capri, 1988 (Lecture Notes in Physics, 343, Springer-Verlag, 1989), 355–66.

25. E. Picasso, *The LEP project*, Proc. 1st European Part. Accel. Conf. (EPAC), Rome 1988 (World Scientific, Singapore 1989), 3–6; for design details, see *LEP Design Report* Vol. 1, collected papers CERN-LEP/TH/83-29, CERN/PS/DL/83-31, CERN/SPS/83-26, LAL/RT/83-09, Vol. 2, CERN-LEP/84-01 (June 1984).

26. B. H. Wiik, *The status of HERA*, 1991 Part. Accel. Conf., San Francisco, May 1991, (IEEE 91 CH 3038-7, 1991), 2905–9.

27. E. Fischer, *Residual gas scattering, beam intensity and interaction rate in proton storage rings*, CERN ISR-VA/67-16.

28. O. Gröbner and R. S. Calder, *Beam induced gas desorption in the CERN Intersecting Storage Rings*, Proc. 1973 Part. Accel. Conf., San Francisco, *IEEE Trans. Nucl. Sci.*, Vol. **NS-20**, No. 3 (June 1973), 760–4.

29. G. E. Fischer and R. A. Mack, *Vacuum design problems of high current electron storage rings*, 11th Annual Symposium of the American Vacuum Society (30 Sept.–2 Oct. 1964), Chicago, Illinois.

30. M. Bernardini and L. Malter, *Vacuum problems of electron and positron storage rings*, 11th Annual Symposium of the American Vacuum Society (30 Sept.–2 Oct. 1964), Chicago, Illinois.

31. O. Gröbner, J. M. Laurent and A. G. Mathewson, *The behaviour of the synchrotron radiation induced gas desorption during the first running period of LEP*, Proc. 2nd European Part. Accel. Conf. (EPAC), Nice 1990 (Editions Frontières 1991), 1341–3.

32. Y. Baconnier, *Neutralisation of accelerator beams by ionisation of the residual gas*, Proc. CERN Accelerator School General Accelerator Physics, Gif-sur-Yvette, 1984, CERN 85-19 (Nov. 1985), 267–300.

33. G. Brianti, *The stability of ions in bunched-beam machines*, Proc. CERN Accelerator School Antiprotons for Colliding Beam Facilities, Geneva, 1983, CERN 84-15 (Dec. 1984), 369–83.

34. A. Poncet, *Ion trapping and clearing*, Proc. CERN Accelerator School Third Advanced Accelerator Physics Course, Uppsala, 1989, CERN 90-04 (April 1990), 74–98.

35. C. Bernardini, G. F. Corazza, G. Di Giugno, G. Ghigo, J. Haissinski, P. Marin, R. Querzoli and B. Touschek, Lifetime and beam size in a storage ring, *Phys. Rev. Let.*, **10** (Jan./June 1963), 407–9.

36. H. Bruck and J. LeDuff, *Beam enlargement in storage rings by multiple Coulomb scattering*, Proc. 5th Int. Conf. on High Energy Accel., Frascati, 1966 (CNEN, Rome, 1966), 284–6.

37. A. Piwinski, *Intra-beam-scattering*, Proc. 9th Int. Conf. on High Energy Accel., Stanford, 1974 (SLAC, 1974), 405–9; and *Intra-beam scattering*, Proc. CERN Accelerator School Advanced Accelerator Physics, Oxford, 1985, CERN 87-03 (April 1987), 402–15.

38. D. Habs, *Crystallization of particle beams*, in *Frontiers of particle beams*, Proc. Joint US–CERN School on Part. Accel., South Padre Island (Lecture Notes in Physics No. 296, Springer-Verlag), 310–37.

39. M. Martini, *Intrabeam scattering in the ACOL-AA machines*, CERN PS/84-7(AA), CERN PS/84-9(AA) (May 1984).

40. A. H. Sørensen, *Introduction to intrabeam scattering*, Proc. CERN Accelerator School 2nd General Accelerator Physics Course, Aarhus, 1986, CERN 87-10 (July 1987), 135–51.

41. H. Hahn, *The relativistic heavy ion collider at Brookhaven*, Proc. 1st European Part. Accel. Conf. (EPAC), Rome 1988 (World Scientific, Singapore 1989), 109–11; for design details, see *RHIC Proposals*, BNL 51801 (Aug. 1984).

42. *Nonlinear dynamics and the beam–beam interaction*, AIP Conf. Proc. No. 57 (1979).

43. *Nonlinear dynamics aspects of particle accelerators*, Proc. US–CERN School on Part. Accel., Sardinia, 1985 (Lecture Notes in Physics 247, Springer-Verlag, 1986).

44. K. Cornelis, M. Meddahi and R. Schmidt, *Experiments on the beam–beam effect in the CERN-SPS in the 1989 collider run*, CERN SPS/AMS/Note 89-13 (Dec. 1989).

45. G. Guignard, *Review of the investigation of the beam–beam interactions at the ISR*, paper contained in Ref. 40, pp. 69–93.

46. L. Z. Rivkin and S. V. Milton, *Disruption limitation to the beam–beam tune shift in circular colliders*, Proc. 2nd European Part. Accel. Conf. (EPAC), Nice 1990 (Editions Frontières, 1991), 255–7.

Appendix A

1. H. Goldstein, *Classical mechanics* (Addison-Wesley, 1980).
2. J. D. Jackson, *Classical electrodynamics* (John Wiley & Sons, 2nd edition, 1975).
3. J. S. Bell, *Hamiltonian mechanics*, Proc. CERN Accelerator School Advanced Accelerator Physics, Oxford, 1985, CERN 87-03 (April 1987), 5–40.
4. B. W. Montague, *Single particle dynamics – Hamiltonian formulation*, Proc. Int. School of Part. Accel. Ettore Majorana, Centre for Scientific Culture, Erice, 1976, CERN 77-13 (July 1977), 37–51.
5. H. S. W. Massey and H. Kestelmann, *Ancillary mathematics* (Pitman, 1959), 865.
6. H. G. Hereward and K. Johnsen, *On the phase equation for synchrotrons*, CERN-PS/HGH-KJ-1 (March 1957).

Appendix B

1. W. H. Lamb and R. J. Lari, *Proc. of the int. symp. on magnet technology*, ed. H. Brechna and H. S. Gordon, SLAC, Sept. 1965 (US Atomic Energy Commission), 487–96.

350

2. W. C. Elmore and M. W. Garrett, Measurement of two-dimensional fields, Part 1: Theory, *Rev. Sci. Instr.*, **25**, No. 5 (1954), 480–5 and I. E. Dayton, F. C. Shoemaker and R. F. Mozley, The measurement of two-dimensional fields, Part 2: Study of a quadrupole magnet, *Rev. Sci. Instr.*, **25**, No. 5 (1954), 485–9.
3. R. Perin, *End effects compensation in the auxiliary magnets of the CERN Intersecting Storage Rings*, CERN ISR-MA/70-16.
4. J. Billan, K. Henrichsen and L. Walckiers, *Atomkernenergie, Kerntechnik*, **46**, No. 3 (1985), 193–7.
5. M. Wilson, *Superconducting magnets* (Oxford, Clarendon Press, 1983).
6. H. Brechna, *Superconducting magnet systems* (Springer-Verlag, 1973), 36–46.
7. K.-H. Mess and P. Schmüser, Superconducting accelerator magnets, *Proc. CERN Accelerator School*, Hamburg, 1988, CERN 89-04 (March 1989), 87–148.
8. J. P. Blewett, 200 GeV intersecting storage accelerators, *8th Int. Conf. on High-Energy Accel.*, CERN, 1971, 501–4.
9. E. J. Bleser, Superconducting magnets for the CBA project, *Nucl. Instr. Methods in Phys. Research*, **A235** (1985), 435–463; see footnote p. 435.
10. *CBA Brookhaven colliding beam accelerator*, Newsletter No. 2 (Nov. 1982), 27–31.
11. *Report of the Task Force on SSC Operations*, SSC-SR-1005 (July 1985).
12. R. Perin, First results of the high-field magnet development for the Large Hadron Collider, *11th Int. Conf. on Magnet Technology*, Tsukuba, Japan, Sept. 1989; alternatively, see R. Perin CERN/SPS/89-34 (EMA).

Appendix C

1. A. A. Kolomensky and A. N. Lebedev, *Theory of cyclic accelerators* (North-Holland Publishing Company, Amsterdam, 1966), 47–8.
2. H. Goldstein, *Classical mechanics*, 2nd edition (Addison-Wesley, 1980), 35–7.

Appendix D

1. H. G. Hereward and K. Johnsen, *On the phase equation for synchrotrons*, CERN-PS/HGH-KJ-1 (March 1957).
2. H. G. Hereward, *What are the equations for the phase oscillations in a synchrotron?*, CERN 66-6 (Feb. 1966).
3. D. Bohm and L. Foldy, The theory of the synchrotron, *Phys. Rev.*, **70**, Nos. 5 & 6 (Sept. 1946), 249–58.
4. J. A. MacLachlan, *Differential equations for longitudinal motion in a synchrotron*, Fermilab report, FN-532 (Jan. 1990).
5. T. Suzuki, Hamiltonian formulation for synchrotron oscillations and Sacherer's integral equations, *Particle Accelerators*, **12** (1982), 237–46.
6. C. J. A. Corsten and H. L. Hagedoorn, Simultaneous treatment of betatron and synchrotron motions in circular machines, *Nucl. Instrum. Methods*, **212** (1983), 37–46.
7. K. R. Symon and A. M. Sessler, *Methods of radio frequency acceleration in fixed field accelerators with applications to high current and intersecting beam accelerators*, Proc. CERN Symposium on High Energy Accelerators and Pion Physics (CERN 1956), 44–58.

Appendix E

1. A. A. Vlasov, On the kinetic theory of an assembly of particles with collective interaction, *J. Phys. USSR*, **9** (1945), 25.
2. J. D. Lawson, The physics of charged-particle beams (Clarendon Press Oxford, 1977), 172–8.
3. F. Ruggiero, *Kinetic theory of charged particle beams*, Proc. CERN Accelerator School 3rd Advance Accelerator Physics Course, Uppsala, 1989, CERN 90-04 (April, 1990), 52–66.
4. D. Habs, *Crystallisation of particle beams*, in *Frontiers of particle beams*, Proc. Joint US–CERN School on Part. Accels. (Lecture Notes in Physics No. 296, Springer-Verlag), 310–37.

INDEX

353